Alcohol

Methods and Protocols

METHODS IN MOLECULAR BIOLOGY™

John Walker, SERIES EDITOR

Alcohol

Methods and Protocols

Laura E. Nagy

Departments of Gastroenterology and Pathobiology, Cleveland Clinic;
Department of Nutrition, Case Western Reserve University, Cleveland, OH

Editor

 Humana Press

Editor:
Laura E. Nagy
Department of Gastroenterology and Pathobiology
Cleveland Clinic
and
Department of Nutrition
Case Western Reserve University
Cleveland, OH 44195
laura.nagy@case.edu

Series Editor:
John Walker
28 Selwyn Avenue
Hatfield, Hertfordshire
AL10 9NP, UK

ISBN 978-1-58829-906-2   e-ISBN 978-1-59745-242-7
ISSN 1064-3745

Library of Congress Control Number: 2007938049

Cover illustrations: Isolation of neuroepithelial cells (Chap. 12, Fig. 3; see complete caption on p. 162). (Background) Neurospheres adherent to culture dish in mitogen-withdrawal, ECM matrix addition paradigm (see complete figure on p. 163).

Printed on acid-free paper

9 8 7 6 5 4 3 2 1

springer.com

Preface

Alcohol consumption is often characterized as an environmental stress to the organism. In response to the stress of alcohol exposure, complex cellular and organismal adaptations occur to manage this insult. In most individuals, modest alcohol consumption over the course of a lifetime does not result in substantive health risks, and in the pathophysiology of some diseases, such as type 2 diabetes, modest alcohol consumption may actually be protective. Yet, chronic and heavy alcohol consumption poses significant health risks. Scientists who study the effects of acute and chronic alcohol consumption know little about what marks the transition from a benign or even protective effect of alcohol to its pathophysiological effect in the development of tissue injury and disease.

Understanding the transition to injury is an important clinical and public health issue, as excessive alcohol consumption can impact nearly every tissue in the body, contributing to more than 60 different medical conditions. The effects of excessive alcohol occur both in the developing organism as well as in the adult. Although the liver, as the site of ethanol metabolism, is particularly sensitive to chronic alcohol exposure, alcohol consumption also leads to damage of other tissues, including brain, heart, and cardiovascular system, as well as disruptions in regulation of endocrine and immune systems. The contribution of alcohol to the development of chronic diseases, such as osteoporosis, heart disease, and diabetes, is particularly relevant today, given the increased incidence of these diseases in our aging population.

Recent advances in understanding the pleiotropic effects of ethanol have been possible because of the development of relevant and rigorously controlled animal and cell culture models of acute and chronic ethanol exposure. Although each of the various models for ethanol exposure may not model perfectly the exposure of humans to alcohol, many model systems have now been developed that can mimic particular conditions of ethanol exposure in target tissues and organs. One of the primary goals of this volume is to provide detailed procedures for several of the more common models of acute and chronic ethanol exposure, enabling studies on the effects of ethanol in both the developing organism and in the adult. Use of these clearly defined models of ethanol exposure, presented in the first section of this volume, will allow for comparison of results among different laboratories, as well as among multiple tissue and organ targets of acute and chronic ethanol exposure.

One of the themes arising in recent studies that investigate the mechanisms of ethanol action on target tissues is the commonality in the impact of ethanol on regulation of cellular metabolism. Thus, in addition to the effects of acute and chronic alcohol on the complex physiology of the intact organism, alcohol exposure also has a profound impact on the biology of individual cells. As with studies in whole animals, investigations to study the impact of ethanol on cellular biology must be rigorously controlled and designed. Recent advances in the development of specific methodologies to mimic the impact of ethanol metabolism in cultured cells, detailed in the second section of this volume, have furthered our understanding of the molecular mechanisms by which ethanol disrupts cellular function.

Although there are common mechanisms of ethanol action on a variety of cell types, studies of the effects of ethanol on cellular function must also take into consideration the complex differentiated function of individual cells and tissues. Thus, expertise in the use of models of ethanol exposure, as well as in the design and analysis of experiments to ascertain the effects of ethanol on the highly regulated function of each differentiated cell and tissue type, must be combined to finely dissect the mechanisms of ethanol action. Therefore, an additional theme of this volume embraces the methodologies to investigate a variety of cells and tissues that are known to be disrupted by ethanol, from intestinal epithelial cells, to cells in the liver, including hepatocytes and Kupffer cells, to cells in the periphery, including skeletal muscle, adipose and bone. Specific methodologies to investigate the effects of ethanol on neuronal function, including the use of neuronal cell lines and organotypic cultures, are also presented.

It is likely that the effects of ethanol on cell, tissue, and organismal function are fundamentally based on the impact of ethanol on transcriptional and post-transcriptional regulation of gene expression. Novel methodologies to study the molecular mechanisms of ethanol action include the use of gene arrays, as well as proteomic analysis of the post-translational modifications of proteins in organelles and cells exposed to ethanol. Chapters providing the specific expertise required for the design and analysis of gene array and proteomic studies are included in this volume to enable investigators new to these data-rich approaches to successfully "mine" the vast amount of data that can be obtained by these approaches.

In the final analysis, studies into the molecular mechanisms for ethanol action not only result in a further understanding of the pathophysiology of ethanol-induced injury, but also contribute to our understanding of the fundamental mechanisms by which organisms have adapted to subtle changes in their environment. Although excessive alcohol consumption can result in profound impairments in the ability of the organism to develop and function, most organisms can readily handle the subtle insults associated with moderate alcohol consumption. Understanding the genetic, molecular, cellular, and physiological responses to ethanol that "tip the balance" from an adaptive response to a maladaptive/pathological response is critical to the development of therapeutic strategies for the intervention and/or prevention of the effects of ethanol on development and tissue injury. I hope that the very detailed and specific methods presented in this volume will further spur investigators to delve into the complex and fascinating story of the adaptive and maladaptive responses humans have developed to the consumption of alcohol.

Laura E. Nagy, PhD
Cleveland Clinic

Acknowledgments

I would like to thank Joyce Nolan for her careful and thorough editing and formatting of the chapters in this volume. This work was supported in part by US Public Health Service Grants RO1 AA11975 and RO1 AA11876.

Contents

Part III   Cell Culture Approaches to Studying Ethanol Exposure

Part IV   Regulation of Specific Organ Systems
in Response to Ethanol

Contributors

Carla-Maria A. Alexander, BS
Interdisciplinary Graduate Program in Immunology, University of Iowa, Iowa City, IA

Kelly K. Andringa, PhD
Department of Environmental Health Sciences, University of Alabama Birmingham, Birmingham, AL

Gavin E. Arteel, PhD
Department of Pharmacology and Toxicology and the James Graham Brown-Cancer Center, University of Louisville Health Sciences Center, Louisville, KY

Shannon M. Bailey, PhD
Department of Environmental Health Sciences, University of Alabama at Birmingham, Birmingham, AL

Zuhair K. Ballas, M.D
The Iowa City VA Medical Center and the Department of Internal Medicine, The Roy J. and Lucille A. Carver College of Medicine, University of Iowa, Iowa City, IA

Alan Cahill, PhD
Department of Pathology, Anatomy, and Cell Biology, Thomas Jefferson University, Philadelphia, PA

Cynthia Camarillo, PhD
Department of Neuroscience and Experimental Therapeutics, College of Medicine, Texas A&M Health Science Center, College Station, TX

Michael J. Carlson, PhD
Interdisciplinary Graduate Program in Immunology, University of Iowa, Iowa City, IA

Arthur I. Cederbaum, PhD
Department of Pharmacology and Biological Chemistry, Mount Sinai School
of Medicine, New York, NY

Xiaocong Chen, BS
Department Pathobiology, Cleveland Clinic; Department of Nutrition,
Case Western Reserve University, Cleveland, OH

Ruth A. Coleman, BS
Department of Pathology, Carver College of Medicine, University of Iowa,
Iowa City, IA

Robert T. Cook, MD, PhD
Department of Pathology, Carver College of Medicine, University of Iowa,
Iowa City, IA

Victor M. Darley-Usmar, PhD
Department of Pathology, University of Alabama at Birmingham, Birmingham, AL

Douglas Dohrman, PhD
Department of Neuroscience and Experimental Therapeutics, College of
Medicine, Texas A&M Health Science Center, College Station, TX

Hector D. Dominguez, PhD
Center for Behavioral Teratology, Department of Psychology, San Diego State
University, San Diego, CA

Alla Dynnyk, DVM
Research Center for Alcoholic Liver and Pancreatic Diseases and Cirrhosis,
Department of Pathology, Keck School of Medicine of the University
of Southern California; Department of Veterans Affairs, Greater Los Angeles
Healthcare System, Los Angeles, CA

Prajwal Gurung, BS
Department of Pathology, Carver College of Medicine, University of Iowa,
Iowa City, IA

Urszula T. Iwaniec, PhD
Department of Nutrition and Exercise Sciences, Oregon State University,
Corvallis, OR

Li Kang, PhD
Department Pathobiology, Cleveland Clinic; Department of Biochemistry,
Case Western Reserve University, Cleveland, OH

Sandra J. Kelly, PhD
Department of Psychology, University of South Carolina, Columbia, SC

Robnet T. Kerns, PhD
Departments of Pharmacology/Toxicology and Neurology, and the Center for Study
of Biological Complexity, Virginia Commonwealth University, Richmond, VA

Lynell W. Klassen, MD
Department of Veterans Affairs Medical Center, Department of Internal Medicine,
University of Nebraska Medical Center, Omaha, NE

Elizabeth J. Kovacs, PhD
Burn and Shock Trauma Institute, Departments of Surgery, Cell Biology,
Neurobiology, and Anatomy, Alcohol Research Program, Loyola University
Medical Center, Maywood, IL

Aimee Landar, PhD
Department of Pathology, University of Alabama at Birmingham, Birmingham, AL

Charles H. Lang, PhD
Department of Cellular and Molecular Physiology, Pennsylvania State University
College of Medicine, Hershey, PA

Charles R. Lawrence, PhD
Department of Psychology, University of South Carolina, Columbia, SC

Kevin L. Legge, PhD
Department of Pathology, University of Iowa Carver College of Medicine,
Iowa City, IA

Tung-Ming Leung, BS
Department of Anatomy, Li Ka Shing Faculty of Medicine, The University
of Hong Kong, Hong Kong, SAR China

Amanda Lindke, BS
Department of Neuroscience and Physiology, SUNY Upstate Medical University,
Developmental Exposure to Alcohol Research Center, Syracuse NY

Emily C. Liong, BS
Department of Anatomy, Li Ka Shing Faculty of Medicine, The University
of Hong Kong, Hong Kong, SAR China

Pranoti Mandrekar, PhD
Department of Medicine, University of Massachusetts Medical School,
Worcester, MA

Megan R. McMullen, BA
Department Pathobiology, Cleveland Clinic, Cleveland, OH

Michael F. Miles, MD, PhD
Departments of Pharmacology/Toxicology and Neurology, and the Center
for Study of Biological Complexity, Virginia Commonwealth University,
Richmond, VA

Michael W. Miller, PhD
Department of Neuroscience and Physiology, SUNY Upstate Medical University,
Developmental Exposure to Alcohol Research Center, Research Service,
Veterans Affairs Medical Center, Syracuse, NY

Rajesh C. Miranda, PhD
Center for Environmental and Rural Health, Texas A&M University College of
Medicine, Department of Neuroscience and Experimental Therapeutics,
Texas A&M Health Science Center, College Station, TX

Hasmik Mkrtchyan, MD
Research Center for Alcoholic Liver and Pancreatic Diseases and Cirrhosis,
Department of Pathology, Keck School of Medicine of the University
of Southern California, Los Angeles, CA

Laura E. Nagy, PhD
Departments of Gastroenterology and Pathobiology, Cleveland Clinic;
Department of Nutrition, Case Western Reserve University, Cleveland, OH

Amin A. Nanji, PhD
Department of Pathology and Laboratory Medicine, Dalhousie University School
of Medicine, Halifax, Canada

Timothy P. Plackett
Burn and Shock Trauma Institute, Departments of Surgery, Cell Biology, Neurobiology,
and Anatomy, Alcohol Research Program, Loyola University Medical Center,
Maywood, IL

Michele T. Pritchard, PhD
Department Pathobiology, Cleveland Clinic, Cleveland, OH

R. K. Rao, PhD
University of Tennessee Health Science Center, Department of Physiology,
Memphis, TN

Mark J. Reimers, PhD
Department of Environmental and Molecular Toxicology, Oregon State University, Corvallis, OR

Daniel R. Santillano, MD, PhD
College of Medicine, Department of Neuroscience and Experimental Therapeutics, Texas A&M Health Science Center, College Station, TX

Annette J. Schlueter, MD, PhD
Department of Pathology, University of Iowa Carver College of Medicine, Iowa City, IA

Becky M. Sebastian, MS
Department Pathobiology, Cleveland Clinic, Cleveland, OH

Michael R. Shey, MS
The Iowa City VA Medical Center and the Department of Internal Medicine, The Roy J. and Lucille A. Carver College of Medicine, University of Iowa, Iowa City, IA, USA

Susan M. Smith, PhD
Department of Nutritional Sciences, University of Wisconsin-Madison, Madison WI

Peter Sykora
Department of Pathology, Anatomy and Cell Biology, Thomas Jefferson University, Philadelphia, PA

Gyongyi Szabo, MD, PhD
Department of Medicine, University of Massachusetts Medical School, Worcester, MA

Robert L. Tanguay, PhD
Department of Environmental and Molecular Toxicology, Oregon State University, Corvallis

Geoffrey M. Thiele, PhD
Department of Veterans Affairs Medical Center, Departments of Internal Medicine, and Microbiology and Pathology, University of Nebraska Medical Center, Omaha, NE

Jennifer D. Thomas, PhD
Center for Behavioral Teratology, Department of Psychology, San Diego State University, San Diego, CA

George L. Tipoe, MD, PhD
Department of Anatomy, Li Ka Shing Faculty of Medicine, The University
of Hong Kong, Hong Kong, SAR China

Barbara Tremper-Wells, PhD
Department of Neuroscience and Physiology, SUNY Upstate Medical University;
Developmental Exposure to Alcohol Research Center, Syracuse NY

Hide Tsukamoto, DVM, PhD
Research Center for Alcoholic Liver and Pancreatic Diseases and Cirrhosis,
Department of Pathology, Keck School of Medicine of the University
of Southern California; Department of Veterans Affairs, Greater Los Angeles
Healthcare System, Los Angeles, CA

Dean J. Tuma, PhD
Department of Veterans Affairs Medical Center; Departments of Internal Medicine
and Biochemistry, University of Nebraska Medical Center, Omaha, NE

Lucas E. Turner, BS
Department of Pathology, Carver College of Medicine, University of Iowa,
Iowa City, IA

Russell T. Turner, PhD
Department of Nutrition and Exercise Sciences, Oregon State University,
Corvallis, OR

Lorraine T. Tygrett
Department of Pathology, Carver College of Medicine, University of Iowa,
Iowa City, IA

Thomas C. Vary, PhD
Department of Cellular and Molecular Physiology, Pennsylvania State University
College of Medicine, Hershey, PA

Shilpi Verma, MS, BSc
Interdisciplinary Graduate Program in Immunology, University of Iowa,
Iowa City, IA

Thomas J. Waldschmidt, PhD
Department of Pathology, Carver College of Medicine, Interdisciplinary Graduate
Program in Immunology, University of Iowa, Iowa City, IA

Susan Wiechert, BA
Department of Pathology, Carver College of Medicine, University of Iowa,
Iowa City, IA

Thomas J. Wronski, PhD
Department of Physiological Sciences, University of Florida, Gainesville, FL

Defeng Wu, PhD
Department of Pharmacology and Biological Chemistry, Mount Sinai School
of Medicine, New York, NY

Betty M. Young, BS
Department of Pathology, Carver College of Medicine, University of Iowa,
Iowa City, IA

Part I
Animal Models of Acute and Chronic Ethanol Exposure

1
Acute Models of Ethanol Exposure to Mice

Timothy P. Plackett and Elizabeth J. Kovacs

Summary Acute alcohol administration has minimal effects on basal immune function. However, when the immune system is challenged, acute alcohol administration alters the immune system's response. In the first 3 h after infection or traumatic injury, the presence of alcohol is associated with a decreased inflammatory response. This defect lasts long after the alcohol is cleared. Conversely, by 48 h after traumatic injury, the presence of alcohol is associated with a heightened inflammatory response. Aside from its in vivo actions, systemic administration of alcohol also alters the ex vivo response of immune cells, resulting in a decreased production of multiple cytokines after stimulation by lipopolysaccharide, concanavilin A, zymosan, and CpG DNA. Here, we describe a standardized model of acute administration of ethanol to mice used to study both the invivo and ex vivo responses of immune cells to ethanol.

Keywords   Ethanol; acute alcohol administration; gavage.

1   Introduction

The interplay between alcohol and the immune system is dependent upon several factors, including the frequency and duration of the alcohol's administration (*1*). The focus of this chapter is the effects of acute alcohol administration, where acute administration

L. E. Nagy (ed.), *Alcohol: Methods and Protocols*
© Humana Press 2008

is defined as a single bolus of alcohol given a single time. This model is designed to mimic the circulation of alcohol in a "social" or "occasional" drinker.

Low concentrations of alcohol have minimal to no effect on baseline in vivo immune function in the absence of an inflammatory insult or infection *(2–4)*. In contrast, ex vivo challenge of splenic macrophages harvested 3 h after systemic alcohol administration results in decreased interleukin (IL)-6 and tumor necrosis factor-alpha (TNF-α) after stimulation by lipopolysaccharide, zymosan, or CpG-DNA *(5)*. Likewise, ex vivo challenge of unfractionated splenocytes 24 hr after systemic ethanol administration results in decreased IL-4 and interferon-gamma (IFN-γ) after stimulation by concanavalin A (con A) *(6)*.

Acute ethanol administration also modifies the in vivo response to several traumatic and systemic injuries. Ninety minutes after lipopolysaccharide injection in conjunction with alcohol administration, there is a suppression of serum TNF-α *(7)*. Similarly, 4 h after traumatic brain injury, the presence of alcohol at the time of injury results in a dose-dependent decrease in central nervous system levels of IL-1β and TNF-α *(3)*. Finally, 24 h after burn injury, the presence of alcohol at the time of injury correlates with a depressed delayed-type hypersensitivity response and an increase in serum IL-6 concentration *(4)*. The same injury is also associated with impaired splenocyte proliferation and IL-4 and IFN-γ production after ex vivo cultivation with con A *(6)*.

2   Materials

2.1   Intraperitoneal Injection

1. 95% denatured ethyl alcohol (Fisher Scientific, Hampton, NH).
2. Sterile water irrigating solution (Baxter, Deerfield, IL).
3. Stericup GV filter unit with 0.22-mm PVDF filter and 150-mL reservoir bottle (Millipore, Billerica, MA)
4. 1-mL Luer-lock syringe (Becton-Dickinson, Franklin Lakes, NJ).
5. 21-gage x 1-inch sterile hypodermic needle with regular wall and regular bevel (Becton-Dickinson, Franklin Lakes, NJ).

2.2   Oral Gavage

1. 95% denatured ethyl alcohol (Fisher Scientific, Hampton, NH)
2. Sterile water irrigating solution (Baxter, Deerfield, IL)
3. Stericup GV filter unit with 0.22-mm PVDF filter and 150-mL reservoir bottle (Millipore, Billerica, MA).
4. 1-mL Luer-lock syringe (Becton-Dickinson, Frankling Lakes, NJ).
5. Stainless-steel 20-gage, 1.5-inch feeding needle (Harvard Apparatus, Holliston, MA).

3 Methods

3.1 Intraperitoneal Injection

1. Prepare a 20% ethanol solution by mixing 210 mL of 95% ethanol with 790 mL of sterile water (*see* **Note 1**).
2. Filter the 20% ethanol through a Stericup GV filter.
3. Calculate the necessary volume of ethanol to inject into each animal (*see* **Note 2**).
   The volume of 20% ethanol needed is based upon the weight of the animal and the desired blood alcohol level (Fig. 1). It is calculated based upon the following formula: volume of ethanol in milliliters= [(weight of the mouse in grams) × (desired blood alcohol level in mg/dL + 31)]/19529.
4. Draw the appropriate volume of ethanol into either a 0.5-mL or 1-mL syringe (*see* **Note 3**).
5. Grasp the mouse by holding the skin at the nape of the neck between the thumb and forefinger, while holding the skin of the dorsal torso between the third and fifth fingers and the thenar eminence. This position should give you firm control of the animal and access to the ventral abdominal surface.
6. Insert the needle into the lower left abdominal quadrant close to the midline at a 30° angle. While advancing the needle, remain within the peritoneal cavity, avoiding injuring any of the intraperitoneal organs.

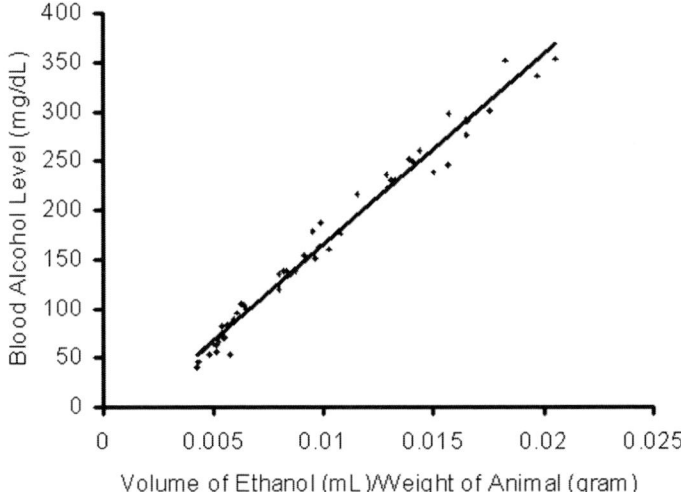

Fig. 1 Blood alcohol levels after intraperitoneal injection. Male and female mice of multiple strains, weights, and ages received intraperitoneal injections of varying volumes of 20% ethanol. Animals were sacrificed 30 min after ethanol injection and serum alcohol levels were determined. Data from individual animals are plotted on a graph and a line of best fit was determined using Microsoft Excel

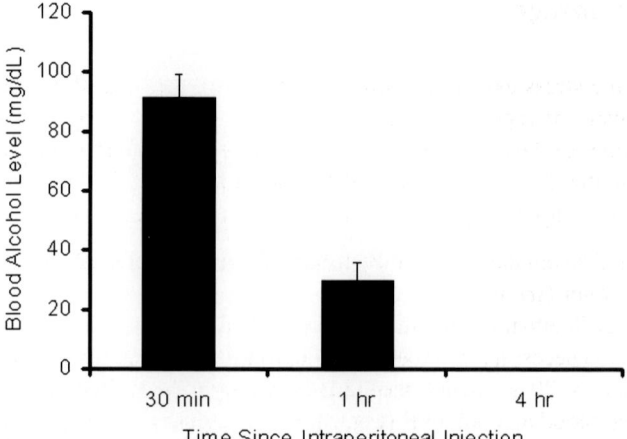

Fig. 2 Time course of blood alcohol levels after intraperitoneal injection. Male and female mice of multiple strains, weights, and ages received 0.15 mL of intraperitoneal injections of 20% ethanol. Animals were sacrificed 30 min, 1 h, and 4 h after intraperitoneal injection and serum alcohol levels were determined. Data are expressed as mean blood alcohol level ± SEM

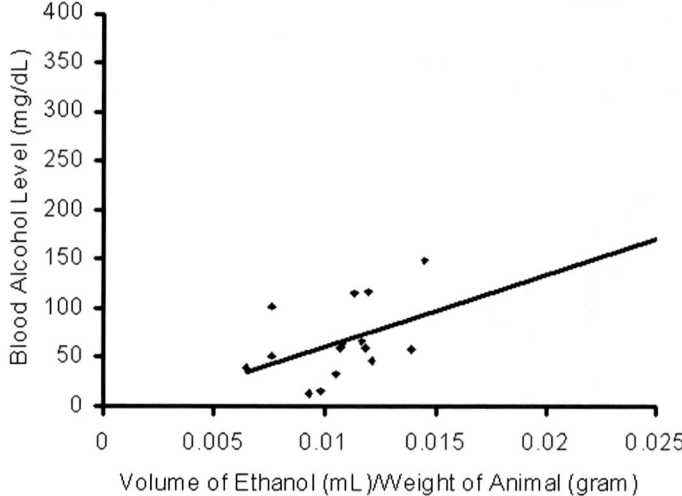

Fig. 3 Blood alcohol levels after oral gavage. Male and female mice of multiple strains, weights, and ages were given varying volumes of 20% ethanol via oral gavage. Animals were sacrificed 30 min after gavage and serum alcohol levels were determined. Data from individual animals are plotted on a graph and a line of best fit was determined using Microsoft Excel

7. Gently aspirate to confirm that the needle is not within a blood vessel or bowel. Then inject the appropriate volume of ethanol.

8. Serum levels of ethanol peak 30 min after intraperitoneal injection (Fig. 2). The ethanol level is decreased by approximately 70% after 1 h and is undetectable after 4 h (*see* **Note 4**).

3.2 Oral Gavage

Because of the stress associated with the procedure, we suggest acclimatizing the animal to gavage by repeating the procedure at least 10 times. This needs to be done at random times of the day to avoid entraining their circadian rhythms. If the animal is not acclimatized, then the stress of the procedure will elevate glucocorticoids which may alter inflammatory and immune responses that are under investigation.

1. Prepare a 20% ethanol solution by mixing 210 mL of 95% ethanol with 790 mL of sterile water (*see* **Note 1**).
2. Filter the 20% ethanol through a Stericup GV filter.
3. Calculate the necessary volume of ethanol to inject into each animal (*see* **Note 5**). The volume of 20% ethanol needed is based upon the weight of the animal and the desired blood alcohol level (Fig. 3). It is calculated based upon the following formula: volume of ethanol in milliliters= [(weight of the mouse in grams) × (desired blood alcohol level in mg/dL + 12)]/7293.6.
4. Draw the appropriate volume of ethanol into either a 0.5-mL or 1-mL syringe.
5. Select the correct length of gavage needle (see Note 6). The maximal length should not exceed the distance between the corner of the mouth and the last rib (which is a rough estimate of the stomach's position).
6. Grasp the mouse by holding the nape of the neck between the thumb forefinger, while holding the skin of the dorsal torso between the third and fifth fingers and the thenar eminence. This position should give you firm control of the animal, especially the neck.

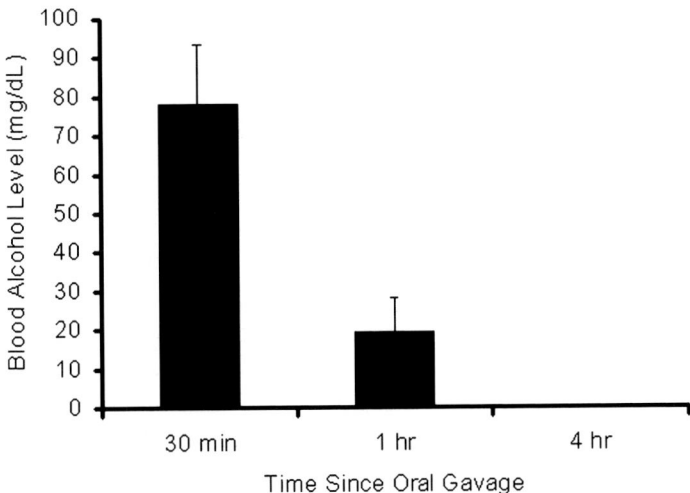

Fig. 4 Time course of blood alcohol levels after oral gavage. Male and female mice of multiple strains, weights, and ages received 0.3-mL intraperitoneal injections of 20% ethanol by oral gavage. The mice were sacrificed at 30 min, 1 h, and 4 h after gavage and serum alcohol levels were determined. Data are expressed as mean blood alcohol level ± SEM

7. Extend the head and neck, so as to position the esophagus in a straight line.
8. Carefully position the needle into the mouth and advance the needle into the esophagus. Do not push if there is resistance; instead, reposition the needle and attempt again. The use of excess force may damage or tear the esophagus (*see* **Note 7**).
9. Administer the appropriate volume of ethanol.
10. Serum levels of ethanol peak 30 min after oral gavage (Fig. 4; *see* **Note 8**). The ethanol level is decreased by approximately 75% after 1 h and is undetectable after 4 h (*see* **Note 4**).

Acknowledgments The authors thank Luis Ramirez and Dr. Mashkoor A. Choudhry for thoughtful discussions. This work was supported by the National Institutes of Health R01 AA12034–04 (EJK), the Ralph and Marion C. Falk Foundation, and the Illinois Excellence in Academic Medicine Grant.

4 Notes

1. True anhydrous alcohol should not be used to prepare 20% ethanol solutions. Anhydrous alcohol is frequently prepared by distilling a small quantity of benzene, a know carcinogen. Instead, 95% ethanol solutions are recommended for preparing 20% ethanol solutions.
2. The determination of the appropriate volume of 20% ethanol to administer via intraperitoneal injection is based upon multiple experiments using mice of various strains, sexes, weights, and ages. For any given volume of ethanol administered, the weight of the animal was the only independent factor that predicted alcohol level 30 minutes after intraperitoneal injection. Determination of the amount of 20% ethanol to administer is based upon a line of best fit from the cumulative results of the aforementioned experiments (Fig. 1). The equation for the slope of the curve is $y=19529x - 30.959$. The accuracy of the resultant equation can not be verified for blood alcohol levels less than 50 mg/dL or greater than 400 mg/dL, nor can the accuracy be verified for species other than mice.
3. In the interest of simplicity, if all the animals are of similar weight, they are all given the same volume of 20% ethanol. Under these circumstances, the volume delivered is based upon the average weight of the mice. For mice receiving an intraperitoneal injection, this results in only small variability in the blood alcohol levels between the animals (standard error of the mean of 4 mg/dL) *(2)*.
4. Experimentation with rats will require higher doses of ethanol per weight than required with mice in order to achieve a similar blood alcohol level *(8)*.
5. The determination of the appropriate volume of 20% ethanol to administer via oral gavage is based upon multiple experiments using male and female mice of various strains and body weights. For any given volume of ethanol administered, the weight of the animal was the only independent factor that predicted alcohol level 30 min after oral gavage. Determination of how much 20% ethanol to administer is based upon a line of best fit from the cumulative results of the aforementioned experiments (Fig. 3). The equation for the slope of the curve is

y=7293.6x – 11.809. The accuracy of the resultant equation can not be verified for blood alcohol levels less than 50 mg/dL or greater than 125 mg/dL, nor can the accuracy be verified for species other than mice.

6. The size of the gavage needle will vary depending on the size of the animal. Suggested sizes for rodents are as follows:

15- to 20-g mice: 22-gage, 1–1.5 inches
20- to 25-g mice: 20-gage, 1–3 inches
25- to 30-g mice: 18-gage, 1.5–3 inches
50- to 100-g rats: 18- to 20-gage, 1–1.5 inches
100- to 200-g rats: 16- to 18-gage, 2.25–3 inches
200- to 350-grats: 14- to 16-gage, 3–4 inches

7. There is a more difficult learning curve with oral gavage. The researcher must take great care to make sure that the gavage needle is in the esophagus and not the trachea. In addition, the unanesthetized animal is fairly difficult to handle while performing oral gavage. However, the use of anesthesia is not recommended with oral gavage, because loss of the gag reflex may occur, increasing the potential for aspiration of ethanol. We recommend practicing the procedure with 10–20 mice before using experimental mice.

8. There was more variability in blood alcohol levels between similarly treated animal subjected to oral gavage than was noted after intraperitoneal injection. This increased variability is likely due to the tendency of mice to partially regurgitate the ethanol.

References

1. Messingham, K. A. N., Faunce, D. E., and Kovacs, E. J. (2002) Alcohol, injury, and cellular immunity. *Alcohol* **28,** 137–149.
2. Faunce, D. E., Gregory, M. S., and Kovacs, E. J. (1997) Effects of acute ethanol on cellular immune responses in a murine model of thermal injury. *J. Leukoc. Biol.* **62,** 733–740.
3. Gottesfeld Z., Moore, A. N., and Dash, P. K. (2002) Acute ethanol intake attenuates inflammatory cytokines after brain injury in rats: a possible role for corticosterone. *J Neurotrauma* **19,** 317–326.
4. Messingham, K. A. N., Heinrich, S. A., Schilling, E. M., and Kovacs, E. J. (2002) Interleukin-4 treatment restores cellular immunity after ethanol exposure and burn injury. *Alcoholism: Clin. Exp. Res.* **26,** 519–526.
5. Goral, J., and Kovacs, E. J. (2005) In vivo ethanol exposure down-regulates TLR2-, TLR4-, and TLR9-mediated macrophage inflammatory response by limiting p38 and ERK1/2 activation. *J. Immunol.* **174,** 456–463.
6. Plackett, T. P., Jarrett, J., Gamelli, R. L., and Kovacs, E. J. (2005) A low blood ethanol level is associated with improved cytokine production in aged mice after traumatic injury. *J Trauma* **59,** 984–989.
7. D'Souza, N. B., Bagby, G. J., Nelson, S., Lang, C. H., and Spitzer, J. J. (1989) Acute alcohol infusion suppresses endotoxin-induced serum tumor necrosis factor. *Alcoholism: Clin. Exp. Res.* **13,** 295–298.
8. Livy, D. J., Parnell, S. E., and West, J. R. (2003) Blood ethanol concentration profiles: a comparison between rats and mice. *Alcohol* **29,** 165–171.

2
A Voluntary Oral-feeding Rat Model for Pathological Alcoholic Liver Injury

George L. Tipoe, Emily C. Liong, Tung-Ming Leung, and Amin A. Nanji

Summary   The variety of animal models used in the study of alcoholic liver disease reflects the formidable task of developing a model that replicates the human disease. We show that oral feeding of fatty acids derived from fish oil and ethanol induces fatty liver, necrosis, inflammation, and fibrosis. Together with the study of oxidative and nitrosative stress markers, cytokines, proteasome function, and protein studies, this model has provided an inexpensive and technically simple method of establishing pathological alcoholic liver injury.

Keywords   Alcoholic liver injury; cytokines; rodents; nitric oxide.

L. E. Nagy (ed.), *Alcohol: Methods and Protocols*
© Humana Press 2008

1   Introduction

The wide variety of animal models used in the study of alcoholic liver disease (ALD) reflects the formidable task of developing an appropriate model that replicates the human prototype. A major advance in the study of ALD pathogenesis is the development of the intragastric ethanol infusion model, which uses an implanted long-term intra-gastric catheter for continuous infusion of ethanol and liquid diets. The epidemiologic association between different types of fat and alcoholic liver disease has prompted the study of different types of fatty acids in the development of ALD in the intragastric feeding rat model. Many of the pathological features of ALD in humans are achieved in rats fed ethanol intragastrically. Women develop ALD more rapidly than men, and liver injury progresses in women with alcoholic hepatitis, even after cessation or reduced drinking. Studies using the intragastric feeding rat model confirmed the obser-vation of enhanced susceptibility of females to alcoholic liver injury. Taking advan-tage of the fact that (1) female rats are more susceptible to alcohol than males and (2) among the different fatty acids, those that are derived from fish oil cause the greatest degree of pathological liver injury, we developed a voluntary oral feeding model for alcoholic liver injury that reproduced many of the pathological and biochemical changes in alcoholic liver injury seen with the intragastric infusion model.

2   Materials

2.1   Animals and Feeding

1. Female Wistar rats weighing approx 175-190 g. Animals housed in cages for rats model no. 1291H (Tecniplast, Italy).
2. Preparation of 1 L of diet: 8% ethanol with 40% energy derived from ethanol, 30% from fat, 23% from protein, and 7% from dextrose. The source of fatty acids in the diet is fish oil (1).
3. The diet components 1 to 8 (Table 1) are added to a blender with 600 mL of water.
4. The components are mixed thoroughly.
5. The suspending agent, component 9 in Table 1, is added to the mixture and mixed.
6. Water is added up to 1 L and mixed.
7. The mixed diet is served inside a water bottle in the absence of chow.

2.2   Assessment of Urine Alcohol

1. Q.E.D. Saliva Alcohol Test (OraSure Technologies, Inc.).
2. 50-mL centrifuge tube (Iwaki Glass Co, Japan).
3. Mineral oil (Sigma-Aldrich Co., St. Louis, MO).

Table 1 Oral Feeding Model Diet Composition

Diet composition	Amount per 100 mL diet
1. Ethanol, absolute (Merck, Germany)	7.24 mL
2. Fish oil (Sigma-Aldrich Co., USA)	3.66 mL
3. Mineral mix (Dyets Inc., USA)	0.93 g
4. Vitamin mix (Dyets Inc., USA)	0.26 g
5. Choline bitartrate (Dyets Inc., USA)	0.05 g
6. Dextrose (Dyets Inc., USA)	1.75 g
7. DL-Methionine (Bio-Serv, USA)	0.11 g
8. Lactalbumin hydrolysate (Bio-Serv, USA)	5.75 g
9. Suspending agent K (Bio-Serv, USA)	0.45 g

2.3 Blood Sample Collection

1. 15-mL centrifuge tube (Iwaki Glass Co, Japan).
2. Swing bucket centrifuge (Sigma-Aldrich Co.).

2.4 Tissue Fixation, Processing, and Histological Staining

1. Chloroform (Merck, Germany).
2. 70%, 80%, 90% 95% and 100% ethanol (Merck, Germany)
3. Neutral buffered formalin saline: 40% Formalin (100 mL), $NaH_2PO_4 \cdot H_2O$ (4 g), Na_2HPO_4 (6.5 g), NaCl (9 g), distilled H_2O (900 mL).
4. Harris's Hematoxylin (Sigma-Aldrich Co.).
5. Eosin (Sigma-Aldrich Co.).
6. Toluene (Merck, Germany).
7. Permount medium (Fisher Scientific).
8. Poly-L-lysine (Sigma Bioscience).

2.5 Sirius Red Staining

1. Sirius Red F3B (Polysciences Inc.).
2. Ethanol (Merck).
3. Picric acid (BDH Laboratory Supplies, UK).
4. Hydrochloric acid (Merck).

2.6 Immunohistochemical Analysis

1. DAKO EnVision™+ System, Peroxidase (DAB) kit (DakoCytomation Denmark A/S, Denmark).
2. Ethanol (Merck).

Table 2 Antibodies Used in Immunohistochemistry

Target protein	Source	Dilution	Manufacturer
Inducible nitric oxide synthase	Rabbit polyclonal	1:100	Transduction Laboratories, USA
Endothelial nitric oxide synthase	Rabbit polyclonal	1:100	Santa Cruz Biotechnology, CA, USA
Cyclooxygenase-2	Rabbit polyclonal	1:200	Cayman Chemical, USA
Nitrotyrosine	Mouse polyclonal	1:200	Zymed Laboratories Inc., USA

3. TBS Buffer: $0.05\,M$ Tris-HCl pH 7.6; $0.15\,M$ NaCl.
4. Trypsin (Sigma Bioscience).
5. Calcium Chloride (Merck).
6. Normal serum (Vector Laboratories).
7. BSA solution: 1% bovine serum albumin (BSA), $0.05\,M$ TBS.
8. Primary antibodies against (Table 2).

2.7 Biochemical Assay of Serum Alanine Aminotransferase (ALT)

1. ALT Substrate: L-alanine ($80\,mM$), nicotinamide adenine dinucleotide (NADH; $0.2\,mM$) and lactate dehydrogenase (LDH; 2 units) in potassium phosphate buffer ($0.1\,M$, pH 7.4)
2. α-ketoglutaric acid ($10\,mM$; Sigma-Aldrich Co).

2.8 Marker of Oxidative Stress (8-Isoprostane)

1. 8-isoprostane Competitive Enzyme Immunoassay kit (Cayman Chemical).

2.9 Electrophoretic Mobility Shift Assay

1. Gel Shift Assay Systems (Promega Corporation).
2. Consensus oligonucleotide (Promega Corporation).

   NF-κB: 5′-AGTTGAGGGGACTTTCCCAGGC-3′
          3′-TCAACTCCCCTGAAAGGGTCCG-5′
   AP-1: 5′-CGCTTGATGAGTCAGCCGGAA-3′
          3′-GCGAACTACTCAGTCGGCCTT-5′

3. [γ^{-32}P]ATP (Amersham Bioscience Ltd, UK).
4. T4 Polynucleotide Kinase (Promega Corporation).
5. TE buffer (10 mM Tris-HCl. pH 8.0; 1 mM ethylene diamine tetraacetic acid [EDTA]).
6. 10× Reaction buffer: 40% glycerol, 200 mM Tris-HCl pH 7.8, 1M NaCl, 50 mM MgCl$_2$, 10 mM EDTA, 50 mM dithiothreitol.
7. Loading dye: 250 mM Tris-HCl, pH 7.5, 0.2% bromophenol blue, 40% glycerol.
8. 4% nondenaturing acrylamide gel prepared in TBE buffer (10× TBE buffer (890 mM Trisbase, 890 mM boric acid, 2 mM disodium EDTA).
9. Aquasol (Packard Instrument Company, Inc.).
10. Antibodies for supershift assay (Table 5).
11. TEMED (Sigma-Aldrich Co.).

2.10   Reverse Transcriptase Polymerase Chain Reaction (RT-PCR)

1. NucleoSpin RNA II extraction kit (Clontech Laboratories, Inc.).
2. β-mercaptoethanol (Sigma-Aldrich Co.).
3. SuperScript™ First-Stand Synthesis System (Life Technologies).
4. AmpliTaq Gold PCR kit (Roche Applied Science).
5. Primers (Invitrogen Inc.; Tables 3 and 4).
6. 0.05 µg/mL ethidium bromide.
7. Bromophenol Blue and Xylene-Scyanol (Sigma-Aldrich Co.).

2.11   Western Blotting

1. Liver homogenizing Buffer 1: 250 mM sucrose, 15 mM NaCl, 5 mM EDTA, 1 mM EGTA, 0.15 mM spermine, 0.5 mM spermidine, 1 mM dithiothreitol,

Table 3 Primers Used in RT-PCR

Target gene	Primer sequence
Inducible nitric oxide synthase	5'-GTGGTGACAAGCACATTTGG-3'
	5'-GGCTGGACTTTTCACTCTGC-3'
Endothelial nitric oxide synthase	5'-GACCCTCACCGCTACAACAT-3'
	5'-CACAGAAGTGGGGGTATGCT-3'
Cyclooxygenase-2	5'-GGAGAGACTATCAAGATAGTGATC-3'
	5'-ATGGTCAGTAGACTTTTACAGCTC-3'
Tumor necrosis factor-α	5'-ATGAGCACAGAAAGCATGATC-3'
	5'-TACAGGCTTGTCACTCGAATT-3'
Glyceraldehydes-3-phosphate dehydrogenase	5'-CCTTCATTGACCTCAACTACATGGT-3'
	5'-TCATTGTCATACCAGGAAATGAGCT-3'

Table 4 Reaction Mixture and Temperature Profile of PCR

Target genes	Reaction mixture (per sample)		Temperature profile	
iNOS	10× PCR buffer	2.5 μL		
	25 mM MgCl$_2$	2.5 μL	95°C	15 min
	10 mM dNTP	1.0 μL	95°C	1 min
	10 μM forward primer	0.5 μL	60°C	1 min
	10 μM reverse primer	0.5 μL	72°C	1 min
	Taq Polymerase	0.25 μL		37 cycles
	Milli-Q H$_2$O	13.75 μL	72°C	10 min
	CDNA	4.0 μL		
ENOS	10× PCR buffer	2.5 μL		
	25 mM MgCl$_2$	2.5 μL	95°C	15 min
	10 mM dNTP	0.5 μL	95°C	1 min
	10 μM forward primer	0.5 μL	60°C	1 min
	10 μM reverse primer	0.5 μL	72°C	1.5 min
	Taq Polymerase	0.125 μL		35 cycles
	Milli-Q H$_2$O	17.375 μL	72°C	10 min
	cDNA	1.0 μL		
COX-2	10× PCR Buffer	2.5 μL		
	25 mM MgCl$_2$	2.5 μL	95°C	15 min
	10 mM dNTP	0.5 μL	95°C	1 min
	10 μM forward primer	0.5 μL	60°C	1 min
	10 μM reverse primer	0.5 μL	72°C	1.5 min
	Taq Polymerase	0.125 μL		37 cycles
	Milli-Q H$_2$O	17.375 μL	72°C	10 min
	cDNA	1.0 μL		
TNF-α	10× PCR buffer	2.5 μL		
	25 mM MgCl$_2$	1.5 μL	95°C	15 min
	10 mM dNTP	0.5 μL	95°C	1 min
	10 μM forward primer	0.5 μL	50°C	1 min
	10 μM reverse primer	0.5 μL	72°C	1 min
	Taq Polymerase	0.125 μL		34 cycles
	Milli-Q H$_2$O	18.375 μL	72°C	10 min
	cDNA	1.0 μL		
GAPDH	10× PCR buffer	2.5 μL		
	25 mM MgCl$_2$	1.5 μL	95°C	15 min
	10 mM dNTP	0.5 μL	95°C	1 min
	10 μM forward primer	1.25 μL	55°C	1 min
	10 μM reverse primer	1.25 μL	72°C	1 min
	Taq Polymerase	0.15 μL		25 cycles
	Milli-Q H$_2$O	16.85 μL	72°C	10 min
	cDNA	1.0 μL		

    15 mM Tris-HCl, pH 7.9, 60 mM KCl, 0.1% leupeptin [Leu], 0.1% aprotinin [Apr]; 0.5% phenylmethylsulfonyl fluoride [PMSF]).

2. Buffer 2: 10 mM Hepes, pH 7.9, 1.5 mM MgCl$_2$, 10 mM KCl, 0.1% Leu, 0.1% Apr; 0.5% PMSF.

3. Buffer 3: 0.5 mM Hepes, pH 7.9, 0.75 mM MgCl$_2$, 0.5 mM EDTA, 0.5 mM KCl, 12.5% glycerol, 0.1% NP-40, 0.1% Leu, 0.1% Apr, 0.5% PMSF.

Table 5 Antibodies Used in Western Blotting

Antibody against	Source	Dilution	Manufacturer
Inducible nitric oxide synthase	Rabbit (Polyclonal)	1:500	Transduction Laboratories, USA
Endothelial nitric oxide synthase	Rabbit (Polyclonal)	1:1000	Santa Cruz Biotechnology, CA, USA
Cyclooxygenase-2	Rabbit (Polyclonal)	1:500	Cayman Chemicals, USA
Nitrotyrosine	Mouse (Monoclonal)	1:1000	Zymed Laboratories Inc, USA
Beta-actin	Mouse (Monoclonal)	1:5000	Sigma Bioscience, USA

4. Buffer 4: 10 mM Tris-HCl, pH 7.9, 5 mM MgCl$_2$, 1 mM EDTA, 10 mM KCl, 20% glycerol, 0.1% NP-40.
5. Bio-Rad Protein Assay Kit (Bio-Rad Laboratories, Inc.).
6. 2× sample buffer: 0.1 M Tris-HCl, pH 6.8, 20% glycerol, 4% sodium dodecyl sulfate [SDS], 0.2% bromophenol blue, 5.25% β-Mercaptoethanol.
7. SDS polyacrylamide gel: for a 10% gel: 0.6 M Tris-HCl, pH 8.8, 10% polyacrylamide, 0.05% SDS, 0.05% ammonium persulfate, 0.1% TEMED.
8. Stacking gel: 0.2 M Tris-HCl, pH 8.8, 4% polyacrylamide, 0.05% SDS, 0.07% ammonium persulfate, 0.1% TEMED.
9. Tank buffer: 0.025 M Tris, 0.192 M glycine, 0.1% SDS, pH 8.3.
10. Transfer buffer: 25 mM Tris, pH 8.3, 192 mM glycine, 20% methanol, 0.1% SDS.
11. Tween Tris-buffered saline (TBST), pH 7.5: 100 mM Tris-HCl, 0.9% NaCl, 0.1% Tween-20.
12. Antibodies (Table 5).
13. Secondary antibodies (1:2000 dilution; DAKO Corporation).
14. ECL™ Western Blotting Detection Reagents (Amersham Pharmacia Biotech).

2.12   *Thiobarbituric Acid Reactive Substances (TBARS)*

1. Tris-KCl buffer: 50 mM Tris-HCl, 154 mM KCl, pH 7.4.
2. Trichloroacetic acid.
3. Thiobarbituric acid.

2.13   *Chymotrypsin (2)*

1. 0.1 M Tris-HCl, pH 7.5.
2. Stop solution: 1% SDS and 1 mL of 0.1 M Tris-HCl, pH 9.1.

2.14   Endotoxin

1. Limulus amoebocyte lysate test (Kinetic-QCL).

2.15   CYP2E1

1. CYP2E1 substrate buffer: 0.45 ml of 0.1 M phosphate buffer, pH 7.4, with 8 mM aniline and 1 mM NADPH.
2. 40% Trichloroacetic acid.
3. 10% Na_2CO_3.
4. 2% Phenol.

3   Methods

3.1   Induction of Fatty Liver, Necrosis, and Inflammation

Prepare ethanol and iso-caloric dextrose control diets fresh daily (*see Note 1*). Female Wistar rats are allowed free access to the ethanol-containing diet or pair-fed isocaloric dextrose control diets for 8 wk (*see Note 2*).

3.1.1   Measurement of Urine Alcohol (*see Note 3*)

1. A urine sample is collected between 2100 and 0900 h in a 50-mL tube with 1 mL of mineral oil to prevent evaporation.
2. The cotton-tipped end of the collector is placed into the urine sample until fully soaked.
3. The collector is inserted into the entry port with gentle and steady pressure until the pink fluid flows past the QA Spot™ at the end of the QED device.
4. After the bar stops moving on the scale bar, the alcohol concentration of the sample can be read.

3.1.2   Assessment and Processing of Blood, Liver Samples, and Histological Sections

3.1.2.1   Processing of Blood, Liver Samples, and Histological Sections

1. Blood samples are incubated at room temperature for 30 min.
2. The samples are centrifuged at 17,000 g for 15 min at 4°C.

3. The upper layer of serum is collected and divided into aliquots and stored at −80°C.
4. The liver tissues are divided into two parts: one portion for snap frozen and stored at −80°C, and another portion is cut into tiny blocks for fixation in formalin.

3.1.2.2   Tissue Fixation

1. Tissue blocks are incubated in 4% neutral buffered formalin for 3 d.
2. The tissues are washed in running water for 3 h.
3. The tissues are immersed in 70% ethanol for overnight.
4. The fixed tissues are dehydrated in ascending percentage of ethanol (80% ethanol for 30 min; 90% ethanol for 30 min; 95% ethanol for 30 min at three exchanges; 100% ethanol for 45 min at three exchanges).
5. The dehydrated tissues are immersed in chloroform overnight at room temperature.
6. Wax is infiltrated into tissues in two different wax-baths with each exchange for 1 h.
7. The tissues are immersed in another wax-bath for another hour under vacuum.
8. The tissues are embedded in paraffin wax.

3.1.2.3   Tissue Sectioning

1. The tissues embedded in paraffin blocks are sectioned at a thickness of 5 μm.
2. The tissue sections are flattened in a 40°C waterbath.
3. The tissue sections are mounted on a clean glass slide coated with poly-L-lysine.
4. Tissue sections are dried in a 60°C oven to melt the excess paraffin wax and transferred to a 37°C oven for drying overnight.

3.1.2.4   Hematoxylin and Eosin (H&E) Staining

1. The paraffin sections are dewaxed and rehydrated by immersing the sections in toluene (twice for 5 min each time).
2. The sections are immersed in a series of descending percentages of ethanol (100% ethanol for 1 min at two exchanges; 95% ethanol for 1 min at two exchanges; 70% ethanol for 1 min).
3. The rehydrated sections are washed in running tap water for 5 min and then immersed in distilled water.
4. The rehydrated sections are stained with Harris's hematoxylin for 5 min and then washed in running tap water for 2 min.
5. The overstained hematoxylin sections are differentiated by a quick bath in acid alcohol followed by Scott's tap water for 1 min.
6. The cytoplasm of the cells in the tissues is stained with 1% eosin solution for 3 min and then is given a quick wash in tap water.

7. The sections are dehydrated in ascending percentages of ethanol (70% ethanol for 10 s; 95% ethanol for 15 s; 95% ethanol for 1 min; 100% ethanol for 2 min at two exchanges; 100% ethanol for 5 min).
8. The slides are transferred to toluene (twice at 5 min each) and then mounted with Permount medium.

3.1.2.5   Quantification of the Necrosis Using Image Analysis

1. The percentage of necrotic area was measured by the Leica QWIN Image Analyzer (Leica Microsystems Ltd., Milton Keynes, UK).
2. The H&E-stained sections were captured with a CCD JVC camera connected to a Zeiss Axiophot microscope at 20× objective.
3. One block per animal and one section per block were chosen for quantification. Five fields were chosen randomly and captured from each section.
4. The percentage of necrosis was expressed as the average of the results obtained from each group by dividing the sum of lightly stained necrotic areas with the sum of the reference field multiplied by 100 *(3)*.
5. The severity of liver pathology was assessed as steatosis (the percentage of liver cells containing fat): 1+, ≤25%; 2+, 26–50%; 3+, 51–75%; 4+, >75%. *(4)*.
6. Necrosis was evaluated as the number of necrotic foci per square millimeter, inflammation was scored as the number of inflammatory cells per square millimeter. At least three different sections were examined per sample of liver *(4)*.

3.1.2.6   Sirius Red Staining and Quantification of Collagen

The collagens are the basic components of connective tissue, and they can be visualized with Sirius Red staining (see **Note 4**).

1. The fixed and sectioned tissues are dewaxed and rehydrated.
2. The tissues are stained with 0.1% picro-Sirius Red in saturated aqueous picric acid for 1 h.
3. The tissue sections are then differentiated in 0.01% hydrochloric acid for 30 min.
4. The stained slides are immediately dehydrated and mounted.

3.2   Biochemical assay of ALT

The degree of necrosis in the liver tissue is assessed by the amount of serum level of ALT. ALT is a biochemical marker of the liver. A high level of circulating ALT indicates a high degree of liver necrosis and injury.

$$\text{Alanine} + \alpha\text{-ketoglutarate} \xrightarrow{\text{ALT}} \text{Glutamate} + \text{Pyruvate}$$

$$\text{Pyruvate} + \text{NADH} \xrightarrow{\text{LDH}} \text{Lactate} + \text{NAD}$$

1. 50 μL of serum is added to a cuvette containing 1 mL of ALT substrate.
2. The mixture is incubated at 37°C for 3 min in a thermostated cuvette compartment.
3. The basal rate of NADH consumption is recorded at the wavelength of 340 nm for 2 min.
4. The rate of utilization of NADH by serum alanine aminotransferase was initiated by adding the α-ketoglutaric acid (10 mM).
5. The change in absorbance was measured for a period of 2 min.
6. The activity of alanine aminotransferase was calculated by applying the following equation:

$$\text{ALT activity (U/L)} = [\text{Abs}_{340nm}/\text{min} \times 1000 \times d \times l] / 6.22 \times 0.05$$

where $\text{Abs}_{340}/\text{min}$ = change of absorbance at 340 nm per minute, 1000 = conversion factor, d = dilution factor, l = light path length of cuvette = 1 cm, 6.22 mM^{-1}cm^{-1} = extinction coefficient of NADH at 340 nm, 0.05 mL = volume of serum sample, and U/L = international unit/liter.

3.3 Markers of Oxidative Stress

3.3.1 Assay on Serum total 8-Isoprostane Level

The serum level of total 8-isoprostane was determined by a Competitive Enzyme Immunoassay kit (*see* **Note 5**).

1. Total (free and esterified) 8-isoprostane is obtained from the serum samples by mixing 3 μL of sample with 12 μL EIA buffer and then a 15 μL of 15% wt/vol potassium hydroxide (KOH) is added.
2. The mixture is incubated in a 40°C water bath for 1 h.
3. 30 μL of potassium dihydrogen phosphate (KH$_2$PO$_4$) (1 M) is added, followed by the addition of 90 μL of EIA buffer to a final dilution of 1:50.
4. To prepare the standards for assay, 100 μL of 8-isoprostane standard is transferred with an ethanol-equilibrated pipet tip to a clean Eppendorf as stock.
5. The stock is then diluted with 900 μL of ultrapure water to a final concentration of 5 ng/mL.
6. A series of eight standards is prepared by transferring 900 μL of EIA buffer to the tube labeled with #1 and 500 μL of EIA buffer to tubes #2 to #8.
7. 100 μL of standard from stock is transferred to tube #1 and mixed thoroughly.

8. A 500-µL aliquot is transferred from tube #1 to #2 and then from tube #2 to tube #3. This step was repeated until tube #8.
9 To wells marked with NSB (nonspecific binding) of the 96-well plate coated with mouse anti-rabbit IgG provided with the kit, 100 µL of EIA buffer is added and 50 µL of EIA buffer is added to wells labeled with B0 (maximum binding).
10. 50 µL of standards and samples are added to the wells accordingly.
11. To the wells labeled with total activity (TA) and blank (Blk), 50 µL of 8-isoprostane AchE tracer and 50 µL of 8-isoprostane antiserum is added, whereas only 50 µL of 8-isoprostane Antiserum is added to the wells labeled NSB.
12. The plate is then covered and incubated at room temperature for 18 h.
13. The content is then emptied and rinsed five times with wash buffer.
14. 200 µL of Ellman's Reagent is added to each well and 5 µL of tracer is added to, T. A.
15. The plate is then covered with a plastic film and allowed to develop in the dark with gentle agitation for 60 min.
16. The final absorbance at 405 nm is measured by a microplate reader.

The amount of tracer detected is inversely proportional to the amount of free 8-isoprostane.

$$\text{Absorbance} \propto [\text{Bound 8-isoprostane Tracer}] \propto 1/[\text{8-isoprostane}]$$

A standard curve is plotted with the percentage of tracer binding ($\%B/B_0$) against the concentration of 8-isoprostane. The percentage of tracer binding is obtained by the following equation:

$$\%B/B_0 = [(x - NSB)/B_0] \times 100$$

where x = Absorbance of individual standards or serum samples, NSB = nonspecific binding, and B_0 = maximum binding.

The concentration of total 8-isoprostane expressed as picogram per milliliter (pg/ml) was calculated by substituting the absorbance of each sample into the equation and determined the concentration by applying the standard curve.

3.3.2 Thiobarbituric Acid Reactive Substances (TBARS)

1. The liver tissue is homogenized in 5 mL of Tris-KCl buffer.
2. 1 mL of 20% trichloroacetic acid and 2 mL of 0.67% (w/v) thiobarbituric acid are added to 2 ml of liver homogenate.
3. The mixture is heated in boiling water bath for 10 min.
4. The mixture is centrifuged to precipitate the protein.

The absorbance of the supernatant at 530 nm is recorded by a spectrophotometer using a molar extinction coefficient of malonaldehyde of $1.56 \times 10^5 \text{cm}^2/\text{mmol}$.

3.3.3 Endotoxin

1. A blood sample is collected in endotoxin-free vials.
2. The blood is centrifuged at $400g$ for 15 min at 4°C.
3. The sample is diluted 1:10 in pyrogen-free water and heated at 75°C for 30 min to remove the inhibitors of endotoxin from plasma.
4. The sample is incubated at 37°C for 10 min with limulus amoebocyte lysate. (Limulus amoebocyte lysate test)
5. The substrate solution is added to the mixture and incubated for 20 min.
6. The reaction is stopped by adding 25% acetic acid.
7. The absorbance of the sample at 410 nm is read by a spectrophotometer.

3.3.4 Chymotrypsin (2)

1. Liver tissue is homogenized to obtain the cytosolic protein fraction. Protein sample of 20 to 200 µg and a peptide substrate concentration of 40 µM are incubated in 0.1 M Tris-HCl (pH 7.5) (total reaction volume is 0.2 mL) at 37°C with continuous shaking for 60 min.
2. The assay is stopped by adding 300 µL of stop solution.
3. The fluorescence of the leaving group, 4-amino-methyl-coumarin (AMC) for N-scuuinyl-Leu-Leu-Val-Tyr-7-amido-4-methylcoumarin (LLVY-AMC) and N-t-Boc-Leu-Ser-Thr-Arg-7-Amido-4-methylcoumarin (LSTR-AMC) is measured at the excitation and emission wavelengths of 390 nm and 440 nm. The fluorescence of β-Naphthylamine (LLE-NA hydrolysis) is measured at the excitation and emission wavelengths of 335 nm and 410 nm.

3.3.5 CYP2E1 (Aniline Hydroxylase Assay) (5,6)

1. Liver microsomes (25 µg) are incubated at 37°C for 1 h in CYP2E1 substrate buffer
2. The reaction is stopped by adding 90 µL of 40% trichloroacetic acid.
3. The samples are incubated on ice for 10 min and then centrifuged for 10 min.
4. 0.36 mL of the supernatant is mixed with 0.24 mL of 10% Na_2CO_3 and 0.36 ml of 2% phenol.
5. The mixture is incubated in the dark for 45 min.
6. The absorbance of the samples is recorded at 630 nm.
7. The activity of the samples is deduced from a standard curve.

3.3.6 Immunohistochemical Analysis

Protein expression of iNOS, eNOS, COX-2, and the formation of nitrotyrosine in the liver sections are detected by immunohistochemistry. A DAKO EnVision™+ System, Peroxidase (DAB) kit (DakoCytomation Denmark A/S, Glostrup, Denmark) is used.

1. The sectioned tissues are dewaxed by immersing the sections in toluene (5 min for two exchanges).
2 The tissue sections are rehydrated in a series of descending percentage of ethanol (100% ethanol for 1 min at two exchanges; 95% ethanol for 1 min at two exchanges; 70% ethanol for 1 min).
3. The sections are rinsed in running tap water.
4. Sections are then rinsed with TBS buffer.
5. Proteolytic digestion is carried out by incubating the sections in $0.05 M$ TBS with 0.1% Trypsin and 0.1% $CaCl_2$ for 5 min at room temperature.
6. Sections are washed with TBS three times.
7. Peroxidase is added to each section and incubated for 5 min and then washed.
8. Blocking solution of 10% v/v normal serum in BSA solution is applied to each section and incubated for 1 h at room temperature.
9. Primary antibody is added to each section except the negative control and incubated at 4°C overnight. (All the primary antibodies were diluted with 1% BSA to the optimum concentration. The primary antibodies and the dilutions used in immunohistochemistry are listed in Table 1).
10. Negative control sections are incubated with normal serum.
11. Sections are washed with TBS for three times, 5 min each.
12. Peroxidase-labeled polymer HRP is added to each section and incubated at room temperature for 30 min.
13. The unbound polymer on the sections is washed with TBS for three times, 5 min each.
14. DAB+ substrate-chromogen solution is added and allowed to develop.
15. The sections are then washed in distilled water and counterstained with Harris's Hematoxylin for 5 s.
16. The stained sections are dehydrated in ascending percentage of ethanol from 70% to 100% and finally in toluene.
17. The slides were mounted with Permount medium.

3.3.7   RT-PCR and Western Blotting of Tissue Samples

3.3.7.1   Semiquantitative Analysis on Gene Expression

3.3.7.1.1   Total RNA Extraction

1. Total RNA is extracted from the liver tissues by using the NucleoSpin RNA II extraction kit.
2. 10 mg of tissue is placed in 400 µl lysis buffer (Buffer RA1) containing 4 µl of β-mercaptoethanol.
3. The tissue is homogenized with a homogenizer on ice.
4. 250 µl of 95% ethanol is added to the lysate and mixed by vortexing and added to a NucleoSpin column.

5. The column with lysate is centrifuged at 13,000 rpm for 1 min and the flow through is discarded.
6. The DNA in the lysate is removed by adding 95 μl of diluted DNase-I reaction mixture and incubated at room temperature for 15 min.
7. The reaction is stopped by adding 500 μl of Buffer RA2 and then centrifuged at 13,000 rpm for 1 min.
8. After the flow through was discarded, the column is washed by adding Buffer RA3 twice (600 μL of RA3, centrifuged at 13,000 rpm for 1 min; 250 μL of RA3, centrifuged at 13,000 rpm for 2 min).
9. The RNA is then eluted twice by adding 30 μL of RNase-free water and stood at room temperature for 15 min.
10. The RNA is collected in the collection tube by centrifuging at 13,000 rpm for 1 min.
11. The concentration of the RNA samples is measured by Gene Quant II Spectrophotometer.
12. The RNA samples are stored at −80°C until use.

3.3.7.1.2 RT-PCR

This is a semiquantitative analysis method to study the expression of messenger RNA level of the genes by amplifying the complementary DNA made from the corresponding RNA. In this part of experiment, total RNA was extracted from the liver tissue and based on the RNA obtained, a complementary DNA was prepared for polymerase chain reaction. The manufacturing of the mRNA of a gene was caused by the trigger of that specific gene and the subsequent making of protein. Thus, the level of expression of mRNA revealed the transcription level of the gene.

3.3.7.1.3 Preparation of Complementary DNA From RNA

1. The complementary DNA is prepared based on the corresponding RNA by SuperScript™ First-Strand Synthesis System for RT-PCR kit.
2. For each sample, 2 μg of RNA is diluted with RNase-free water to a final volume of 10 μL in a microtube.
3. 1 μL of oligo(dT) and 1 μL of 10 mM dNTP are added to the tube.
4. The mixture is incubated at 65°C for 5 min.
5. The mixture is chilled on ice for 3 min.
6. 9 μL of reaction mixture (10× First-Strand RT Buffer, 2 μL; 25 mM MgCl2, 4 μL; 0.1 M DTT, 2 μL; RNaseOUT Recombinant RNase Inhibitor, 1 μL) is added to the tube and mixed thoroughly.
7. The tube is then incubated at 42°C for 2 min.
8. 1 μL (50 units) if SuperScript II RT is added to the mixture and incubated at 42°C for 50 min.
9. The reaction is terminated by incubating at 70°C for 15 min.
10. The final cDNA is stored at −80°C.

3.3.7.1.4 Polymerase Chain Reaction

After obtaining the complementary DNA, the genes of interest are amplified by polymerase chain reaction on the specific primers. Sense and antisense primers for the genes (GAPDH, iNOS, eNOS, COX-2, and tumor necrosis factor-α) are designed by the online software provided by Whitehead Institute for Biomedical Research and are manufactured by Invitrogen Inc. The detailed panel for reaction mixture and thermal cycle for PCR and the specific primers used in RT-PCR are shown in Tables 3 and 4.

The final PCR products are loaded to a 2% agarose gel containing 0.05 µg/mL ethidium bromide. The intensity of the PCR products was quantified by an image analyzer software Image J (National Institutes of Health, Bethesda, MD).

3.3.7.2 Protein Expression Analysis by Western Blotting

3.3.7.2.1 Total Protein Extraction

3.3.7.2.1.1 Cytosolic Protein Extraction

1. Liver tissue (100 mg) is placed in a 5-mL tube containing 500 µL of ice-cold Buffer 1
2. The tissue is then homogenized with a homogenizer and followed by centrifugation at 5,000 rpm for 10 min at 4°C.
3. The pellet is saved for nuclear protein extraction.
4. The supernatant is transferred to a new 1.5 ml tube and centrifuged at 13,000 rpm for 20 min at 4°C.
5. The supernatant is divided into aliquots and stored at −80°C.

3.3.7.2.1.2 Nuclear Protein Extraction

1. The pellet saved from cytosolic protein extraction is resuspended in 500 µL of Buffer 2.
2. The mixture is mixed thoroughly before centrifuged at 6000 rpm for 10 min at 4°C.
3. The pellet is resuspended in 400 µL of nuclear Buffer 3 and mixed.
4. The suspension is then rocked on a rotor for 30 min at 4°C.
5. The samples are then centrifuged at 13,000 rpm for 30 min at 4°C.
6. The supernatant is then transferred to a dialysis tubing and dialyzed against 500 mL of Buffer 4 for three changes with continuous stirring at 4°C overnight.
7. The nuclear protein is collected and stored in aliquots at −80°C.

3.3.7.2.1.3 Protein Assay

The concentration of the cytosolic and nuclear protein samples were measured by Bio-Rad Protein Assay Kit (Bio-Rad Laboratories, Inc., Hercules, CA).

1. Protein standard is prepared by dissolving BSA in distilled water.
2. Serial dilution is performed to give a set of protein standards (0 mg/mL, 0.015625 mg/mL, 0.03125 mg/mL, 0.0625 mg/mL, 0.125 mg/mL, 0.25 mg/mL, 0.5 mg/mL, and 1 mg/mL).
3. The concentrated (5×) protein assay reagent is diluted with distilled water to working concentration (1×) before used.
4. The protein samples are diluted with distilled water to an optimum concentration which lay within the range of the protein standards.
5. 20 μL of standards or samples are added to a well of a 96-well plate in duplicate followed by adding 200 μL of diluted protein assay reagent.
6. The absorbance of each well is measured by a microplate reader at the wavelength of 570 nm.

3.3.7.2.1.4 Measurement of Protein Expression With Western Blotting

Western blotting is the analysis of the expression of a protein by the interaction between the protein and the corresponding antibody.

1. A fixed amount of protein (Table 6) of each sample is diluted with distilled water to a final volume of 10 μL.
2. 10 μL of 2× sample buffer is added to each sample.
3. The mixtures are mixed and then boiled at 99°C for 5 min before chilled on ice.
4. SDS polyacrylamide separation gel is prepared for the separation of different sizes of protein from the total protein extract.
5. A layer of stacking gel is prepared and settled on top of the separating gel.
6. The gel is placed in a tank filled with tank buffer.

Table 6 Conditions for Western Blot Analysis

	iNOS	eNOS	COX-2	Nitrotyrosine
Protein	Cytosolic	Cytosolic	Cytosolic	Cytosolic
Amount of protein (μg)	100	40	50	40
Percentage of acrylamide gel	7%	7%	10%	10%
Blocking time (hr)	1 h	1 h	1 h	6 h
Primary antibody manufacturer	Transduction Laboratories	Transduction Laboratories	Cayman Chemicals	Zymed Laboratories
Primary antibody concentration	1:500	1:1000	1:500	1:1000
Source of primary antibody	Rabbit (Polyclonal)	Rabbit (Polyclonal)	Rabbit (Polyclonal)	Mouse (Monoclonal)
Secondary antibody concentration	1:2000	1:2000	1:2000	1:2000
Film exposure time (min)	10 min	5 min	10 min	5 min

7. The chilled samples are loaded onto the wells on the stacking gel and run down the gel at a constant current of 30 mA.
8. The separated proteins on the gel are transferred onto an Immun-Blot™ PVDF Membrane in transfer buffer at a constant voltage of 20 V at 4°C overnight.
9. The membrane is removed from the transfer set up and washed briefly in TBST and incubated in blocking buffer (5% powdered nonfat milk in TBST).
10. The membrane is washed briefly with TBST followed by incubation with a primary antibody against the specific protein for overnight at 4°C with gentle agitation.
11. The antibody bound membrane is washed thoroughly with TBST for five changes with each change for 5 min.
12. The membrane is then incubated with secondary antibody for 2 h at room temperature with agitation (Table 6).
13. The washing procedure is repeated after incubation with secondary antibody.
14. Excessive washing buffer is removed and 0.5 mL ECL™ Western Blotting Detection Reagents is added to each membrane and incubated at room temperature for 1 min.
15. After removing excessive ECL reagent, the membrane is covered with an X-ray film and exposed in a cassette in the dark room.
16. The exposed X-ray film is developed in developer and fixer solutions.
17. The intensity of the bands were measured and quantified by Image J (National Institutes of Health).

3.3.8 Analysis of the DNA-Binding Activity of the Transcription Factors (NF-κB and AP-1) With EMSA

The transcription of DNA into RNA requires the binding of transcription factors onto the specific site of the DNA. To assess the activity of these transcription factors, EMSA was performed using the Gel Shift Assay Systems.

1. Nuclear protein is obtained by the methods described in Western Blotting and followed by quantification of the amount of nuclear protein in the extract.
2. Specific consensus oligonucleotide of: NF-κB for the binding of the corresponding transcription factor is firstly labeled with $[\gamma^{-32}P]ATP$ through a phosphorylation reaction (2 μl of consensus oligonucleotide (1.75 pmol/μL), 1 μL of T4 Polynucleotide Kinase 10× Buffer, 1 μL of $[\gamma^{-32}P]ATP$ (3000 Ci/mmol at 10 mCi/mL), 5 μL of Nuclease-free water, and 1 μL of T4 Polynucleotide Kinase (5–10 U/μL).
3. The mixture is incubated at 37°C for 10 min.
4. The labeling process is stopped by adding 1 μL of 0.5 M EDTA and diluted with 89 μL of TE buffer.
5. The labeled oligonucleotide is then purified by passing through a Quick Spin Column for radiolabeled DNA purification and the unincorporated nucleotides are removed.
6. The buffer in the column is drained by gravity and then placed in a collection tube.

Table 7 Antibodies Used in Supershift Assay

Antibody	Source	Manufacturer
NF-κB p50	Rabbit (Polyclonal)	Santa Cruz Biotechnology, CA, USA
NF-κB p65	Rabbit (Polyclonal)	Santa Cruz Biotechnology, CA, USA

7. The column is centrifuged at 1100 g in a swing-bucket rotor for 2 min and the eluted buffer is discarded together with the collection tube.
8. The column is then placed in a new collection tube in an upright position.
9. The connection between the two tubes is sealed with a piece of parafilm.
10. The labeled probe is added to the tube carefully to the centre of the column.
11. The column is placed in a 15-mL centrifuge tube and centrifuged at 1100 g for 2 min.
12. The purified probe is collected as the eluate after centrifugation and stored at −20°C afterwards.
13. The radioactivity of the labeled probes is measured by mixing 1 μl labeled oligonucleotide with 1 mL of Aquasol in a vial.
14. The radioactivity is then measured by LS-6500 Multi-purpose Scintillation Counter.
15. DNA binding reaction is then performed by adding 24 μg of each nuclear protein sample with 2 μL of 10× Reaction buffer, 2 μL of salmon sperm DNA (ssDNA), and 1 μL of labeled oligonucleotide (at least 50,000 cpm/μl).
16. The mixture is brought to a brief vortex and incubated at room temperature for 20 min.
17. 2 μL of loading dye is added instead of protein sample as negative control.
18. The mixed samples together with the negative control are added to the wells of a 4% nondenaturing acrylamide gel, 6.8 mL 40% acrylamide solution and 10% ammonium persulgate and 49.7 mL of distilled water, the gel is pre-run at 100 V in 0.5× TBE buffer for 30 min).
19. The samples are run down the gel by applying an electric force at 100 V.
20. Supershift is done to verify the correctness of the binding of the nuclear protein in the experiment.
21. 24 μg of the nuclear protein is mixed with 3 μL of 10× Reaction Buffer, 2 μL of ssDNA, and 1 μL of antibody (NF-κB p50 or NF-κB p65; 200 μg/mL) against the target nuclear protein (Table 7).
22. The whole mixture is incubated at room temperature for 30 min before the addition of labeled oligomer and then incubated for another 20 min.
23. The supershift mixture is loaded to the gel with other samples.
24. The gel is then backed by a filter paper and wrapped with a piece of saran-wrap followed by drying in a gel dryer at 50°C for 30 min.
25. The signal of radioactivity of the binding is detected by exposing the gel to an X-ray film in a cassette at −80°C for 24 h.
26. The film is developed and the expression of the bands is quantified afterwards.

Acknowledgments The authors gratefully acknowledge the support of the National Institutes of Health (grant to A.A.N.) and grants from the Committee of Research and Conference Grants, The University of Hong Kong.

4 Notes

1. The fish oil diet is kept under nitrogen at 4°C to avoid auto-oxidation of the fatty acids. Auto-oxidation causes a foul odor to develop in the diet. To overcome the reluctance of the animals to feed on the fish oil diet, we flavored the diet with chocolate powder. The chocolate had no effect on the biochemical or pathological alterations induced by ethanol. However, other investigators, using the same dietary regimen, have not used the chocolate flavoring and have not encountered any problems.

2. Because ethanol is a dietary nutrient that displaces other nutrients as a source of calories, the nutrient variables must be rigorously controlled. The pair-feeding regimen used in the oral-ethanol feeding model allowed for isocaloric pair-feeding.

3. Blood ethanol levels are monitored by daily measurement of 24-h urine ethanol content. We have ascertained that urine alcohol levels correlate well with blood alcohol levels. There is no cycling of blood or urine alcohol levels (compare intragastric model).

4. The Sirius Red stained collagen was classified into three parameters: total collagen, collagen along the central vein, and collagen in the pericellular area. The collagen deposited along the central vein was that between the endothelium and the adjacent hepatocytes. The pericellular area collagen was the extracellular matrix accumulated in the hepatic sinusoids away from the central vein. One block was chosen per animal. One section was chosen per block. Five fields from each slide were selected randomly and captured with CCD JVC camera with a Zeiss Axiophot microscope. The collagen stained with Sirius Red was quantified using LEICA Qwin Image Analyzer (Leica Microsystems Ltd., Milton Keynes, UK). The percentage of total collagen was calculated by the sum of the Sirius Red positive area divided by the sum of the reference area multiplied by 100. Similar method was used in the evaluation of the percentage of the collagen in the central vein. By subtracting the average value of the two parameters, the percentage of the collagen in the pericellular region was obtained.

5. This assay is based on the competition between the free 8-isoprostane and the 8-isoprostane acetylcholinesterase (AChE) conjugate (Tracer) for a limited amount of isoprostane-specific rabbit antiserum binding sites.

References

1. Nanji, A. A., Zhao S, Sadrzadeh, S. M., Dannenberg, A. J., Tahan, S. R., and Waxman, D. J. (1994) Markedly enhanced cytochrome P450 2E1 induction and lipid peroxidation is associated with severe liver injury in fish oil-ethanol-fed rats. *Alcohol Clin. Exp. Res.* **18,** 1280–1285.

2. Donohue, T. M., Jr., Kharbanda, K. K., Casey, C. A., and Nanji, A. A. (2004) Decreased proteasome activity is associated with increased severity of liver pathology and oxidative stress in experimental alcoholic liver disease. *Alcohol. Clin. Exp. Res.* **28,** 1257–1263.

3. Nanji, A. A., Jokelainen K, Tipoe, G. L., Rahemtulla A, and Dannenberg, A. J. (2001) Dietary saturated fatty acids reverse inflammatory and fibrotic changes in rat liver despite continued ethanol administration. *J. Pharmacol. Exp. Ther.* **299,** 638–644.

4. Tipoe, G. L., Leung, T. M., Liong E, So H, Leung, K. M., Lau, T. Y., Tom, W. M., Fung, M. L., Fan, S. T., and Nanji, A. A. (2006) Inhibitors of inducible nitric oxide (NO) synthase are more effective than an NO donor in reducing carbon-tetrachloride induced acute liver injury. *Histol. Histopathol.* **21,** 1157–1165.

5. Nanji, A. A., Sadrzadeh, S. M., Yang, E. K., Fogt, F., Meydani, M., and Dannenberg, A. J. (1995) Dietary saturated fatty acids: a novel treatment for alcoholic liver disease. Gastroenterology. **109,** 547–554.

6. Imai, Y., Ito, A., and Sato, R. (1966) Evidence for biochemically different types of vesicles in the hepatic microsomal fraction. *J. Biochem. (Tokyo).* **60,** 417–428.

3
Intragastric Ethanol Infusion Model in Rodents

Hide Tsukamoto, Hasmik Mkrtchyan, and Alla Dynnyk

Summary Alcohol-associated life-style disease, as exemplified by alcoholic liver disease (ALD), is multifactorial with intricate interactions among genetic and environmental factors predicating individual predisposition. To experimentally dissect the interfaces of these interactions for better understanding of the pathogenesis, it is essential to have an animal model that provides maximal control over ethanol and dietary intake and that enables a precise addition or deletion analysis for a risk or protective factor of interest. Rodent intragastric ethanol infusion (IEI) model was developed two decades ago to meet this requirement. Work conducted with the model to date demonstrates the importance of both maximal ethanol intake and secondary risk factors in ALD. Mouse IEI model proved to be particularly useful for genetic analysis of the ALD pathogenesis and has the potential of producing synergistic pathologic outcome by combination of risk factors. The model is best used by alcohol researchers through a center-supported core facility and its tissue sharing program.

Keywords Alcoholic liver disease; risk factors; animal model; Tsukamoto-French model.

1   Introduction

An early impetus for creation of the rodent intragastric ethanol infusion (IEI) model was scientific desire to test whether sufficient ethanol intake results in significant alcoholic liver disease (ALD) in rodents that otherwise do not voluntarily consume adequate amounts of ethanol because of their natural aversion. An answer derived from the model for this important question was both yes and no. With maximal intake of ethanol reaching as high as 49% Cal, the rats develop severe hepatic steatosis and elevated ALT levels, even with a diet low in fat that does not cause these changes in the Lieber-DeCarli model *(1)*. An increase in the dietary content of polyunsaturated fat causes further progression to steatohepatitis *(2)*, and the addition of carbonyl iron *(3)* or enteral lipopolysaccharide administration *(4)* potentiates liver fibrosis (*see* **Note 1**). Thus, although maximal ethanol intake indeed cause more accentuated liver injury than that observed in other models, it alone is not sufficient for progression of ALD (*see* **Note 2**).

Daily determination of blood alcohol levels (BALs) in the rat IEI model that achieved maximal ethanol intake also disclosed three other novel findings. The first and most important was demonstration of cyclic BAL with a periodicity of 5 to 6 d and a magnitude between 50 and 450 mg/dL *(5)*. This pattern is observed regardless of whether a constant or variable dose of ethanol was infused. An intriguing aspect of this phenomenon is that it has a switch mechanism with the threshold BAL of approximately 250 mg/dL, above which ethanol clearance is induced and below which the clearance is suppressed *(5)*. The two additional findings were remarkable metabolic and physical tolerance of rats rendered chronic alcoholic by the IEI method. In adult rats, ethanol intake of 13 to 15.5 g/kg/d is shown to be tolerated well and, in growing rats (body weight less than 300 g), the dose can reach 18 g/kg/d. In adult mice, the dosage can be increased from 22.7 to 33 g/kg/d during a 4-wk period. Under these conditions, the animals develop marked physical tolerance that cannot be observed in other models. For instance, after 2~3 wk of ethanol infusion, BALs up to 250 mg/dL may not be associated with any gross physical signs of ethanol intoxication. With BALs between 250 and 350 mg/dL, only slight-to-moderate inhibition of motor activities can be observed. At the peaks of the cyclical BAL (400~500 mg/dL), the animals exhibit appreciably suppressed motor activity, particularly in the lower part of the body. Thus, in the IEI model, BAL between 250 and 350 mg/dL represents slight-to-moderate intoxication, and the levels between 400 and 500 mg/dL cause heavy intoxication. The reported BAL in the voluntary ethanol consumption models (150~200 mg/dL) are the concentrations that do not achieve any physical sign of intoxication or liver pathology in the IEI model.

These fundamental differences are very important in trying to set up or use the rodent IEI models. In the past, the authors have encountered several examples of unsuccessful reproduction of the liver pathology in the model by other investigators who in fact failed to challenge rats with sufficient alcohol intake because they believed that their measured BAL of ~250 mg/dL were sufficiently intoxicating.

The development of mouse IEI model was an important turning point in ALD research because it opened a door to genetic studies on ALD (*see* **Note 3**). Early work by the late Thurman's laboratory set a stage for this new era. They demonstrated the importance of endotoxin-stimulated Kupffer cells, their NADPH oxidase activation and generation of ROS, subsequent TNF-α induction, and proinflammatory responses in the ALD pathogenesis by using mice deficient in TNF receptor I, CD14, Toll-like receptor 4, LBP, p47phox, and ICAM-1 for the IEI model (see ref. **6** for review).

The primary purpose of this chapter is to provide methodological details concerning the rat and mouse IEI models. The original IEI method was described by the author in 1984 and it also included a description for venous catheterization for simultaneous repetitive blood sampling in addition to the intragastric infusion (**7**). It also reported the model's basic physiological parameters, including stress response. Since then, the IEI model has evolved in terms of both its method and applications, and a more comprehensive review of the IEI model for studies on ALD has previously been published (**8**). For the IEI model to be successfully established and used (*see* **Note 4**), the following three technical components need to be executed well: 1) catheter construction and surgery; 2) preparation of diet and ethanol/dextrose solution; and 3) diet/ethanol infusion and animal monitoring (*see* **Note 5**). The chapter will cover these components in a manner as detailed as possible within the allocated space.

2 Materials

2.1 *Intragastric Catheters and Infusion Set-up*

Materials required for construction of rat and mouse intragastric (gastrostomy) catheters are listed in Tables 1 and 2, including of their potential sources and catalog numbers. Similarly, materials needed for an infusion set-up are listed in Tables 1 and 2. For easy reference as to where these listed materials are used in the catheters and infusion set-up, the items are also identified by lower case alphabets in the last column of Tables 1 and 2 and three drawings for the rat and mouse catheters and infusion set-up in Fig. 1A-C. Photographs of the mouse IEI model and its set-up are also shown in Fig. 2A,B.

2.2 *Diet*

Lactalbumin hydrolysate is purchased from Invitrogen (cat. no. 11800–42). All other dietary components listed for low-, medium-, and high-fat diets in Table 3 are purchased from Sigma Chemical Co. (St. Louis, MO) except for the trace mineral mix (Dyets Inc., cat. no. 210090), vitamin mix (Dyets Inc., AIN-76

Table 1 Rat Intragastric Catheter and Infusion System: Materials and Vendors

Material	Description	Vendor (cat. no.)	Figure 1 A	C
Silasic® silicone tubing	ID0.030' × OD0.065"	Fisher Sci. (11–189–15C)	c	
Dacron felt		Boston felt (54–6-032)	e	
Rodent plastic swivel	21 gauge	Lomir Biomed. (RSP3)	a	a
Rodent Swivel tether connector		Lomir Biomed. (RST01)	b	
Stainless steel anchoring button assembly		Instech Lb. (625ss)	d,f	
Tubing connector	18 gauge	Small Parts (B-TC-18/3)		c
Intramedic® Luer stub adaptor	20 gauge	VWR (63019–841)		e
Monoject® 60-mL syringes	With Luer lock tips, 100/cs	VWR (82–002–318)		f
Tygon® plastic tubing	ID0.030' × OD0.090'	Fisher Sci. (14–170–15C)		b,d
RTV silicone rubber	Dow Corning (3110 RTV)	Motion Industries		
RTV catalyst #4	Dow Corning	Motion Industries		

Table 2 Mouse Intragastric Catheter and Infusion System: Materials and Vendors

Material	Description	Vendor (cat. no.)	Figure 1 B	C
Silasic® silicone tubing	ID0.025" × OD0.047"	Fisher Sci. (11–189–15B)	f	
	ID0.030' × OD0.065"	Fisher Sci. (11–189–15C)	c	
Dacron felt		Boston felt (54–6-032)	e	
Tygon® plastic tubing	ID0.020" × OD0.060"	Fisher Sci. (14–170–15B)	b	b
	ID0.030" × OD0.090"	Fisher Sci. (14–170–15C)		d
Stainless-steel wire T-340V	0.011" diameter × 30" length	Small Parts (GWX-0110–30)	d	
Precision swivel	23 gauge	Instech Lb. (375–22p)	a	a
Intramedic® Luer stub adaptor	20 gauge	VWR (63019–841)		e
Terumo® 10-mL syringe	With Luer lock tip	Fisher Sci. (22–272315)		f
Tubing connector	18 gauge	Small Parts (STCY-18–10)		c
RTV silicone rubber	Dow Corning (3110 RTV)	Motion Industries		
RTV catalyst #4	Dow Corning	Motion Industries		
Nylon tubing for catheter protection	ID3/32" × OD1/8"	Small Parts (3814–2)		

Fig. 1 (**A**) A drawing of a rat intragstric catheter and its components. Items a~f are identified in the last column of **Table 1**. g, skin of the dorsal neck; h, abdominal skin; i, abdominal muscles. (**B**) A drawing of a mouse intragastric catheter and its components. Items a~f are all identified in the last column of **Table 2**. (**C**) A drawing of infusion set-up. Items a~ d are identified in **Tables 1** and **2**

A.

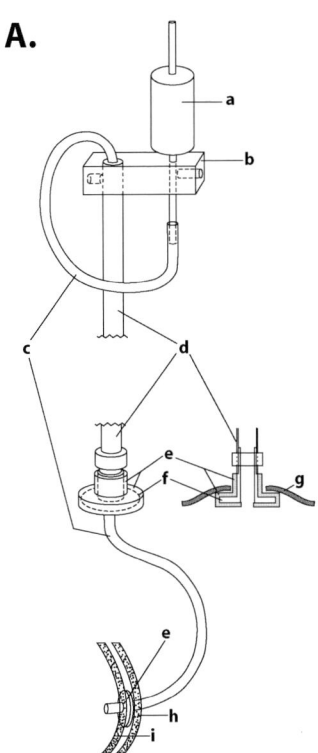

B.

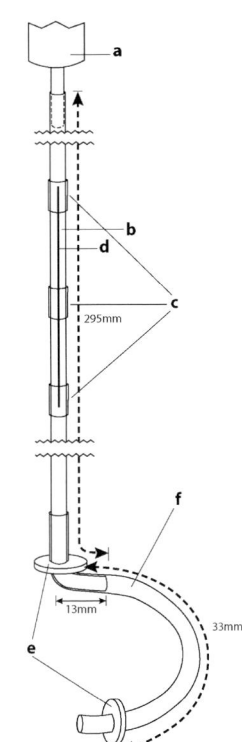

C.

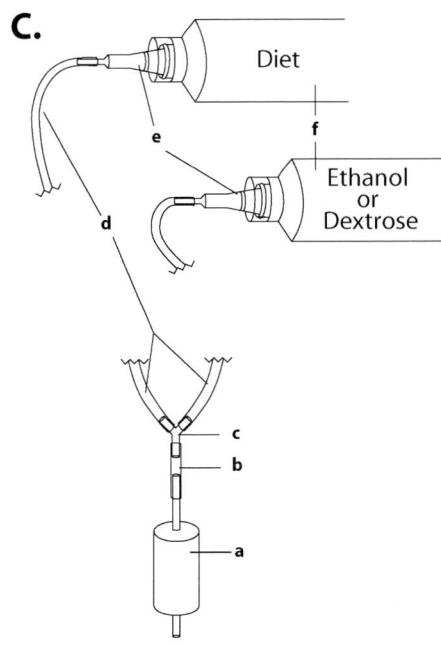

Fig. 2 (A) Overview of mouse IEI model set-up. **(B)** A close-up picture of an IEI mouse in a cage. The lower vertical portion of the catheter from the dorsal neck is protected by a nylon tubing segment as shown

Table 3 Diet Preparation

	For 3 L		
	Low fat	Medium fat	High fat
	(8%)	(25%)	(35%)
(A)			
Millipore water	1500 mL	1500 mL	1500 mL
Lactalbumin hydrolysate	276 g	276 g	276 g
(B)			
Millipore water	390 mL	390 mL	600 mL
Heat to dissolve dextrose			
Dextrose (D-glucose)	379.8 g	191.0 g	57.8 g
Citric Acid	6 g	6 g	6 g
Mineral			
1) KH_2PO_4 (17.325 g/dL)	72 mL	72 mL	72 mL
2) $CaCl_2$ $2H_2O$ (43.39 g/dL)	36 mL	36 mL	36 mL
3) NaCl (19.535 g/dL)	24 mL	24 mL	24 mL
4) $MgSO_4$ $7H_2O$ (8.88 g/dL)	12 mL	12 mL	12 mL
5) $MnSO_4$ (0.605 g/dL)	12 mL	12 mL	12 mL
6) KI (21.5 mg/dL)	12 mL	12 mL	12 mL
7) $(NH_4)_6Mo_7O_{24}4H_2O$ (12.5 mg/dL)	12 mL	12 mL	12 mL
8) $CuSO_4$ $5H_2O$ (1.22 g/dL)	12 mL	12 mL	12 mL
9) Ferric ammonium citrate (3.115 g/dL)	12 mL	12 mL	12 mL
10) ZnCl (255 mg/dL)	12 mL	12 mL	12 mL
11) Trace mineral mix (Dyet)	1.435 g	1.435 g	1.435 g
Cool down to dissolve vitamin			
Vitamin mix (Dyet AIN-76)	13.25 g	13.25 g	13.25 g
Choline chloride	2.65 g	2.65 g	2.65 g
(C)			
Corn oil	41.4 g	130.2 g	193.2 g
Xanthan Gum	4 g	6.1 g	8.5 g

cat. no. 300050), choline chloride (Dyets Inc., Bethlehem, PA cat. no. 400775), and 100% pure ethanol from Proof 200 (Proof 200, Gold Shield Chemical Co., Hayward, CA). In this chapter, corn oil (Mazola) is described as the source of polyunsaturated fat, which can be replaced with different types of fat. Macro- and micro-nutrient compositions of high fat diet are listed in Table 4.

3 Methods

3.1 Catheter Construction

3.1.1 Rat Intragastric Catheter

1. Cut silicone rubber (Silastic) tubing (ID0.03" × OD0.065", "c" in Fig. 1A) to a length of 60 cm.

Table 4 Diet Composition of High-Fat Diet

Lactalbumin hydrolysate	92 g/L
Dextrose	19.3 g/L
Corn oil	64.4 g/L
Citric acid	2 g/L
Choline chloride	0.88 g/L
Vitamin mix (AIN 76)	4.42 g/L
Mineral	
Calcium	1.422 g/L
Phosphate	0.948 g/L
Sodium	0.615 g/L
Magnesium	0.035 g/L
Manganese	7.87×10^{-3} g/L
Iodine	7.64×10^{-4} g/L
Molybdenum	3.88×10^{-5} g/L
Cupper	0.0124 g/L
Iron	~0.0247 g/L
Zinc	4.89×10^{-3} g/L
Trace element mix	0.957 g/L
Selenium	6.46×10^{-5} g/L
Silicon	4.12×10^{-3} g/L
Chromium	4.30×10^{-4} g/L
Lithium	4.30×10^{-5} g/L
Boron	2.15×10^{-4} g/L
Fluoride	4.33×10^{-4} g/L
Nickel	1.65×10^{-4} g/L
Vanadium	4.33×10^{-5} g/L

2. Cut a Dacron felt disk with a 6-mm diameter ("e" in Fig. 1A). Make a small hole in its center, insert the above Silastic tubing, and glue it at 5~7 mm from the distal tip of the tubing using RTV silicone rubber and catalyst #4.
3. Cut Dacron felt in corresponding shapes with slightly larger sizes to cover both sides of the flange and the stem of a stainless steal anchoring button ("f" in Fig. 1A). Glue the felt together with RTV silicone rubber and catalyst #4.
4. Connect a protective spring coil ("d" in Fig. 1A) to the anchoring button at its distal end and to the swivel tether connector ("b" in Fig. 1A) at its proximal end. Connect a flow-through swivel ("a") to the connector.
5. Pass the proximal end of the Silastic tubing through the Dacron felt-covered button, the spring coil, and the connector and connect it to the swivel as shown in Fig. 1A.

3.1.2 Mouse Intragastric Catheter

1. Cut Tygon plastic tubing (ID0.020" × OD0.060": "b" in Fig. 1B) to a length of 295 mm.

2. Place one end of the Tygon tubing into boiling water to bend it at 13 mm from the end with an angle of 90~100°.
3. Cut Silastic tubing (ID0.025" × OD0.047": "f" in Fig. 1B) to a length of 48 mm.
4. Expand one end of the Silastic tubing by dipping in chloroform for ~15 s and insert one end of the Tygon tubing into the expanded lumen of the Silastic tubing so that the length of the Silastic tubing is 33 mm between its distal tip and the angle of the catheter (*see* Fig. 1B). As chloroform evaporates, the Silastic tubing shrinks to its original size resulting in a tight connection. Note this 33-mm length is for a 8~10-wk-old, 25-g mouse and needs to be adjusted according to the animal size.
5. Cut stainless-steel stiffening wire to a length of 140 mm ("d" in Fig. 1B) and cut three 8 mm-length segments of Silastic tubing (ID0.030" × OD0.065" ("c" in Fig. 1B).
6. Expand the three Silastic tubing segments in chloroform and slip them over the Tygon tubing from its proximal end so that they are positioned in the middle of the long vertical portion of the catheter. Slide the stiffening wire between the Tygon tubing and the three Silastic tubing segments as shown in Fig. 1B.
7. Cut one Dacron felt disk with a 7-mm diameter and another with a 5-mm diameter ("e" in Fig. 1B). Make a small hole in their centers and insert the distal end of the catheter so that the larger disk is positioned just above the angle and another disk at 2~3 mm from the distal tip. Glue both disks at the respective locations using the silicone rubber and catalyst.
8. Cut 5 cm of nylon tubing (ID3/32" × OD1/8") and slide it over the catheter from the proximal end. This tubing serves to protect the lower vertical portion of the catheter from chewing.
9. Connect the proximal end of the catheter to the swivel ("a" in Fig. 1B).

3.2 Infusion Set-up

1. Connect the swivel to the 18-gauge Y-tube connector ("c" in Fig. 1C) using a Tygon tubing ("b" in Fig. 1C). This Tygon tubing has a smaller diameter for the 18-gauge Y connector but it will expand to fit. If needed, dip it briefly in boiling water.
2. Use 60-mL and 10-mL syringes ("f" in Fig. 1C) for infusion for rat and mouse models, respectively.
3. Connect two ends of the Y connector to diet and ethanol/dextrose syringes by the Tygon tubing ("d" in Fig. 1C) and 20-gauge Intramedic luer stub adaptors ("e" in Fig. 1C).
4. Set the diet and ethanol/dextrose syringes onto infusion pumps. We use the PHD 2000 infusion pump from Harvard Apparatus. Other pumps will suffice as long as they deliver steady and accurate infusion.

3.3   Diet Preparation

For the rodent IEI models, soluble and particulate-free diet is required. For this reason, we use highly soluble lactalbumin hydrolysate and dextrose as protein and carbohydrate sources, respectively as shown in Table 3. We also prefer to use "homemade" mineral stock solutions to maximize the solubility, but a commercially available highly soluble mineral mix (Dyets Inc. cat. no. 210061) also may be used. As examples, diets containing three different levels of polyunsaturated fat as corn oil are shown in Table 3. Corn oil can be replaced with a different type of fat as long as it does not result in a solidifying effect at the room temperature. Macro-nutrient, vitamin and mineral compositions of the high fat diet are shown in Table 4.

1. In a beaker A, dissolve 276 g of lactalbumin hydrolysate with 1500 mL of Millipore water on a stirrer.
2. In a beaker B, dissolve 379.8, 191.0, or 57.8 g of dextrose with 390, 390, or 600 mL of Millipore water for low-, medium-, and high-fat diet using low to medium heat.
3. Dissolve 6 g of citric acid in the beaker, B.
4. Make 10 mineral stock solutions as listed in Table 3 and add indicated volumes of these stock solutions to the beaker B in a sequence shown in the table as being stirred on a stirrer. Add the tracer mineral mix and observe all minerals dissolved completely.
5. Cool down the beaker B solution to 38°C. Dissolve vitamin mix and choline chloride. The binder in the vitamin mix may not dissolve completely. This will be filtered out before blending.
6. Place both beaker A and B in a refrigerator overnight to observe no precipitations.
7. Weigh 41.4, 130.2, or 193.2 g of corn oil (100% corn oil, cholesterol free) and 4, 6.1, or 8.5 g of Xanthan Gum for the low-, medium-, or high-fat diet, respectively.
8. Measure the total volume of the solution A and, B. As you measure the solution B, filter it using a 70 μm of Spectra nylon mesh.
9. Subtract the volumes of A and B and the volume of corn oil (shown in Table 3) from 3000 to calculate a volume of additional water needed to be added.
10. Blend together the solution A, additional water, and corn oil in a commercial-use blender (Waring Products. Model 38BL61, Dynamic Corporation of America, New Hartford) at a high speed for 30 s.
11. Add approximately one third of Xanthan Gum and blend for 30 s, and another one third for 30 s.
12. Add the solution B and the last third of Xanthan Gum into the blender and blend for 1 min.
13. Slowly vacuum-filter the blended diet through a 70-μm Spectra nylon mesh.
14. Refrigerate the diet overnight, aspirate and discard a top form layer. Aliquot the diet into dated bottles for immediate freezing and storage at −20°C.

3.4   Surgical Procedures

3.4.1   Rat Surgery

1. Anesthetize a male Wistar rat (body weight 350–375 g) by intraperitoneal injection of ketamine (80 mg/kg) and xylazine (4 mg/kg).
2. Prepare surgical areas (mid-abdomen and dorsal neck) for aspectic surgery by clipping hair and swabbing with iodine solution. Place the animal on the heating pad to prevent hypothermia. Protect the surgical field by wrapping the animal with Steri-Drape™ (3 M). Use autoclaved surgical instruments, towels, and gowns. Gas-sterilize the catheters. Wear a sterile gown, a cap, a mask, and sterile gloves.
4. Make a ventral midline incision in the skin of the abdomen from the xiphoid cartilage extending to the midabdomen.
5. Make another skin incision on the dorsal midline from the base of the skull to the midscapular region.
6. Use a hemostat to create a subcutaneous tunnel from the ventral abdominal incision to the dorsal neck incision. Use the same hemostat to bring the distal tip of the catheter from the dorsal neck to the abdominal incision through the created tunnel.
7. Make a ventral midline incision on the linea alba to open the abdominal cavity.
8. Make a small incision in the abdominal muscle at 2 cm to the right of the midline incision. Through this incision, bring the catheter into the abdominal cavity.
9. Expose the stomach, retract the great curvature of the forestomach portion, and block it with gauze wet with saline.
10. Make a small hole through the forestomach wall and insert the catheter tip into the level of the Dacron disk.
11. Use 5-O silk to make several stitches to secure the Dacron disk to the stomach wall.
12. Replace the stomach to the original position in the abdominal cavity as the catheter is pulled from the right side of the abdominal wall.
13. Close the peritoneal cavity with 3-O chromic gut and the skin with 3-O nonabsorbable suture such as Braunamid (B Braun).
14. Place the rat in a prone position. Suture the anchoring button to the muscles of the dorsal cervix and close the skin using 3-O Braunamid.
15. Inject Buprenex (0.04 mg/kg, subcutaneously) as an analgesic postoperatively and monitor the recovery from anesthesia.
16. Place the rat in shoe-box or metabolic cage and clamp the swivel above the cage to allow free movement of the animal.

3.4.2   Mouse Surgery

The surgical procedure for a mouse is essentially same as that described for the rat except for the following differences:

1. Use the ketamine and xylazine at the dosage of 80 mg/kg and 10 mg/kg, respectively.
2. Use 6-O silk for all sutures, including stitches for the Dacron disks, for closing the peritoneal cavity, and for skin incisions.

3.5   Diet and Ethanol Infusion

3.5.1   Rat Intragastric Infusion

1. Provide drinking water but remove chow.
2. Set the diet and ethanol/dextrose infusion rate at 120 mL/kg/day.
3. Start diet and dextrose (13.5% w/v) a day after the surgery and continue this infusion for at least 4 d before initiating ethanol infusion.
4. Make stock 95% ethanol solution from 100% pure ethanol to make different concentrations of ethanol.
5. Increase the concentration of ethanol solution to achieve the incrementally increasing dose of ethanol as summarized in Table 5A. The concentrations of dextrose solutions to achieve parallel pair-feeding are also shown in the same table.
6. Transfer an aliquot bottle of diet from a −20°C freezer to a refrigerator 24 h before feeding to allow slow thawing. Thaw out the amount of diet just enough for each day so that freshly thawed diet is always fed.

Table 5A Ethanol and Dextrose Dosage Schedule for Rat IEI Model

Ethanol dose (g/kg/day)	95% ethanol concentration[a] (% v/v)	Dextrose doncentration[b] (% w/v)	Days/Week
9	9.3	13.5	Day 1
9.5	9.8	14.2	Day 2
10	10.3	15	Day 3
10.5	10.8	15.7	Day 4–6
11	11.3	16.5	Day 7–9
11.5	11.8	17.2	Day 10–12
12	12.3	18	Day 13–15
12.5	12.9	18.7	Day 16–21
13	13.4	19.5	Week 4
13.5	13.9	20.2	Week 5
14	14.4	21	Week 6
14.5	14.9	21.7	Week 7
15.0	15.4	22.5	Week 8
15.5	15.9	23.2	Week 9[c]

Based on infusion rate of 120 mL/kg/day

[a] Ethanol Conc = ethanol dose (g/kg/day)/120 (mL/kg/day)/0.810 (density of 95% ethanol at 25°C) × 100 = ethanol dose × 1.0288.

[b] Dextrose Conc = ethanol dose (g/kg/day) × 7.2 (Cal/g for ethanol)/120 (mL/kg/day)/4 (Cal/g for dextrose) × 100 = ethanol dose × 1.5.

[c] Ethanol dose to be adjusted according to the tolerance of individual animals.

7. Fill a 60-mL syringe with 55 mL of diet and another 60 mL of syringe with 55 mL of ethanol or glucose solution. Set them on a same infusion pump set to infuse 120 mL/kg/day.

8. The next day, observe whether the anticipated volume of diet and ethanol/glucose has been infused. Record the remaining amounts and time.

9. Disconnect the Intramedic stub connector from the diet syringe, connect a 3-mL syringe filled with saline, temporally occlude the ethanol/glucose line by clamping it with a hemostat, and slowly flush the diet line with ~1 mL of saline. Remove clamping of the ethanol/glucose line.

10. Refill both the diet and ethanol/glucose syringes, connect them to the lines, and start infusion.

11. Repeat **steps 7~9** for each rat.

3.5.2 Mouse Intragastric Infusion

For mice, we use the diet and ethanol infusion rate of 8 mL/20 g body weight/day. The concentrations of ethanol and dextrose solutions to be used for different dosage levels are shown in Table 5B. The feeding technique is identical except for:

1. Feed regular chow for at least 2 d from the morning of the day after the surgery and begin infusing diet and dextrose from the third day to allow the smooth recovery of mice from a relatively larger loss of body weight (as compared with a rat) and stress associated with surgery.

2. Use a 10-mL syringe for both diet and ethanol/dextrose infusion.

3. Use a 1-mL syringe to slowly flush the diet line with ~0.3 mL of saline before filling the syringe with fresh diet.

3.6 Animal Monitoring

Monitoring animals is a very important technical component of the model, particularly for assessing the degree of intoxication. One may monitor daily urine alcohol concentration if an animal is housed in a metabolic cage (*see* **Note 6**). The authors have occasional observed adverse effects of long-term housing of the IEI models in the metabolic cage such as urinary bladder infection. For this reason, if daily assessment of urine alcohol levels is not absolutely required, the animals should be house in a regular housing condition with a bedding material which supports their normal behavior. The dosage schedules shown in Table 5A and 5B will keep animals within a desired range of alcohol intoxication. However, there is always individual variability in their metabolic and physical tolerance to alcohol during chronic feeding even among same litter mates. For this reason, careful assessment of animals is required in the morning and late afternoon. Because of the very high metabolic rate of mice, changes in their ethanol metabolism and intoxication take

Table 5B Ethanol and Dextrose Dosage Schedule for Mouse IEI Model

Ethanol dose (g/kg/day)	95% Ethanol concentration[a] (% v/v)	Dextrose concentration[b] (% w/v)	Days/Week
22.7	7.0	9.4	Day 1 and 2
24.3	7.5	10.1	Day 3 and 4
26.0	8.0	10.8	Day 5 and 6
27.5	8.5	11.5	Day 7 and 8
29.2	9.0	12.2	Day 9~14
30.9	9.5	12.9	Week 3
33.0	10.2	14.8	Week 4
35.0	10.8	15.7	Week 5***

Based on infusion rate of 8 mL/20 g BW/day.
[a] Ethanol Conc = ethanol dose (g/kg/day)/400(mL/kg/day)/0.810 (density of 95% ethanol at 25°C) × 100 = ethanol dose × 0.3086.
[b] Dextrose Conc = ethanol dose (g/kg/day) × 7.2 (Cal/g for ethanol)/400 (mL/kg/day)/4 (Cal/g for dextrose) × 100 = ethanol dose × 0.45.
[c] Ethanol dose to be adjusted according to the tolerance of individual animals.

place more dramatically in this species than rats, necessitating more frequent monitoring. Daily alcohol determination in urine samples can de performed by using a commercially available kit. Our center's animal core utilizes an automated analyzer for alcohol determination that requires only 5 μL of sample (Analox Instruments, Lunenburg, MA). For general physical assessment of the animal's state of alcohol intoxication during chronic ethanol feeding, the following criteria and manipulations may be helpful:

1. Intoxication Degree 0: Try to touch and pull a tail. An animal swiftly tries to escape you and assumes a defensive position toward you. BAL <200 mg/dL.
2. Intoxication Degree 1+: An animal is slightly slow to respond with slightly diminished motor coordination in lower extremities. The animal's cage may smell of alcohol. 50 mg/dL < BAL < 300 mg/dL.
3. Intoxication Degree 2+: Motor coordination of lower body is obviously impaired, but the mouse still can move around in a cage. 300 mg/dL < BAL < 400 mg/dL.
4. Intoxication Degree 3+: The mouse moves only its head and arms. 400 mg/dL < BAL < 450 mg/dL.
5. Intoxication Degree 4+: Anesthetic stage accompanied by a loss of whole body motor coordination. BAL > 450 mg/dL.
6. Intoxication Degree -1: Withdrawal stage. An animal is jumpy and sensitive to noise with a raised tail. BAL < 50~75 mg/dL.
7. One may remove ethanol solution from a syringe by 1~4 mL for a rat, 0.2~0.6 mL for a mouse for 2+~3+ intoxication.
8. For 4+ intoxication, stop both diet and ethanol infusion and withdrawal gastric content via the catheter using a 3 ml or 1 ml syringe for a rat or mouse, respectively.
9. For -1 intoxication (withdrawal), slowly inject 0.5 mL or 0.1 mL of ethanol solution for a rat or mouse, respectively. Repeat it 2~3 times with 15-min intervals until physical sings of withdrawal disappear.

10. If any manipulations of the ethanol infusion rate are needed, as described previously, treat corresponding control animals exactly same (remove or inject exactly same amounts of dextrose solution) to maintain perfect pair-feeding.

Acknowledgments The authors are thankful for the technical support provided by all animal core personnel in the past. This work was supported by P50AA11999.

4 Notes

1. The IEI model has also been used for studies on alcohol-induced damage of other organs such as kidney *(9)* and pancreas *(10)*. Lung, brain, heart, and bones are other organs that deserved to be investigated.
2. The IEI model can be used to induce hypercaloric alimentation and consequent obesity. The mouse IEI model has recently been used to produce obesity, insulin resistance, and non-alcoholic steatohepatitis *(11)*. The IEI model can easily be used to test synergistic effects of alcohol (ASH) and obesity (NASH) on liver pathology, as well as be adapted to combine both voluntary and forced feeding.
3. Continued use of the mouse IEI model for genetic studies on experimental ALD will be a major activity of our center's animal core that serves center members at USC, UCLA, and UCSD as well as investigators across the nation.
4. The tissue-sharing program of our animal core facilitates analysis of multiple parameters by different investigators in the same samples or in the same IEI model.
5. The center also provides hands-on training on the IEI model that usually entails spending 3~5 d to cover the aspects covered by this chapter.
6. A microsensor for alcohol determination, which is under development by several companies, may be incorporated into the model in near future to continuously monitor tissue alcohol concentrations.

References

1. Tsukamoto, H., French, S. W., Benson, N., Rao, G. A., Larkin, E. C., and Largman, C. (1985) Severe and progressive steatosis and focal necrosis in rat liver induced by continuous intragastric infusion of ethanol and low fat diet. *Hepatology* **5,** 224–232.
2. Tsukamoto, H., Towner, S. J., Ciofalo, L. M., and French, S. W. (1986) Ethanol-induced liver fibrosis in rats fed high fat diet. *Hepatology* **6,** 814–822.
3. Tsukamoto, H., Horne, W., Kamimura, S., Niemelä, O., Parkkila, S., Ylä-Herttuala, and Brittenham, G. M. (1995) Induction of liver cirrhosis in rats fed alcohol and iron. *J. Clin. Invest.* **96,** 620–630.
4. Mathurin, P., Deng Q-G., Keshavarian, A., Choudary, S., Holmes, E. W., French, S. W., and Tsukamoto, H. (2000) Exacerbation of alcoholic liver injury by enteral endotoxin. *Hepatology* **32,** 1008–1017.
5. Tsukamoto, H., French, S. W., Reidelberger, R., and Largman, C. (1985) Cyclical pattern of blood alcohol levels during continuous intragastric infusion in rats. *Alcoholism Clin. Exp. Res.* **9,** 31–37.

6. Wheeler, M. D., Kono H, Yin, M., Nakagami, M., Uesugi, T., Arteel, G. E., Gabele, E., Rusyn, I., Yamanashi, S., Froh, M., et al. (2001) The role of Kupffer cell oxidant production in early ethanol-induced liver disease. *Free Radic. Biol. Med.* **31,** 1544–1549.
7. Tsukamoto, H., Reidelberger, R. D., French, S. W., and Largman, C. (1984) Long-term cannulation model for blood sampling and intragastric infusion in the rat. *Am. J. Physiol.* **147,** R595–R599.
8. Tsukamoto, H. (1998) Animal models of alcoholic liver disease, in McCullough AJ (ed). *Clinics in Liver Disease, Alcoholic Liver Disease.* WB Saunders Co, Philadelphia, pp. 739–751.
9. Smith, S. M., Yu, G. S. M., and Tsukamoto, H. (1990) IgA nephropathy in alcohol abuse. An animal model. *Lab Invest* **62,** 179–184.
10. Pandol, S. J., Periskic, S., Gukovky, I., Zaninovic, V., Jung, Y. Zong, Y., Solomon, T. E., Gukovskaya, A. S., and Tsukamoto, H. (1999) Ethanol diet increases the sensitivity of rats to pancreatitis induced by cholecystokinin octapeptide. *Gastrotenterology* **117,** 706–716.
11. Deng Q-G., She, H., Xiong, S., French, S. W., Koop, D., and Tsukamoto, H. (2005) Non-alcoholic steatohepatitis induced by overfeeding in mice. *Hepatology* **42,** 905–914.

4
A Practical Method of Chronic Ethanol Administration in Mice

Ruth A. Coleman, Betty M. Young, Lucas E. Turner, and Robert T. Cook

Summary Mice provide a useful model for the study of immune deficiency caused by chronic alcohol abuse. Their suitability is related to several factors, including in particular the extensive knowledge base in the immunology of mice already existing in the literature. Specific modeling of the immunodeficiency of the chronic human alcoholic requires that ethanol must be administered to the model for a significant portion of its life span. In mice, it has proven to be necessary to administer ethanol daily for up to 32 wk or longer to observe all the immune abnormalities that occur in middle-aged alcoholic humans. Such time spans are problematic with many of the common protocols for ethanol administration. It has been shown by others and confirmed by our group that the most practical way of accomplishing such long protocols is by administering ethanol in water as the only choice of water. Details of management of the chronic ethanol mouse colony are described here that are necessary for the success of such studies, including methods for initiating ethanol administration, maintenance of barrier protection, monitoring weight gain, strain differences and fetal alcohol exposure.

L. E. Nagy (ed.), *Alcohol: Methods and Protocols*
© Humana Press 2008

Keywords Alcohol; ethanol; mice; chronic alcohol abuse; immune deficiency, fetal alcohol exposure.

1   Introduction

It is well known that chronic alcoholics become immune deficient. They have up to a fourfold increased incidence of pneumonia with a resultant enormous medical cost and lost productivity *(1–3)*. Other infectious diseases also are increased, including tuberculosis (TB), which can be up to 15-fold increased, and a number of others, including hepatitis, C., bacterial peritonitis, diphtheria, meningitis, schistosome infections, bronchitis, salmonellosis, bacterial endocarditis, and listeriosis *(1,2)*. These increases are not confined to any one culture or ethnic group *(2)* and are clearly an accompaniment of long-term alcohol abuse. There is also an increased severity of the infections in these individuals, and mortality is slightly to greatly higher in the various infections.

Rodent models of the increased infectious disease of the alcoholic have produced data clearly demonstrating that under specific conditions of diet and length and level of exposures, alcohol also creates an increased mortality in rodents. Bacterial pneumonia, TB, listeriosis, salmonellosis, and sepsis (the latter after combined burn-alcohol exposures) have all been demonstrated *(2,4–10)*. Mice have several desirable characteristics of a useful experimental model of alcohol abuse, including the availability of many genetic variants and a well-studied immune system. However, in the chronic alcoholic, it is typical to require 20–30 yr of consistent abuse before the development of clinically apparent immunodeficiency. Therefore, it would be most desirable to have a murine model of ethanol administration that allows substantial and consistent ethanol exposure for a proportionate length of the life span. This time span of exposure in mice has proven difficult or impossible to attain with fully liquid diets containing ethanol, or by repeated ethanol gavage or injections. The use of ethanol in water coupled with normal chow diets is the most practical solution for long-term ethanol exposure in mice. It has been demonstrated by others *(11,12)* and confirmed by us *(15)* that this method of ethanol administration causes flattening of the normal diurnal variation of serum corticosterone. However, the method does not cause net elevation of corticosterone, and most importantly does not cause depletion of the corticoid-sensitive T- and B-cell precursor populations *(15)*. It is therefore demonstrably free of the systemic stress effects observed with other administration protocols, especially during acute and short-term exposures. In the following chapter, we provide a detailed methodology for the use of the ethanol in water protocol in mice.

2   Materials

1. Source of mice: Common mouse strains, including C57BL/6, Balb/c, and C3H, are obtained from the National Cancer Institute (NCI-Frederick, MD) or commercial sources as specific pathogen-free (SPF) mice. Most congenic, mutant, and knockout mice and their appropriate controls are obtained from commercial suppliers. In some cases, it is more economical or reliable to maintain an in-house breeding colony for strains that are used with some frequency but are expensive or difficult to obtain from commercial sources in sufficient numbers at one time.
2. Ethanol: Pharmaceutical grade ethyl alcohol (95%) is purchased from AAPER Alcohol and Chemical Company (Shelbyville, KY). Highly purified, 18 mOhm water is produced in house for diluting the ethanol and is the sole water source for control mice.
3. Feed and bedding: Feed and bedding are provided by the animal care research facility: Standard pelleted rodent chow (Harlan-Teklad NIH-31 Modified Mouse/Rat Diet-Irradiated) and Cellu-Dri cage bedding, recycled cellulose fiber (Shepherd Specialty Papers, Gurnee, IL) have been used in most of the authors' studies.
4. Cages and bottles: Equipment that has improved animal health and made the delivery of drinking water containing ethanol more reliable includes a caging system that allows for maximum housing density with excellent air filtration and accessibility (e.g., Thoren™ Caging System, Thoren Caging Systems, Inc., Hazelton, PA). Delivery of a liquid with reduced surface tension, such as 20% w/v ethanol in water requires a bottle with a pinhole opening to reduce leaking and the potential for ethanol-soaked bedding and hypothermic mice. The Thoren 9-oz. "ink" bottle with stainless-steel cap is satisfactory for this purpose.

3   Methods

3.1   Overview of Ethanol Administration Methods in Mice

Ethanol dosing can be accomplished by several methods, including gavage, intraperitoneal injection, surgical implantation of catheters for continuous delivery, use of liquid diets containing ethanol with carefully adjusted and matched nutritional components, or various concentrations of ethanol in water given as the only water source. This last method is the one that will be described in detail as a way of producing a chronic mouse model consuming up to 20% weight per volume (w/v) ethanol for periods in excess of 52 wk or, approximately 50% of the normal mouse life span.

The maintenance of progressively immunodeficient mice chronically consuming 10 to 20% weight/volume (w/v) of ethanol requires good animal care infrastructure. This includes infection control or barrier housing (*see* **Note 1 (13)**), well-trained animal care personnel for routine mouse care, in-house veterinarians, and one or two laboratory research staffers, whose responsibilities include daily attention to the animal colony. It is necessary to have someone from the research laboratory responsible for ordering the mice, starting and increasing ethanol dosages, checking animal weights, and in general observing the overall health of the mice on a daily basis. If the mouse population is intended for use by multiple investigators, laboratory personnel must also keep an inventory and the ethanol concentrations tolerated by various wild-type, congenic and mutant strains of mice.

3.2   Animal Management for Immunology Studies

3.2.1   Animals: Housing and Maintenance

Male and female SPF mice at 6–7 wk of age are obtained from one of several licensed commercial sources, the National Cancer Institute, or bred in-house. Use of female mice allows housing of four or five animals per cage in shoebox style caging in barrier facilities (these facilities will be discussed later). When male mice are used, only one or two can be housed together, because dominant males will often kill their inebriated and/or smaller cage mates.

Upon arrival, mice are placed on high resistance, distilled, deionized water demonstrated to be free of significant lipopolysaccharide contamination. Water from the same source is used to dilute the 95% ethanol for the experimental animals. Mice are acclimated for a period of 1-3 wk, allowing them to adjust to the facilities, water, and the 12-h light/dark cycle (6 am to 6 pm light/6 pm to 6 am dark) maintained in their rooms.

All mice during acclimation, and later in both water and ethanol experimental groups, receive a sterilizable pelleted rodent chow ad libum (Harlan/Teklad Mouse/Rat Irradiated Diet #7913), with a guaranteed analysis of not less than 18% crude protein, 6% crude fat, and 5% crude fiber. Food pellets are placed in the bedding as well as in the overhead bins to facilitate feeding. Very small or very drunk mice have problems with balance and may become too weak to eat from the overhead feeders. A few food pellets in the bedding allow these mice to feed, resulting in greatly improved weight gain and survival rates.

3.2.2   Troubleshooting

As mentioned previously, the low surface tension of ethanol can cause two potential problems. First, leaking or steady dripping will soak not only the bedding, but also the mice. It saturates the fur almost immediately and produces hypothermia that is very difficult to counteract. Use of the "ink" bottles and pinhole tops when possible, remedies this problem. If sipper tubes are all that are available, use ones with the smallest openings possible.

The second problem occurs with both water and ethanol, but results are magnified with ethanol for the aforementioned reasons. Mice have a natural tendency to pile and push their bedding around while building a nest. The absorbent bedding comes in contact with the bottle and the contents of the bottle are wicked out very quickly. In this case, the remedy is to use a deeper style of shoebox cage with only 1–2 inches of bedding in the bottom.

3.3   Barrier Facility

3.3.1   Immunodeficiency and Ethanol

Because of the immunodeficiency caused by consistent excessive ethanol ingestion, measures to minimize exposure to infectious agents are imperative. Therefore, all mice are housed in barrier facilities. All materials entering this barrier area must be autoclaved or sprayed with a disinfectant. Personnel entering this area are required to wear a gown, hair covering, mask, gloves and shoe covers. Any handling of mice for purposes such as cage changes, weighings, and injections, must be done in a biosafety hood. Once mice are removed from these barrier rooms, they may not be returned. Instead, they must go into a secondary clean area until euthanized.

3.3.2   Use of Infectious Disease Models

Separate infectious barrier areas are required for ethanol mice that have received infectious agents. For infectious barrier areas, all personal protective gowning must be removed and placed in a waste receptacle before leaving the room. Bottles, cages, and bedding must be wrapped before leaving the infectious room and taken to be autoclaved before being emptied. Waste is then ready for disposal; cages and bottles are ready for washing, drying, sterilizing, and returning to the individual rooms. Cages and bottles from infectious barrier rooms are specially marked and are never used in the clean barrier area or the general population rooms.

3.4   Chronic Administration of Ethanol to Mice

3.4.1   Labeling

Although many labeling schemes can be envisioned, it is imperative for long-term studies with many groups and cages to have sufficient details on the cage to allow laboratory personnel to rapidly determine the origin and exact experimental conditions of each group. Typical cage cards provide the following information for each cage of mice: supplier, date of birth, date received, mouse strain, room number, account numbers, principal investigator, contact person, and a phone number. In a facility handling groups of mice for several different investigators or for differing lengths of ethanol exposure or other conditions, cages need to be additionally labeled to distinguish different study groups. The following has worked well for many years in the authors' studies:

MS#___/FC__   OR   MS#___/ME__

Where MS# = Mouse Study #123 and F or M for male or female, and C or E for water control or ethanol Therefore, this label, MS#100/FC26, indicates the mice in this cage are female, controls in cage #26, part of Study #100. To further identify individual cages, color-coded labels for water and different concentrations of ethanol are affixed to each cage card. Blue indicates water, yellow indicates 10% w/v ethanol, green indicates 15% w/v ethanol, red indicates 20% w/v ethanol, and so on. This system simplifies administration of the proper fluids to several hundred cages of study mice and permits investigators from collaborative groups to identify their own conditioned mice as previously assigned on a secure web site.

If one laboratory oversees and supplies all water control and ethanol mice for several laboratory groups, a secure web site is an essential tool for dissemination of information about the mouse colony. The web site maintained by the authors contains such information as mouse strain, age, weeks on ethanol, number of mice in each group and location of mice. An additional section entitled "Delivery Schedule" gives detailed information for individual labs and has very specific information for picking up mice, and other more general information for all users.

A second log-in record is created to collate additional information. It is generated when an order for mice is placed. The MS# is assigned and all available information such as number of mice, strain name, investigator IDs, and expected arrival date is recorded. When mice arrive, the sheet is updated with date of birth, age on arrival, and "special" instructions/requirements of individual investigators are written in and highlighted. Other sections include dates for starting and ramping-up ethanol, final target ethanol concentration and, finally, dates that will correspond to the specific target number of weeks of ethanol for which this particular group of mice is scheduled.

This simple sheet and the order/billing statements from the animal care facility allows the mouse colony manager to break down mouse usage numbers by individual investigator and to have rapid access to details of any one of a dozen or more concurrently running studies.

3.4.2   Ethanol Preparation and Administration

Mice at 6-7 wk of age are placed on distilled water and acclimated to their surroundings for a minimum of 1 wk. At 8-12 wk of age, depending upon the strain, mice are weighed individually or by cage groupings and divided into groups of approximately equal weight mice.

Table 1 Ethanol (EtOH) Preparation for Mouse Consumption

EtOH % W/V →	5%	10%	12%	15%	18%	20%
MLs of 95% EtOH	67	134	161	201	241	268
MLs of 18 mOhm H_2O	933	866	839	799	759	732
Total volume	1000	1000	1000	1000	1000	1000
MLs of 95% EtOH	134	268	322	402	482	535
MLs of 18 mOhm H_2O	1866	1732	1678	1598	1518	1465
Total volume	2000	2000	2000	2000	2000	2000
MLs of 95% EtOH	201	402	482	602	723	803
MLs of 18 mOhm H_2O	2799	2598	2518	2398	2277	2197
Total volume	3000	3000	3000	3000	3000	3000

Ethanol for the mice is calculated as a percent weight per volume or number of grams of ethanol per ml of the final fluid mixture. Pharmaceutical grade 95% ethanol is used as the source of ethanol to avoid the hygroscopic properties of absolute ethanol. It also avoids the possibility of contamination by drying agents used to produce absolute ethanol. Taking the density of absolute ethanol as 0.789, Table 1 was created and is posted for use by laboratory personnel preparing 1, 2, or 3 L of the most commonly used concentrations of ethanol. The volumes of 95% ethanol shown are rounded to the nearest whole ml and produce calculated concentrations within 0.1% of the stated value.

Administration of ethanol begins with a 7-d gradual increase in the percent of ethanol (phase-in or ramp-up). When mice are of the appropriate size, they receive 10% ethanol (w/v) for 2 d, 15% ethanol for 5 d, and then on day 8, they are increased to their maximum concentration. This final concentration is based on the ethanol tolerance of various mouse strains in the protocol at the time. If a previously untested mouse strain is introduced to ethanol, it is best to obtain a small pilot group and stretch the ramp-up phase out to 2 wk. During this time, the mice are kept at each level of ethanol for at least 4 or 5 d.

Observing and initially weighing the mice daily will help detect signs that the mice are not tolerating a certain percentage of ethanol. A mouse that has lost weight or has unkempt fur and a hunched back needs to have its ethanol concentration reduced for several days. Mouse growth rates may slow until they adjust to the taste and smell of the alcohol, but actual weight loss is rare. By carefully increasing the ethanol concentration, it is possible to find a level at which most mouse strains can be maintained in good physical condition.

3.5 Response to Ethanol in Several Mouse Strains

3.5.1 Weight Gain and Practical Starting Weights

As stated in the previous section, successfully administering ethanol to mice depends upon many factors, including strain, weight, and age. When first starting to work with chronic ethanol mice, it is important to weigh the animals at regular intervals. It is typical for most strains to slow or stop gaining weight during the first 2 or 3 wk of ethanol exposure, but weight gain will then resume. C57BL/6 will usually catch up to their control counterparts and may outweigh them after 6-8 wk on ethanol (Fig. 1). Balb/c will resume gaining weight at a rate parallel to the controls but will not completely make up the difference. C3H are most unusual in that the discrepancy between controls and ethanol mice is greater than for other strains. C3H tend to stay within 1–2 g of their starting weight even though their water control counterparts continue a robust weight increase. Despite the lower weights than controls these ethanol C3H mice remain vigorous and well groomed and actually appear more robust than the controls which become obese and tend to be inactive. A 2-wk pilot study successfully kept the water and ethanol C3H groups matched in

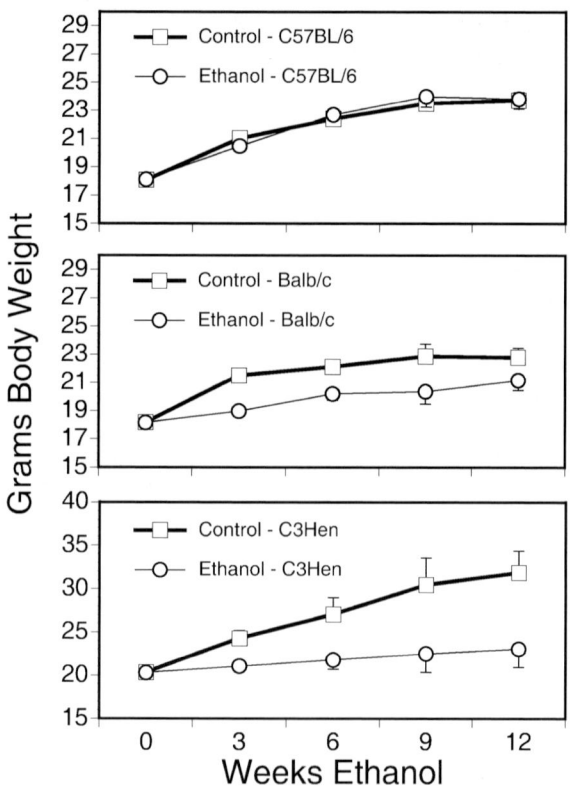

Fig. 1 Weight gains in three mouse strains ingesting 20% (w/v) ethanol as the only water source. The weights shown were obtained from large numbers of studies in which initial numbers were in the several hundreds (C57BL/6), and as groups were evaluated at different times, the ending numbers were substantially smaller. Ending numbers (12 wk ethanol) were C57Bl/6, 45 water and 35 ethanol; BALB/c, water 18, ethanol, 17; C3H, water 20, ethanol 26. C57BL/6 water vs ethanol was never significantly different in weight. Water vs ethanol mouse weights are significantly different in both BALB/c and C3H at various time points. See also Song et al. *(14)*

Table 2 Strain Tolerance for Maximum Sustained Ethanol Ingestion and Recommended Successful
Starting Weights

Strain	% W/V EtOH Maximum	Phase-up starting weight
C57BL/6 w/t (Both CD45.1 & CD45.2)	20%	17 g
B:6-Scid	20%	18 g
B:6-IFNγ-KO	18%	18 g
Balb/c w/t	18%	18–20 g
Balb/c-Scid	18%	19–20 g
C3Hen	20%	20+ g
C3HeJ	15%	20+ g

weight by pair feeding the water mice only that weight of chow eaten by the ethanol
mice on the previous day. Daily food weighing and adjustment for a longer chronic
study has not been evaluated. Table 2 shows the starting weights of several mouse
strains tested in our laboratory and the maximum ethanol concentrations practical
to achieve while maintaining good health of the strain.

3.6 Fetal Alcohol Exposure

Fetal alcohol exposure throughout gestation can be problematic in the mouse
because of fetal loss or abnormality when using high ethanol concentrations. In our
hands, a protocol that is manageable and produces some immune system abnormalities is as follows. Female breeders are kept on 5% (w/v) ethanol at least 2 wk
before breeding. When a male is placed in with the female, the concentration of
ethanol in water is increased to 10%. Because these matings are for timed pregnan-
cies, the males are removed at either 24 or 48 h (*see* **Note 2**). The 10% ethanol is
maintained throughout the pregnancy and when pups are born, the ethanol in the
mother's water is increased to 12% and maintained at this level until the pups are
weaned. When the ethanol females' pups are weaned those females are placed back
on 5% ethanol until the breeding cycle begins again. At the same time ethanol mice
are mated, an equal number of water control mice are mated. When pups are born,
some pups are euthanized, as necessary, to match litter sizes with the matched
water control females and their litters so that approximately equal nutrition is
obtained by ethanol pups and controls.

3.7 Long-Term Indicators of Health

3.7.1 Weight Gains

As indicated earlier, it is quite normal for mice to slow or stop gaining weight in
the first 1 or 2 wk of 20% ethanol after the phase up period. Significant actual
weight loss, however, requires modification until a well tolerated level is found.

3.7.2   Inebriation

Because mice are nocturnal animals, it is not unusual to find inebriated mice in the early morning hours. Mice will exhibit behaviors such as staggering, inability to stand on hind legs at the feeder, and even apparent loss of consciousness. At the end of the nocturnal feeding period (0600 h) blood alcohol levels may be as high as 400 mg/dL *(14)*. The mouse metabolism of ethanol is quite rapid, and it is not usually necessary to do anything special. If, however, cage mates are attacking the most inebriated mouse (especially in the case of males), they should be separated temporarily or permanently. In the case of one dominant or excessively aggressive animal, isolate the aggressor.

3.7.3   Barbering

Natural habits of mice may become magnified/intensified while on ethanol. C57Bl/6 are known to "barber" each other. This exaggerated grooming can lead to bald spots and in severe cases, open sores. If this happens, the injured mouse must be immediately isolated as the blood and open sore will invite more aggressive attacks, often leading to the death of the injured animal. Once the wound heals and hair regrows, the mouse can be returned to a group cage. Three barbered mice and one unbarbered mouse can be resolved by removing the dominant unbarbered mouse.

3.7.4   Changes in Response

Finally, it is important to be alert to sudden changes in a specific strain's response to ethanol. Suppliers occasionally change their source of breeding colonies for routine shipments and despite apparently equivalent genetics, mice of the same nominal strain from different colonies do not always respond in the same way to a given ethanol concentration. Resolution of this problem requires adequate source records to assure the exact breeding colony is known (*see* **Note 3**).

Acknowledgments   Work in the authors' laboratory has been supported by grant numbers AA09598 and AA014405 from the NIH, and the Department of Pathology of the University of Iowa.

4   Notes

1. It is recommended facilities be humidity controlled at $45 \pm 5\%$ relative humidity and $22 \pm 2°C$ temperature *(13)*.
2. Once a male has been exposed to ethanol, it is only used to mate with ethanol females. This eliminates concerns about a solely paternal effect of ethanol exposure transmitted through normal female matings at subsequent times.

3. This laboratory has experienced on one occasion a sudden and substantial deterioration of ethanol tolerance in new shipments of a mouse strain that had been used successfully for several years. Examination of the detailed shipping records kept for all mice in our laboratory permitted the observation that the supplier had changed the location of the breeding colony from which the mice were received. Restoration of tolerance was immediately achieved when the source was changed to the original breeding colony. It is likely that subtle genetic differences, of possible interest in their own right, exist between different breeding colonies from the same supplier, and requires vigilance by the laboratory group as to the suitability of the mice from different locations.

References

1. MacGregor, R. R., Louria, D. B. (1997) Alcohol and infection [review]. Curr. Clin. Top. Infect. Dis. 17, 291–315.
2. Cook, R. T. (1998) Alcohol abuse, alcoholism, and damage to the immune system—a review. Alcohol Clin. Exp. Res. 22, 1927–1942.
3. Happel, K. I., and Nelson, S. (2005) Alcohol, immunosuppression, and the lung. Proc. Am. Thorac. Soc. 2, 428–432.
4. Jerrells, T., and Sibley, D. (1995) Effects of ethanol on cellular immunity to facultative intracellular bacteria. Alcohol Clin. Exp. Res. 19, 11–16.
5. Nelson, S., Summer, W., Bagby, G., et al. (1991) Granulocyte colony-stimulating factor enhances pulmonary host defenses in normal and ethanol-treated rats. J. Infect. Dis. 164, 901–906.
6. Lister, P. D., Gentry, M. J., and Preheim, L. C. (1993) Granulocyte colony stimulating factor protects control rats but not ethanol-fed rats from fatal pneumococcal pnneumonia. J. Infect. Dis. 168, 922–926.
7. Bermudez, L. E., Petrofsky, M., Kolonoski, P., and Young, L. S. (1992) An animal model of Mycobacterium avium complex disseminated infection after colonization of the intestinal tract. J. Infect. Dis. 165, 75–79.
8. Mendenhall, C. L., Grossman, C. J., and Roselle, G. A. (1990) Host response to Mycobacterium infection in the alcoholic rat. Gastroenterology 99, 1723–1726.
9. Saad, A. J., Domiati-Saad, R., and Jerrells, T. R. (1993) Ethanol ingestion increases susceptibility of mice to Listeria monocytogenes. Alcohol Clin. Exp. Res. 17, 75–85.
10. Messingham, K. A., Faunce, D. E., and Kovacs, E. J. (2002) Alcohol, injury, and cellular immunity. Alcohol 28, 137–149.
11. Kakihana, R., and Moore, J. A. (1976) Circadian rhythm of corticosterone in mice: The effect of chronic consumption of alcohol. Psychopharmacologia 46, 301–305.
12. Sipp, T. L., Blank, S. E., Lee, E. G., and Meadows, G. G. (1993) Plasma corticosterone response to chronic ethanol consumption and exercise stress. Proc. Soc. Exp. Biol. Med. 204, 184–190.
13. Reeb-Whitaker, C. K., Paigen, B., Beamer, W. G., et al. (2001) The impact of reduced frequency of cage changes on the health of mice housed in ventilated cages. Lab. Animals 35, 58–73.
14. Song, K., Coleman, R. A., Zhu, X., et al. (2002) Chronic ethanol consumption by mice results in activated splenic T cells. J. Leuk. Biol. 72, 1109–1116.
15. Cook, R. T., Schluetter, A. J., Coleman, R. A., et al. (2007) Thymocytes, pre-B cells and organ changes in a mouse model of chronic ethanol ingestion. Absence of subset specific glucoeorticoid-induced immune cell loss. Alcohol Clin. Exp. Res. 31, 1746–1758.

Part II
Animal Models of Ethanol Exposure During Development

5
Analysis of Ethanol Developmental Toxicity in Zebrafish

Robert L. Tanguay and Mark J. Reimers

Summary It is largely accepted that vertebrates are more susceptible to chemical insult during the early life stage. It is implied that if a chemical such as ethanol is developmentally toxic, it must interfere with, or modulate, critical signaling pathways. The probable molecular explanation for increased embryonic susceptibility is that collectively there is no other period of an animal's lifespan when the full repertoire of molecular signaling is active. Understanding the mechanism by which ethanol exposure disrupts vertebrate embryonic development is enormously challenging; it requires a thorough understanding of the normal molecular program to understand how transient ethanol exposure disrupts signaling and results in detrimental long-lasting effects. During the past several years, investigators have recognized the advantages of the zebrafish model to discover the signaling events that choreograph embryonic development. External development coupled with the numerous molecular and genetic methods make this model a valuable tool to unravel the mechanisms by which ethanol disrupts embryonic development. In this chapter we describe procedures used to evaluate and define the morphological, cellular and molecular responses to ethanol in zebrafish.

Keywords Zebrafish; development; embryo; apoptosis; gene expression.

1   Introduction

Although the teratogenic properties of ethanol have been firmly established, the underlying mechanism(s) of toxicity remain unclear. Numerous model systems have been developed to help unravel the mechanisms by which embryonic ethanol exposure disrupts development, including rodent models *(1)*, larger mammalian models such as sheep *(2,3)*, and non-mammalian models, including the chick *(4,5)*, *Drosophila melanogaster* **(6,7)**, *Caenorhabditis elegans* **(8,9)**, and zebrafish *(10,11)*. Although each model system has unique advantages and disadvantages, zebrafish have properties that make them an important part of an overall integrated approach to study ethanol-mediated developmental toxicity. Like many models, much of the anatomy and physiology of fish is highly homologous to humans *(12,13)*. Zebrafish also possess all of the classical sense modalities, including vision, olfaction, taste, touch, balance, and hearing, and their sensory pathways share an overall homology with humans. Cognitive behavioral tests suggest that anatomic substrates of cognitive behavior also are conserved between fish and other vertebrates. Thus, similar to observations of hippocampal lesions in mammals, lesions of the structural homolog of the hippocampus in fish selectively impair spatial memory *(14)*.

Another major advantage of zebrafish is that the embryos develop externally and are optically transparent. Thus, using simple microscopic techniques, it is possible to resolve individual cells in vivo across a broad range of developmental stages. Resolution is increased by using transgenic zebrafish models that express fluorescent reporter genes *(15,16)*. Several additional features of zebrafish biology, including small size, rapid embryonic development, and short life cycle (reviewed in *(13,17,18)*), make this model system logistically attractive for developmental studies. Such features make genetic screens feasible to identify genes involved in normal development and in disease susceptibility *(19–21)*. High-throughput chemical screens also have proven feasible in zebrafish to identify novel developmentally active small molecules *(22–25)*. In addition, transient gene repression techniques using morpholinos (Gene Tools, Philomath, OR) are routinely used to rapidly evaluate the role of developmentally expressed genes *(26,27)*. Finally, because the zebrafish genome has been completely sequenced, it is now possible to integrate new discoveries at the biochemical, cellular, and molecular levels with observations at the structural and functional level.

Studies in zebrafish demonstrate that developmental ethanol exposure leads to craniofacial abnormalities, cardiac and structural malformations, and developmental delay similar to results observed in mammals *(11,28–32)*. The importance of ethanol biotransformation in these ethanol-dependent responses in zebrafish remains unknown; however, the alcohol metabolism enzymes that are expressed in mammals are also conserved in zebrafish *(33,34)*. Although the utility of zebrafish for developmental studies is well appreciated, few laboratories have exploited the full potential of zebrafish to help understand how ethanol exposure interferes with normal vertebrate development. In this chapter, we provide techniques that we have developed to evaluate the morphologic and cellular responses to ethanol in zebrafish embryos. Emphasis is placed on methods used to determine the ethanol dose and on evaluation of gene expression and cellular death.

2 Materials

1. Fish water: 0.3 g/L Instant Ocean salts (Aquatic Ecosystems Apopka, FL), in reverse osmosis water.
2. Incubator set at 28 ± 0.1°C.
3. Compound stereo microscope for viewing embryos (0.7X and 10X).
4. Dumont forceps No. 5 to manually remove chorions.
5. Inverted microscope with rhodamine and FITC filter sets equipped with digital camera (5–40X).
6. 3-aminobenzoate ethyl ester methanesulfonate salt (tricaine) anesthetic (Sigma A-5040) prepared at 4 mg/mL in water, pH adjusted to 7.0 with Tris-HCl, pH 9.0.
7. Alcohol dehydrogenase from *Saccharomyces cerevisiae*.
8. β-nicotinamide adenine dinucleotide (NAD, Sigma N6754; 1 mg/mL in 0.5 M Tris-HCl buffer, pH 8.8.)
9. Absolute ethyl alcohol USP, 200 proof.
10. 5 μg/mL acridine orange in fish water.
11. Glass exposure vials with Teflon-lined lids (~20 mL).
12. 0.7 % w/v low-melt agarose for mounting embryos.
13. 4% w/v paraformaldehyde in 1X phosphate-buffered saline (PBS).
14. Proteinase K stock solution 10 mg/mL in water (Roche cat. no. 3115836).
15. PBSTX: 1X PBS, 0.5% Triton X-100.
16. PBST: 1X PBS + 0.1% TWEEN-20.
17. 1X thioltransferase (TTase) equilibration solution: 1X TTase buffer, 2 mM CoCl$_2$, and dH$_2$O.
18. Terminal deoxynucleotidyl transferase (TdT)-mediated dUTP nick end labeling (TUNEL) reaction mixture (50-μL solution): 1X TTase buffer, 2 mM CoCl$_2$, 0.2 μL FL-dUTP (Roche, 11-373-242-910), 0.25 μL TTase enzyme (Roche, 3-333-566), and dH$_2$O.
19. Normal calf serum (NCS, Sigma, C-5280).
20. Phenol Red (0.2%) 0.1 g of phenol red (Sigma P5530) in 5 mL of dH$_2$O. Filter through 0.22-micron filter to remove debris that can clog micropipettes.
21. Penicillin (Sigma, P3032, and Streptomycin (S-9137).
22. Danieau's solution: 58 mM NaCl, 0.7 mM KCl, 0.4 mM MgSO$_4$, 0.6 mM Ca(NO$_3$)$_2$, 5 mM HEPES, pH 7.6.

3 Methods

This section describes how zebrafish embryos are exposed to ethanol during defined developmental windows. Because this model is a relatively new one for ethanol-mediated toxicity investigations, assays are described that broadly define the effects of ethanol exposure on cellular and embryonic morphology.

3.1   Zebrafish Husbandry

Adult AB strain zebrafish (*Danio rerio*) are raised and housed according to Institutional Animal Care and Use Committee protocols. Zebrafish are typically reared in 2.0-liter polycarbonate tanks on recirculating systems in which the water is maintained at 28 ± 1°C at pH of 7.0 ± 0.2. The fish are fed twice daily with either crushed TetraMin® Tropical Flake or live artemia from INVE (Salt Lake City, UT). We have found that a diverse diet improves egg production and quality. Adult male and female zebrafish are placed into spawning baskets in polycarbonate tanks the day before the embryos are needed. The following morning, newly fertilized eggs are collected, and embryos are rinsed several times in fish water. Normally dividing and spherical embryos between the 256-cell stage (2.5 hours post fertilization; hpf) and the oblong stage (3.7 hpf) are selected and utilized for all of the ethanol exposure studies.

3.2   Embryonic Ethanol Exposure and Endpoint Assessments

Embryos are waterborne-exposed to ethanol using a static method in which 30 embryos are exposed to ethanol in 20-mL glass vials sealed with Teflon®-lined lids to prevent losses by volatilization. The length and developmental stage of ethanol exposure can be carefully controlled as the embryos develop externally. At the end of the experimentally defined exposure period, the embryos are washed several times with warm (28°C) water and allowed to develop at 28°C in chemical-free water until endpoint assessments. For the acute toxicity studies described here, the daily noninvasively monitored developmental parameters include yolk sac edema, pericardial edema, axial edema/blistering, axial malformations, eye diameter, axial length, and overall developmental progression (Fig. 1).

3.3   Embryonic Ethanol Dose Determination

Because aquatic ethanol exposures are routinely done by adding ethanol directly to the water, it is important to determine the embryonic ethanol concentration after a given waterborne concentration. This information is necessary to permit comparison to dosing regiments in other model systems. To determine the embryonic ethanol dose, an enzymatic assay was developed that uses purified alcohol dehydrogenase (ADH) to convert whole embryonic tissue-derived ethanol as the substrate. The reduction of NAD is used to estimate the ethanol concentration.

1. Transfer 30 embryos (equals ~ 50 μL volume) into 1.5-mL microcentrifuge tubes on ice.
2. Remove ethanol solution and quickly wash embryos 2X with 150 μL of cold 3.5% v/v perchloric acid to remove residual ethanol.
3. Add ~150 μL of 3.5% perchloric acid.

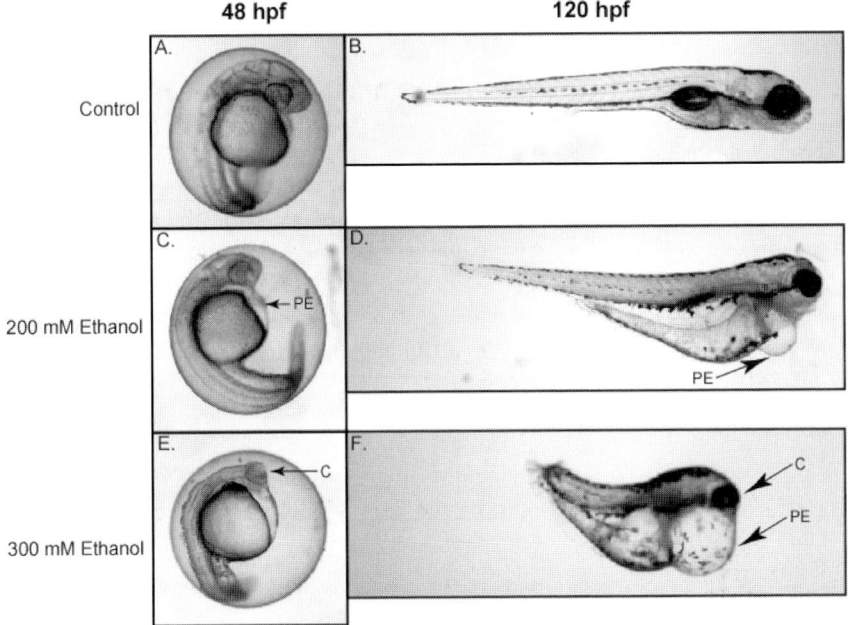

Fig. 1 Gross morphological responses to an ethanol exposure. Embryos were exposed to ethanol from 4 hpf until 24 hpf to the indicated water concentrations of ethanol. Images were acquired at 48 and 120 hpf (Cyclopia, C; Pericardial edema, PE)

4. Homogenize embryos with a disposable microcentrifuge pestle on ice.
5. Centrifuge the samples at 4°C for 10 min at 12,000 g.
6. Transfer the supernatant to a 1.5-mL microcentrifuge tube on ice.
7. Prepare ethanol standards 0 mg% to 1200 mg %.
8. Make a fresh yeast alcohol dehydrogenase solution (0.75 mg/mL in water).
9. The total reaction volume for this assay is 200 μL conveniently run in 96-well plates.
10. Add 174 μL of NAD solution to each well.
11. Add 8.7 μL of either a sample or standard to each well.
12. Initiate the reaction by adding 17.3 μL of a 0.75 mg/mL ADH solution to the well.
13. Incubate the reaction for 10 min at 37°C.
14. After the 10-min incubation, the production of NADH is measured @ 340 nM using a 96-well plate spectrophotometer.
16. Calculate the embryonic dose of ethanol using the standard curves.

3.4 Cell Death Assays

3.4.1 In Vivo Cell Death

We have recently demonstrated that embryonic ethanol exposure produces a concentration-dependent increase in embryonic cellular death (*32*). To visualize in vivo cell death, control and ethanol-exposed embryos are incubated with acridine orange, which preferentially intercalates the DNA of cells undergoing cellular death. This permits the fluorescent detection of cellular death in the live embryo. Because the acridine orange assay is not absolutely specific, as non apoptotic cells can be labeled, it is necessary to confirm the results using at least one additional method such as TUNEL which is described below.

1. After ethanol exposure, manually remove chorions using No. 5 Dumont forceps.
2. Incubated embryos with 5 µg/mL acridine orange in small glass vials for 1 h in the dark (*see* **Note 1**).
3. Remove the acridine orange solution and properly dispose this toxic solution as hazardous waste.
4. Rinse the embryos several times with fish water.
5. Euthanize the embryos with tricaine (0.04 mg/mL) before mounting
6. Mount stained embryos in by adding add a drop of 42°C 0.7% w/v low melt agarose to prewarmed microscope slides.
7. Using a wide-bore transfer pipet, place embryos on top of the agarose.
8. Position embryos using a fine-tip probe.
9. To quickly solidify the agarose, carefully move the microscope slide to a cooled surface.
10. Acridine orange positive cells are visualized using rhodamine filter sets.

Examples of ethanol-induced cellular death in live 24 hpf embryos are illustrated in Fig. 2.

3.4.2 Apoptotic Cell Death With TUNEL Analysis

The TUNEL analysis is more specific for the detection of apoptotic cellular death in whole zebrafish embryos.

Acridine Orange Staining

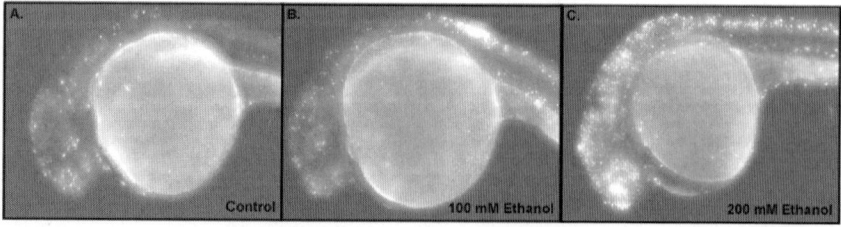

Fig. 2 In vivo cell death detection. Zebrafish embryos were exposed to the indicated concentrations of ethanol from 3 hpf to 24 hpf before acridine orange staining. In the absence of ethanol exposure (**A**), acridine orange-positive cells are detected. Ethanol exposure leads to an increase in acridine orange positive cells (**B** and **C**)

1. After ethanol exposure, manually remove chorions using No. 5 Dumont forceps.
2. Rapidly anesthetize with tricaine (0.04 mg/mL).
3. Fix embryos in 4% paraformaldehyde overnight at 4°C on rocker.
4. Place 5-10 embryos in 1.5-mL tubes and rinse 3X in PBSTx.
5. Incubate in 500 µL of proteinase K (1 µg/mL) at 37°C. Incubation time depends on the embryonic stage: 30 min for embryos 24-36 hpf, 45 min 48 hpf, and 60 min 60-72 hpf.
6. Briefly wash in PBSTx and then fix the permeabolized embryos in 4% paraformaldehyde for 10 min. This step helps to protect fragile tissues and completely inactivates the proteinase K.
7. Wash in PBSTx at least six times and let stand in PBSTx for 20 min.
8. Remove most of the PBSTx. Equilibrate embryos in 25 µL of 1X TTase buffer + CoCl$_2$+dH$_2$O solution for 15 min at room temperature (*see* **Note 2**).
9. Remove liquid and then incubate embryos in 25 µL of reaction solution on ice, in the dark, for 60 min then incubate an addition 60 min, in the dark, at 37°C.
10. Wash three times in PBSTx if mounting and viewing immediately. For longer storage, wash in PBST and store in PBST. Keep samples in the dark at all times.
11. Mount in 50% glycerol for immediate fluorescence microscopic imaging using fluorescein isothiocyanate (FITC) filter sets.

Examples of ethanol-induced apoptosis in 24 hpf embryos are illustrated in Fig. 3.

3.5 Global Gene Expression Analysis

With the completion of the zebrafish genome sequence, a number of commercial microarray platforms have been developed. We have used custom and commercial arrays to evaluate the impact of chemical and drug exposure on zebrafish

TUNEL Staining

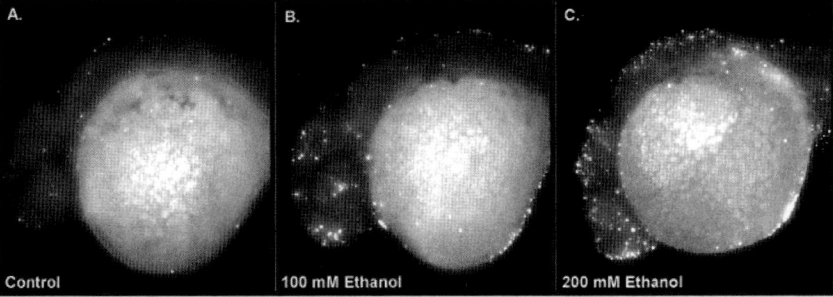

Fig. 3 Ethanol-induced apoptotic cellular death. Zebrafish embryos were exposed to the indicated concentrations of ethanol from 3 hpf to 24 hpf before *in situ* TUNEL detection of apoptotic cells

embryonic gene expression. The methods after RNA isolation used for microarray analysis are identical, regardless of the RNA source, so we will focus this section on the specific methods we used to collect total embryonic RNA for global gene expression studies. First, because embryos are small (approximately 1 mg total wet weight @24 hpf) it is necessary to pool embryos at each time point. It is critical to use developmental progression landmarks prior to pooling to avoid mixing embryos from different developmental stages. For developmental time points between 24 and 120 hpf, we routinely pool 30–40 embryos.

1. Transfer pooled embryos to 1.5-mL microcentrifuge tubes using wide-bore Pasteur pipets.
2. Rapidly anesthetize with tricaine (0.04 mg/mL).
3. Remove water and add 250 μL of Tri-Reagent (Molecular Research Center, Cincinnati, OH)
4. Immediately homogenize using a disposable plastic pestle for approximately 1 min.
5. Add an additional 750 μL of Tri reagent, mix and briefly sonicate using a probe sonicator.
6. Homogenates can either be stored @ −80°C or the RNA can be immediately isolated following the manufacturer's recommendations.
6. Resuspend the RNA pellets in 25 μL of nuclease-free water and quantify RNA.
7. Using this method we routinely obtain at least a 10 μg/RNA per sample, which is sufficient to quantify and generate hybridization probes (*see* **Note 3**).

After data analysis, the gene expression levels can be validated using standard methods such as northern blotting, quantitative PCR (qRT-PCR), RNAse protection assays, or western blotting. Because of the nearly transparent embryo, gene expression can also be temporally and spatially assessed using *in situ* hybridization and immunohistochemical methods.

3.6   *Temporal and Spatial Gene Expression Analysis*

The small size of the zebrafish embryo is a distinct advantage, as gene expression can be rapidly evaluated in the whole animal. Thus, investigators can assess the temporal and spatial distribution of cellular ethanol targets. mRNA expression patterns can be rapidly determined using *in situ* mRNA hybridization methods *(35)*. The Zebrafish Information Network is an outstanding resource for mRNA expression data as all reported zebrafish *in situ* hybridization data are collected and displayed (http://zfin.org/cgi-bin/webdriver?MIval=aa-ZDB_home.apg). It is, of course, preferable to measure the protein expression; however, this is only feasible when antibodies are available that recognize zebrafish proteins. To follow is a detailed method that we use to define protein expression patterns in embryonic zebrafish. For each wash step, we use 1 mL of the indicated reagent and place the tubes on their sides while on the rocker platform.

1. After ethanol exposure, manually remove chorions using No. 5 Dumont forceps.
2. Rapidly anesthetize with tricaine (0.04 mg/mL).
3. Fix embryos in 4% paraformaldehyde overnight at 4°C on rocker.
4. Place no more than 10 embryos in 1.5 mL.
5. Wash fixed embryos 2X, 20 min each with PBST rocker @ room temperature.
6. Wash with dH$_2$O 1h on a rocker @ room temperature.
7. Remove water and wash with −20°C 100 % Acetone for 15 min @ −20°C.
8. Remove acetone, and wash again with water.
9. Further permeabalized by adding freshly prepared collagenase (1 mg/mL, Sigma cat. no. C9891). The length of collagenase digestion is dependent on the developmental stage: 24 hpf, 10 min; 48 hpf, 30 min; 72 hpf, 45 min; 96 hpf, 60 min; and 120 hpf, 75 min.
10. Wash permeabolized embryos rinse 3X, 10 min each, in PBST.
11. Block embryos for 1 h in 10% NCS in PBST on a rocker @ room temperature.
12. Remove blocking reagent and add primary antibody diluted in 10% NCS in PBST. (The dilution factor must be empirically determined for each antibody).
13. Incubate overnight @ 4°C on a rocker.
14. Wash 4X, 30 min each in PBST @ room temperature.
15. Add fluorescently labeled secondary antibodies diluted in 10% NCS in PBST. For most supplied commercial antibodies such as Alexa-546 (Invitrogen, Carlsbad, CA) we use 1:500 to 1:2000 dilutions.
16. Incubate for 5 h @ room temperature or overnight @ 4°C on a rocker.
17. Wash 4X, 30 min each, with PBST.
18. Mount in 50% glycerol for fluorescence microscopic imaging using appropriate filter sets.

3.7   Transient Gene Repression Techniques

A powerful and rapid method to determine the role of a gene in a biological response, such as exposure to ethanol, is to block its developmental expression. This is conveniently accomplished using morpholinos (MO) which are short (typically 25) morpholino subunits. Each subunit is composed of a nucleic acid base, a morpholine ring, and a nonionic phosphorodiamidate intersubunit linkage. Morpholinos (Gene Tools, Philomath, OR) are designed to block translation initiation or RNA splicing. The key is to deliver the antisense reagent as early as possible following fertilization by microinjection, as homogenous distribution of the MO is necessary for effective expression knockdown. We only inject one- to four-cell stage embryos. The microinjection technique is identical to methods described for transgenic animal production (36) and will not be detailed here. An illustration of the MO technique is depicted in Fig. 4.

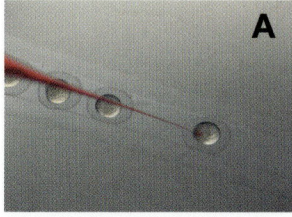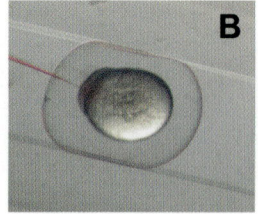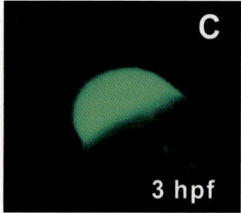

Fig. 4 Morpholino injection. (**A**) Single-cell stage embryos placed in channels before microinjection of morpholino with finely pulled needles. (**B**) Close-up image of a single one-cell embryo just after single injection with morpholino. (**C**) Typical fluorescent pattern in embryo 3 h after MO microinjection

1. Design and order targeting and control MO from Gene Tools.
2. We prefer to order 3'-fluorescein-tagged MO to permit assessment of microinjection success.
3. Dilute MO stocks at 25 mg/mL in 1X Danieau's solution.
4. MO injection solutions are prepared by adding 3 μL of 2% phenol red into 3 μl of the appropriate concentration of MO.
5. Load embryos into the agarose injection troughs as described *(36)*.
6. Inject approx 1–3 nL through the chorion directly into the cell or into the top of the yolk, just under the cells (yolk streaming will pull the MO into the dividing cells).
7. Remove embryos from injection plate and incubate @28°C in fish water containing penicillin (100 μg/mL) and streptomycin (84 μg/mL).
8. Screen embryos at 3 hpf for uniform MO distribution using fluorescence microscopy.

4   Notes

1. Incubate the embryos in glass vials and in the dark during the acridine orange treatment. The embryos become fragile during the treatment and adhere less to the glass than to the walls of plastic microcentrifuge tubes.
2. Make the equilibration and reaction buffers at the same time. However, do not add the FL-dUTP and enzyme to the solution until equilibration step is completed.
3. For microarray studies, it is strongly recommended that at least three independent biological replicates are performed for the RNA isolation (i.e., separate pools of embryos), which improves the confidence of the observed gene expression level determinations.

References

1. Becker, H. C., Diaz-Granados, J. L., and Randall, C. L. (1996) Teratogenic actions of ethanol in the mouse: a minireview. *Pharmacol. Biochem. Behav.* **55,** 501–513.

2. Brien, J. F., Clarke, D. W., Richardson, B., and Patrick, J. (1985) Disposition of ethanol in maternal blood, fetal blood, and amniotic fluid of third-trimester pregnant ewes. *Am. J. Obstet. Gynecol.* **152,** 583–590.
3. Richardson, B. S., Patrick, J. E., Bousquet, J., Homan, J., and Brien, J. F. (1985) Cerebral metabolism in fetal lamb after maternal infusion of ethanol. *Am. J. Physiol.* **249,** R505–R509.
4. Bupp Becker, S. R., and Shibley, I. A., Jr. (1998) Teratogenicicty of ethanol in different chicken strains. *Alcohol* **33,** 457–464.
5. Su, B., Debelak, K. A., Tessmer, L. L., Cartwright, M. M., and Smith, S. M. (2001) Genetic influences on craniofacial outcome in an avian model of prenatal alcohol exposure. *Alcohol Clin. Exp. Res.* **25,** 60–69.
6. Ranganathan, S., Davis, D. G., and Hood, R. D. (1987) Developmental toxicity of ethanol in *Drosophila melanogaster. Teratology* **36,** 45–49.
7. Ranganathan, S., Davis, D. G., Leeper, J. D., and Hood, R. D. (1987) Effects of differential alcohol dehydrogenase activity on the developmental toxicity of ethanol in *Drosophila melanogaster. Teratology* **36,** 329–334.
8. Dhawan, R., Dusenbery, D. B., and Williams, P. L. (1999) Comparison of lethality, reproduction, and behavior as toxicological endpoints in the nematode *Caenorhabditis elegans. J. Toxicol. Environ. Health A* **58,** 451–462.
9. Thompson, G. and de Pomerai, D. I. (2005) Toxicity of short-chain alcohols to the nematode Caenorhabditis elegans: A comparison of endpoints. *J. Biochem. Mol. Toxicol.* **19,** 87–95.
10. Laale, H. W. (1971) Ethanol induced notochord and spinal cord duplications in the embryo of the zebrafish, *Brachydanio rerio. J. Exp. Zool.* **177,** 51–64.
11. Reimers, M. J., Flockton, A. R., and Tanguay, R. L. (2004) Ethanol and acetaldehyde-mediated developmental toxicity in zebrafish. *Neurotoxicol. Teratol.* **26,** 769–781.
12. Ackermann, G. E., and Paw, B. H. (2003) Zebrafish: a genetic model for vertebrate organogenesis and human disorders. *Front. Biosci.* **8,** d1227–d1253.
13. Rubinstein, A. L. (2003) Zebrafish: from disease modeling to drug discovery. *Curr. Opin. Drug Discov. Devel.* **6,** 218–23.
14. Rodriguez, F., Lopez, J. C., Vargas, J. P., Broglio, C., Gomez, Y., and Salas, C. (2002) Spatial memory and hippocampal pallium through vertebrate evolution: Insights from reptiles and teleost fish. *Brain Res. Bull.* **57,** 499–503.
15. Amsterdam, A., S. Lin, L.G. Moss, and N. Hopkins (1996) Requirements for green fluorescent protein detection in transgenic zebrafish embryos. *Gene* **173,** 99–103.
16. Higashijima, S., Okamoto, H., Ueno, N., Hotta, Y., and Eguchi, G. (1997) High-frequency generation of transgenic zebrafish which reliably express GFP in whole muscles or the whole body by using promoters of zebrafish origin. *Dev. Biol.* **192,** 289–299.
17. Wixon, J. (2000) Featured organism: *Danio rerio*, the zebrafish. *Yeast* **17,** 225–31.
18. Dodd, A., Curtis, P. M., Williams, L. C., and Love, D. R. (2000) Zebrafish: Bridging the gap between development and disease. *Hum. Mol. Genet.* **9,** 2443–2449.
19. Haffter, P., Granato, M., Brand, M., Mullins, M. C., Hammerschmidt, M., Kane, D. A., Odenthal, J., van Eeden, F. J., Jiang, Y. P., Heisenberg, C. P., et al. (1996) The identification of genes with unique and essential functions in the development of the zebrafish, *Danio rerio. Development* **123,** 1–36.
20. Mullins, M. C., Hammerschmidt, M., Haffter, P., and Nusslein-Volhard, C. (1994) Large-scale mutagenesis in the zebrafish: in search of genes controlling development in a vertebrate. *Curr. Biol.* **4,** 189–202.
21. Amsterdam, A. and Hopkins, N. (2006) Mutagenesis strategies in zebrafish for identifying genes involved in development and disease. *Trends Genet.* **22,** 473–478.
22. Peterson, R. T., Link, B. A., Dowling, J. E., and Schreiber, S. L. (2000) Small molecule developmental screens reveal the logic and timing of vertebrate development. *Proc. Natl. Acad. Sci. USA* **97,**12965–12969.
23. Peterson, R. T., Shaw, S. Y., Peterson, T. A., Milan, D. J. Zhong, T. P., Schreiber, S. L., MacRae, C. A., and Fishman, M. C. (2004) Chemical suppression of a genetic mutation in a zebrafish model of aortic coarctation. *Nat. Biotechnol.* **22,** 595–599.

24. Shafizadeh, E., Peterson, R. T., and Lin, S. (2004) Induction of reversible hemolytic anemia in living zebrafish using a novel small molecule. *Comp. Biochem. Physiol. C Toxicol. Pharmacol.* **138**, 245–249.
25. Mathew, L. K., Andreasen, E. A., Sengupta, S., Peterson, R. T., and Tanguay, R. L. (nd) Chemical genetics to probe regenerative biology. *J. Biol. Chem.* **282**, 35202–35210.
26. Nasevicius, A., and Ekker, S. C. (2001) The zebrafish as a novel system for functional genomics and therapeutic development applications. *Curr. Opin. Mol. Ther.* **3**, 224–228.
27. Nasevicius, A., and Ekker, S. C. (2000) Effective targeted gene 'knockdown' in zebrafish. *Nat. Genet.* **26**, 216–220.
28. Baumann, M., and Sander, K. (1984) Bipartite axiation follows incomplete epiboly in zebrafish embryos treated with chemical teratogens. *J. Exp. Zool.* **230**, 363–376.
29. Carvan, M. J., 3rd, Loucks, E., Weber, D. N., and Williams, F. E. (2004) Ethanol effects on the developing zebrafish: Neurobehavior and skeletal morphogenesis. *Neurotoxicol. Teratol.* 26, 757–768.
30. Laale, H. W. (1977) The biology and use of zebrafish, *Brachydanio rerio*, in fisheries research. A literature review. *J. Fish Biol.* **10**, 121–173.
31. Loucks, E. and Carvan, M. J., 3rd. (2004) Strain-dependent effects of developmental ethanol exposure in zebrafish. *Neurotoxicol. Teratol.* **26**, 745–755.
32. Reimers, M. J., La Du, J. K., Periera, C. B., Giovanini, J., and Tanguay, R. L. (2006) Ethanol-dependent toxicity in zebrafish is partially attenuated by antioxidants. *Neurotoxicol. Teratol.* **28**, 497–508.
33. Reimers, M. J., Hahn, M. E., and Tanguay, R. L. (2004) Two zebrafish alcohol dehydrogenases share common ancestry with mammalian class I, II, IV, and V alcohol dehydrogenase genes but have distinct functional characteristics. *J. Biol. Chem.* **279**, 38303–38312.
34. Lassen, N., Estey, T., Tanguay, R. L., Pappa, A., Reimers, M. J., and Vasiliou, V. (2005) Molecular cloning, baculovirus expression, and tissue distribution of the zebrafish aldehyde dehydrogenase 2. *Drug Metab. Dispos.* **33**, 649–656.
35. Jowett, T., and Lettice, L. (1994) Whole-mount in situ hybridizations on zebrafish embryos using a mixture of digoxigenin- and fluorescein-labelled probes. *Trends Genet.* **10**, 73–74.
36. Linney, E., and Udvadia, A. J. (2004) Construction and detection of fluorescent, germline transgenic zebrafish. *Methods Mol. Biol.* **254**, 271–288.

6
The Avian Embryo in Fetal Alcohol Research

Susan M. Smith

Summary The avian embryo has proven utility for studying ethanol's damaging effects upon the embryo. Chicken and quail are long-established models for developmental biology research; much of what we know regarding limb, craniofacial, neural crest, hindbrain, and cardiac morphogenesis was first established with avian models. These models also are for popular mechanistic studies of teratogens, including ethanol. Avian models have been used to explore ethanol's effects on neurogenesis, cardiogenesis, intracellular signaling, neurobehavior, and apoptosis. Presented here are several of these methodologies for adaptation by interested researchers.

Keywords Fetal alcohol syndrome; chick embryogenesis; apoptosis; calcium signaling; electroporation; ratiometric imaging.

L. E. Nagy (ed.), *Alcohol: Methods and Protocols*
© Humana Press 2008

1   Introduction

The avian embryo has proven utility for studying ethanol's damaging effects upon the embryo. Chicken and quail are long-established models for developmental biology research; much of what we know regarding limb, craniofacial, neural crest, hindbrain, and cardiac morphogenesis was first established with avian models. They are also popular for mechanistic studies of teratogens, including ethanol. Avian models also have been used to explore ethanol's effects on neurogenesis, cardiogenesis, intracellular signaling, neurobehavior, and apoptosis (e.g., *(1–6)*). Presented here are several of these methodologies for adaptation by interested researchers.

Avian models offer several advantages for developmental research. Fertile eggs are inexpensive, commercially available, and require only an incubator to develop. The shell is easily windowed to directly view or manipulate the embryo and easily resealed to continue development. Significant genetic and molecular resources are available (www.chicken-genome.org; www.ncbi.nlm.nih.gov/projects/genome/guide/chicken), and developmental processes are strongly conserved between avian and mammalian embryos.

2   Materials

2.1   Egg Handling

1. Unincubated fertile chicken eggs (*see* **Note 1**).
2. A refrigerator set at 15°C for egg storage.
3. Forced air egg incubator (e.g., Humidaire, New Madison OH; Model 1502, G.Q.F. Manufacturing Co., Savannah, GA) set to 38°C and 95% relative humidity, and ventilated to ambient air (*see* **Note 2**). Humidity is maintained using a pan of distilled water.
4. Incubator racks that hold eggs horizontal. These can be made inexpensively from egg cartons, 1-inch thick Styrofoam sheets, or 65-mm Syracuse glass dishes holding a plastic ring, nest of cotton or modeling clay to form an egg rest.
5. Stereoscopic microscope (e.g., Wild M5A) equipped with ×1 objective, ×10 eyepieces, and low-intensity fiber optic illumination to visualize the *in ovo* embryo.

2.2   Ethanol Treatment

1. 100% ethanol, USP grade stored, and handled in glass, not plastic.
2. Sterile water.
3. 10X phosphate-buffered saline (PBS): $30 \, mM$ KCl, $1.3 \, mM$ NaCl, $20 \, mM$ K_2HPO_4, $80 \, mM$ Na_2HPO_4, pH 6.4.

2.3 Pharmacologic Manipulation of Embryos With Microbead Implants

1. Column filtration microbeads of 75–150 μm in diameter. We use SM-2 (20–50 mesh) for hydrophobic compounds, AG1-X2 (200–400 mesh) for anions, AG50-X2 (200–400 mesh) for cations, and AffyGel Blue for proteins (all from BioRad, Hercules, CA). Beads should be washed extensively according to manufacturer's directions before use, and stored in sterile water at 4°C.
2. Chemical agents for microbead implantation, diluted in dimethyl sulfoxide if hydrophobic and 0.9% saline if hydrophilic. Generally, the loading concentration is 100- to 1000-fold greater than the concentration used for direct exposure, and must be determined experimentally *(3,11)*.
3. Medium 199 (Gibco/BRL, Bethesda, MD) or 0.05% Neutral Red in 0.9% saline, sterile filtered and warmed to 38°C.
4. Egg white albumin, ~1–2 mL for each egg that is operated upon. This can be taken from the infertile eggs of the clutch.
5. Scotch brand Super 33+ black electrical tape (3 M, Minneapolis, MN).

2.4 Apoptosis Assessment

1. 1X Tyrode's buffer with calcium (TWC): 137 mM NaCl, 2.7 mM KCl, 1.36 mM CaCl$_2$, 0.5 mM MgCl$_2$, 0.3 mM Na$_2$HPO$_4$, 12 mM NaHCO$_3$, 5.6 mM glucose, pH 7.8–8.2.
2. Acridine orange (5 μg/mL in TWC; Sigma/Aldrich) or LysoTracker Red (0.5 μM in TWC; Molecular Probes). Acridine orange is a carcinogen and appropriate handling precautions should be taken.
3. Fluorescent microscope (epi-illumination or inverted) outfitted with FITC (acridine orange) or Texas Red (LysoTracker Red) filter set, ×10 objectives, and a digital camera with imaging software.
4. Paper rings cut from Whatman #1 filter paper, outer diameter 1.5 mm, inner diameter 5–6 mm.

2.5 Ratiometric Imaging of Ca$_i^{2+}$ With Fura-2

1. TWC containing 25 uM Fura-2-AM and 0.1% Pluronic F-127 (both from Molecular Probes).
2. 100% ethanol, diluted to the desired concentration in 1X TWC.
3. 0.1 mM ionomycin (Calbiochem) in 1X TWC.
4. 2 M MnCl$_2$ in 1X TWC.
5. A inverted fluorescence microscopy apparatus equipped with: ×10 Fluor-S lens, Xenon lamp, a filter wheel (Lambda 10–2, Sutter Instruments, Novato, CA)

equipped with a Fura-2 filter set and a wide band emission filter (Chroma Technology, Brattleboro VT); a black and white digital camera (e.g., CoolSnap-ES, Photometrics, Tucson AZ) capable of capturing images at least every 1 s; a computing system equipped with software for ratiometric imaging (e.g., MetaFluor, Universal Imaging, West Chester PA). Spreadsheet program such as Excel for data analysis.

6. Slide chamber for holding the embryo. We use custom-cut, silicon rubber rings (10-mm inner diameter, 2-mm thick, McMaster) mounted on a glass coverslip. Soak rings overnight in distilled water and adhere them to the coverslip using surface tension. Avoid chemical sealants as these can leach into the medium and perturb calcium signaling.

2.6   Genetic Manipulation of the Ethanol-Exposed Avian Brain

1. cDNA construct diluted to 0.8 µg/mL in TWC; tint the solution with a small amount of Fast Green (4 µL of Fast Green solution in 100 µL of DNA) to monitor its injection. Vectors that offer robust expression in the chick embryo include pCAGGS (12), pCAX (13), RCAS (14), or the bi-cistronic construct pMES (13). If the construct is not epitope-tagged, or a bi-cistronic plasmid is not used, then express enhanced green fluorescent protein (eGFP) in the same vector, and inject the two constructs as a mixture at an experimentally determined ratio. We find 3:1 to be suitable (e.g., 1.2 µg/µL test cDNA to 0.4 µg/µL eGFP).

2. Sterile saline.

3. Pulled glass micropipets. Start with capillary pipets of 1.0 mm diameter and 3 inches long (cat. no. 1B100–3, World Precision Instruments, Sarasota, FL). For targeting the neural tube lumen, prepare pulled pipettes of 5–10 µm in diameter using a microprocessor-controlled vertical pipette puller adjusted to Setting 2 (PUL-100, World Precision Instruments). Store pipettes in a closed glass dish layered with Kimwipes.

4. Stereomicroscope (e.g., Wild M5A) equipped with a ×1 objective and micromanipulator apparatus to stabilize the electrodes and pipettor.

5. 1-mm platinum electrodes with a fixed 4-mm gap (Protech International, San Antonio, TX).

6. Electroporator. The CUY-21 (Protech International) offers an exceptionally stable signal that has low risk of damaging the embryo (reviewed in (13)).

3   Methods

3.1   Egg Handling

Remove eggs from 15°C refrigerator and bring to room temperature (2–4 h). Transfer to flat trays and incubate at 38°C, 95% humidity. At the desired time of development, open three to four eggs to confirm the correct stage has been achieved. Use the criteria of Hamburger and Hamilton *(15)* to establish the embryo's developmental stage. Incubate eggs horizontally rather than blunt end up, to position the embryo for developmental manipulations. The embryo resides at the uppermost position directly beneath the shell. It is harvested by carefully cracking the egg into a 100-mm Petri dish, or visualized by using small dissecting scissors to cut a small hole in the shell (*see* Subheading 3.3.).

3.2 In Ovo Ethanol Exposure

1. Prepare solutions just before injection. For controls, prepare 1× phosphate-buffered saline in sterile water. For experimentals, dilute ethanol to 10% in sterile water containing a final concentration of 1× phosphate-buffered saline. Store and handle all solutions in glass, not plastic.
2. Remove a dozen eggs from incubator. To target neural crest, use embryos between HH4 (gastrulation) and HH10 (10 somites, e.g., between 18h to 36h of incubation). Position eggs horizontally, blunt end outward on a Styrofoam tray or on dishes. Rotate each egg 180° along its horizontal axis to dislodge the embryo from the overlying shell. Clean the injection site by wiping the egg's blunt end with a tissue soaked in 95% ethanol. Using a metal dissecting probe (Fisher cat. no. 08–965A) or 16-gauge needle, through the shell's blunt end, pierce a small hole that is only large enough to accommodate the injection needle. Insert the probe no deeper than 5mm such that only the air sac is pierced.
3. Using a 2.5-mL glass syringe (e.g., Hamilton) fitted with a 1.5-inch 20-gage sterile needle, insert all but one-quarter inch of the needle into the hole, bevel side down and held directly horizontal. This places the injection into the approximate yolk center; affirm the technique with test injections of India ink. Slowly (over 3–5 seconds) inject 250 μL of saline or ethanol solution into each egg; because the air sac volume is 350 μL, larger injection volumes will crush the embryo and should be avoided. Rotate the egg 180° along its horizontal axis. Seal the injection hole with a small piece of cellophane tape, label the treatment on the shell using pencil, and immediately return eggs to incubator (*see* **Note 3**).

3.3 Pharmacologic Manipulation of Embryos With Microbead Implants

1. Prepare beads for implantation. Transfer ~5–10 μL of beads to a 1.5-mL Eppendorf tube. Spin briefly, decant. Add 50 μL of the desired agent to beads. Mix vigorously on a benchtop shaker (Vortex Genie 2, speed 5) for 20 min. Spin briefly in microfuge. Decant (*see* **Note 4**).
2. Add 1 mL of 0.05% Neutral Red in 0.9% saline to stain beads for subsequent visualization. Shake for 10 min. Spin. Decant.

3. Add 1 mL of 0.9% saline. Shake for 10 min. Spin. Decant.

4. Repeat **step 3** twice more. Transfer beads to sterile culture dish.

5. Remove from the incubator four eggs at the appropriate stage of development. Rotate each egg 180° horizontally to dislodge the embryo from the overlying shell. Using a 12-mL syringe, withdraw ~1 mL of albumin through the injection hole at the blunt end; insert the needle downward to avoid damaging the yolk or embryo. Using a metal probe, chip a small hole directly above the embryo, no larger than 5–10 mm diameter. Gently tease out the underlying membrane. The embryo should immediately drop into the egg; if not, repierce the air sac. Rock the egg slightly to center the embryo beneath the hole. If not immediately using the egg, loosely seal the hole with a small piece of low-tack cellophane tape.

6. Remove the cellophane tape and view the opened egg under the stereomicroscope. Add one drop (30 µL) of warmed Medium 199 atop embryo to visualize it; wait 10–20" for dye uptake.

7. Using a mouth capillary pipettor, hand-held pipettor (5 µL), or similar device, pick up a bead and place it in the desired location of the embryo. Let the embryo rest for 30–60 seconds; a slight drying of the embryo's surface is usually sufficient to hold the bead in place. For larger structures (e.g., limb bud) a small hole can be teased in the structure using a fine tungsten needle, and the bead tucked within it. Discard eggs with bleeding or a pierced yolk as they will not survive.

8. Gently refill the egg with reserved albumin so that the embryo rises to the opening. Seal the egg with a small piece of Super 33+ electrical tape. Rotate the egg 30° along the horizontal axis such that the embryo is under shell rather than tape. Notate the treatment on shell using pencil. Reincubate the egg to the desired developmental stage. Do not rock opened eggs because the surgery hole may leak.

9. Ethanol or saline injections are given 1–2 h after bead implantation, so that cells can uptake the compound prior to ethanol delivery. Three hours after injection, the electrical tape can be gentle removed using a fingernail and the bead removed using gentle aspiration from a hand-held pipettor (5 µL), to reduce the agent's effects on downstream events. Reseal the egg with fresh electrical tape and reincubate. Keep the tape and hole size as small as possible, to facilitate the embryo's development.

3.4   *Apoptosis Assessment*

1. Prewarm all solutions to 38°C.

2. Gently crack egg into a 100-mm Petri dish, keeping the embryo-side of the egg upward. Immediately drop a filter paper ring over embryo. Using small scissors, quickly cut around the ring, freeing the embryo from its membranes. Using forceps or a spatula, transfer the embryo and ring to a Petri dish containing prewarmed TWC.

3. Using a Pasteur pipet, gently swish solution beneath the embryo to remove any adhering yolk, which has significant autofluorescence and will interfere with the analysis. Hold embryos at 38°C until ready to use. With practice, it should take no longer than 5 min to harvest one dozen embryos.

4. Transfer the embryos, with or without their paper rings, to a fresh, prewarmed dish containing 5 µg/mL acridine orange in TWC. Incubate 5 min 38°C with gentle rocking (*see* **Note 5**).

5. Transfer embryos to fresh TWC at 38°C. Destain embryos 15 min 38°C with gentle rocking.

6. Using forceps or a spatula, quickly transfer an embryo to a clean glass slide and immediately visualize and photograph the fluorescent apoptotic cells using a microscope equipped with a FITC filter set. Keep all yolk and extraembryonic membranes out of the field of view, as their strong autofluoresence will disrupt the exposure.

7. Once all embryos are photographed, code the images and score the level of apoptosis using previously established criteria. We use a 1–5 scale (1 being endogenous apoptosis, 5 being most severe) to quantify the apoptosis severity within the hindbrain and migrating neural crest populations *(7)*. Imaging software (NIH Image, Metamorph) can used to quantify the signal.

3.5 Ratiometric Imaging of Ca_i^{2+} With Fura-2

1. Harvest embryos using the paper ring method and hold them in 38°C TWC. Harvest no more than can be processed in half an hour. Use a Spoonula™ (cat. no. 14–375–10, Fisher Scientific) to carefully transfer an embryo, dorsal side down, to a flat-bottom slide chamber containing 150 µL of TWC. Replace the solution with 150 µL of the Fura-2/Pluronic F-127 solution and incubate for 1 h at 38°C, protected from light. Gently rinse embryos with two exchanges of TWC. Embryos can now be imaged for Ca_i^{2+}, or treated with pharmacological agents (**step 2**; *see* **Note 6**).

2. Replace TWC with the desired pharmacologic agent diluted in TWC and incubate 20 min, 37°C. The precise loading time depends upon the compound and the presence of intracellular enzymes, such as deesterases, needed to activate the compound. Wash with two exchanges of 120 µL of TWC.

3. Position embryo in slide chamber under microscope objectives. Gently remove 100 µl TWC (final volume 20 µL) and focus on desired region using DIC or bright field optics; for chick embryos, use ×10 objectives. Using MetaFluor software and a Fura-2 filter set, define 7–10 regions for analysis, and then collect a succession of images (1 second apart over 15–20 seconds) to establish background signal. Gently overlay the embryo with 20 µL of ethanol diluted in TWC and again collect images 1 second apart for 30–45 seconds, or until the ethanol signal returns to baseline.

4. Apply 0.1 mM ionomycin (20 µL) to the embryo to affirm its viability and determine R_{max}; exclude nonresponding embryos from further analysis. Once the signal plateaus (usually by 30 seconds after addition), challenge embryo with 2M MnCl$_2$ (20 µl) to quench Fura-2 and determine R_{min}. Generally, we collect images for 30–60 seconds per treatment to ascertain R, R_{max}, and R_{min}.

5. Download the ratiometric data (ratio of fluorescent intensity at 510 nm emission, following dual excitation at 340 and 380 nm) for each imaged region into a

spreadsheet such as Excel. For each region calculate Ca_i^{2+} release using the equation $Ca_i^{2+} = K_d \times [(R - R_{min}) / (R_{max} - R)] \times (F_{max}^{380} / F_{min}^{380})$ *(18,19)*. The terms K_d, F_{max}^{380}, and F_{min}^{380} are cancelled if the ethanol response relative to baseline (no ethanol addition) is calculated. We calculate a mean response per embryo from 7 to 10 defined regions and generally analyze 5 to 10 embryos per treatment to generate an overall mean response.

3.6   Genetic Manipulation of the Ethanol-Exposed Avian Brain

1. Incubate chick embryo to the desired developmental stage. Cut a small window in the egg directly above the embryo (*see* Subheading 3.3.). Visualize the embryo under the stereomicroscope optics (*see* **Note 7**).
2. Position a pair of 1-mm platinum electrodes with a fixed 4-mm gap such that they straddle the embryo, parallel with its long axis. Apply slight downward pressure such that the embryo is sandwiched directly between them. Add 2–3 drops of saline (~75–100 µL) to immerse the embryo and electrodes.
3. Gently insert the pulled glass pipet into the lumen of the neural tube. Position the pipette relative to the target population; for hindbrain neural crest, insert the pipet at the level of somite 3 and point it toward the cranial direction. Using an autoinjector or oral pipetting device, slowly inject 2 µL of DNA into the lumen. In a successful injection, the Fast Green should fill the entire cranial neural tube lumen.
4. Using a foot pedal control, immediately (within 5 seconds of injection) electroporate the embryo using experimentally defined conditions. For the HH8–10 embryo and the CUY-21 electroporator, we use 20V/100 ms on, 999.9 ms off, 5 pulses. To avoid killing the embryo, keep the amperage <1.0 ohm. Remove the electrodes and gently wipe them clean with a Kimwipe.
5. Implant a microbead, if desired, as described in Subheading 3.3. Otherwise, refill the egg with albumin, reseal using Scotch 33+ electrical tape, rotate the egg such that the embryo is not beneath the tape, and reincubate.
6. Monitor expression of the eGFP construct at times thereafter using a fluorescent microscope equipped with FITC filters. The embryo can be viewed under brightfield optics, and apoptosis detected using LysoTracker Red, as described in Subheading 3.4. The transfected material can be detected using numerous approaches, including epitope tagging, direct immunostain, or monitoring of eGFP co-expression. In our hands, the pCAGGS construct offers robust neural expression by 6h after electroporation. Because electroporation affects only one side of the neural tube, the contralateral side serves as in internal control.

4   Notes

1. Fertile eggs can be obtained from commercial vendors (e.g., SPAFAS), or from a local layer facility. Most studies utilize layer strains due to high egg productivity, but broiler strains are also acceptable. Always indicate the strain used, as

their alcohol responses can vary *(1,7,8)*. Store eggs at 15°C to avoid killing the embryo, and use eggs within 1 wk of arrival. For studies of early embryogenesis, the freshest eggs are best. Careful attention should be paid to storage and incubation temperatures to assure synchronous development. Additional details on egg handling are found in several excellent reviews *(9,10)*.

2. Cell culture or closed air incubators are unsuitable because the embryos must receive fresh, circulating air to prevent suffocation. Unopened eggs should be rotated using an automated rocker, or manually twice daily, to prevent embryos from adhering to the overlying shell. Opened eggs should not be rotated because of the potential for leakage. Because temperature dictates the rate of development, do not overcrowd the incubator to assure even air circulation. Distribute eggs randomly throughout the interior, rather than clustered by treatment. Maintain humidity using a tray of distilled water. Sterilize the tray weekly to prevent bacterial or mold growth, which can kill the embryos.

3. This acute ethanol exposure model achieves a peak alcohol level of 50–60 mM at 30–90 min after injection *(7)*. Ethanol then evenly distributes between embryo, yolk, and white, and is sustained at ~9 mM thereafter, as the embryos lack appreciable alcohol metabolic activity, and evaporation losses are minimal. As an alternate *in ovo* exposure route, position the egg vertically and inject the ethanol into the air sac through a hole pierced in the shell's blunt end. Because liquids placed directly on the embryo can be damaging, especially at early stages *(16)*, we recommend using the yolk-injection method.

4. The precise concentration used is determined experimentally, and is generally 100- to 1000-fold greater than that used for direct exposure. Washed beads release the adsorbed compound following steady-state kinetics for ~18–24 h; a detailed discussion of these kinetics is found in Eichele et al. *(7)*. The bead loading time can be increased to 1 h, depending on the compound of interest. Beads must be used immediately upon preparation. For studying ethanol effects on neural crest apoptosis, place the bead immediately lateral to the presumptive hindbrain of a HH8- embryo 1–2 h before ethanol challenge. Remove the bead 3 h after ethanol injection to reduce its effects on later development. If the embryo is going to develop longer than 24 h, one to two drops of antibiotic (penicillin-streptomycin) can be applied to minimize bacterial growth; egg white lysozyme has modest antibacterial properties.

5. Embryos must be maintained at 38°C because cooling reduces their ability to export the dye, and background fluorescence will rise significantly. Because avian embryos significantly autofluoresce in the FITC range, LysoTracker Red offers a cleaner background signal. Its use is identical to that for acridine orange, except that staining is for 30 min. Protect reagents and embryos from light. It may be necessary to tease away the cephalic membranes using fine forceps to fully expose the head for staining and imaging.

6. Ca_i^{2+} imaging can also be performed using Fluo-3, as described *(17)*; however, this reagent is less sensitive than Fura-2, it requires an external standard to normalized the signal, and it displays significant quenching and cell leakage. Both epiillumination and inverted microscope systems can be used for fluorescent Ca_i^{2+} imaging. However, the inverted microscope offers superior focal capability, and the embryo moves less upon ethanol challenge. A small, wire harp (Warner Instruments) can be laid across the embryo to reduce its movement.

7. Electroporation and embryo treatment can also be performed *ex ovo*, using a whole embryo culture system such as described *(9)*.

References

1. Bruyere, H. J., and Stith, C. E. (1993) Strain-dependent effect of ethanol on ventricular septal defect frequency in White Leghorn chick embryos. *Teratology* **48**, 299–303.
2. Fang, T-T., Bruyere, H. J., Kargas, S. A., Nishikawa, T., Takagi, Y., and Gilbert, E. F. (1987) Ethyl alcohol-induced cardiovascular malformations in the chick embryo. *Teratology* **35**, 95–103.
3. Garic-Stankovic, A., Hernandez, M., Chiang, P. J., Armant, D. R., Debelak-Kragtorp, K. A., and Smith, S. M. (2005) Ethanol selectively triggers neural crest apoptosis thru its activation of a pertussis toxin-sensitive G-protein and a phospholipase Cβ-dependent Ca2+ transient. *Alcohol Clin. Exp. Res.* **29**, 1237–1246.
4. Kentroti, S., and Vernadakis, A. (1996) Ethanol neurotoxicity in culture: selective loss of cholinergic neurons. *J. Neurosci. Res.* **44**, 577–585.
5. Means, L. W., McDaniel, K., and Pennington, S. N. (1989) Embryonic ethanol exposure impairs detour learning in the chick. *Alcohol* **6**, 327–330.
6. Rovasio, R. A., and Battiato, N. L. (2002) Ethanol induces morphological and dynamic changes on in vivo and in vitro neural crest cells. *Alcohol Clin. Exp. Res.* **26**, 1286–1298.
7. Debelak, K. A., and Smith, S. M. (2000) Avian genetic background modulates the neural crest apoptosis induced by ethanol exposure. *Alcohol. Clin. Exp. Res.* **24**, 307–314.
8. Bupp-Becker, S. R., and Shibley, I. A., Jr. (1998) Teratogenicity of ethanol in different chicken strains. *Alcohol Alcoholism.* **33**, 457–464.
9. Darnell, D. K., and Schoenwolf, G. C. (2000) Culture of avian embryos, in *Methods in Molecular Biology. Vol 135: Developmental Biology Protocols, v.1.* (Tuan, R. S., and CW Lo, C. W., eds.). Humana Press, Totowa, NJ. pp. 31–38.
10. Ros, M. A., Simandl, B. K., Clark, A. W., and Fallon, J. F. (2000) Methods for manipulating the chick limb bud to study gene expression, tissue interactions, and patterning, in *Methods in Molecular Biology. Vol 135: Developmental Biology Protocols, v.1.* (Tuan, R. S., and CW Lo, C. W., eds.). Humana Press, Totowa, NJ, pp. 245–266.
11. Eichele, G., Tickle, C., and Alberts, B. M. (1984) Microcontrolled release of biologically active compounds in chick embryos: beads of 200-microns diameter for the local release of retinoids. *Anal. Biochem.* **142**, 542–555.
12. Niwa, H., Yamamura, K., and Miyazaki, J. (1991) Efficient selection for high-expression transfectants with a novel eukaryotic vector. *Gene* **108**, 193–200.
13. Krull, C. E. (2004) A primer on using in ovo electroporation to analyze gene expression. *Dev. Dyn.* **229**, 433–439.
14. Fekete, D. M., and Cepko, C. L. (1993) Replication-competent retroviral vectors encoding alkaline phosphatase reveal spatial restriction of viral gene expression/transduction in the chick embryo. *Mol. Cell Biol.* **13**, 2604–2613.
15. Hamburger, V., and Hamilton, H. L. (1951) A series of normal stages in the development of the chick embryo. *J. Morphol.* **88**, 49–92.
16. Drake, V. J., Koprowski, S. L., Lough, J. W., and Smith, S. M. (2006) The gastrulating chick embryo as a model for evaluating teratogenicity: a comparison of three approaches. *Birth Defects Res. A.* **76**, 66–71.
17. Debelak-Kragtorp, K. A., Armant, D. R., and Smith, S. M. (2003) Ethanol-induced cephalic apoptosis requires phospholipase C-dependent intracellular calcium signaling. *Alcohol Clin. Exp. Res.* **27**, 515–523.
18. Grynkiewicz, G., Poenie, M., and Tsien, R. Y. (1985) A new generation of Ca2+ indicators with greatly improved fluorescence properties. *J. Biol. Chem.* **260**, 3440–3450.
19. Kao, J. P. Y. (1994) Practical aspects of measuring [Ca2+] with fluorescent indicators. *Meth. Cell Biol.* **40**, 155–181.

7
Artificial Rearing

Hector D. Dominguez and Jennifer D. Thomas

Summary Prenatal alcohol exposure disrupts development, leading to a range of effects referred to as fetal alcohol spectrum disorders (FASD). FASDs include physical, central nervous system, and behavioral alterations. Animal model systems are used to study the relationship between alcohol-related central nervous system damage and behavioral alterations, risk factors for FASD, mechanisms of alcohol-induced damage, as well as treatments and interventions. When using a rodent model, it is important to recognize that the timing of brain development relative to birth differs between humans and rodents. Thus, to model alcohol exposure during the third trimester equivalent, rats must be exposed during early postnatal development (postnatal days 4–9). Artificial rearing is one experimental paradigm that is used to expose neonatal rats to alcohol during this period of brain development. Neonatal rat pups are housed in an artificial rearing environment and automatically fed a milk diet substitute via an intragastric cannula to ensure adequate growth during the treatment period. Alcohol is delivered in the milk diet. This chapter provides a description of the methods needed for this administrative technique, including preparation of the artificial rearing environment, gastrostomy surgery, and care of artificially reared rat pups.

Keywords Fetal alcohol syndrome; central nervous system; development; intra-gastric infusion; ethanol.

1   Introduction

Fetal alcohol spectrum disorders (FASDs) include a range of physical, neuro-logical, and behavioral alterations associated with prenatal alcohol exposure. Studies using animal model systems are conducted to better identify risk factors associated with increased vulnerability to prenatal alcohol, mechanisms by which alcohol disrupts development, and potential interventions/treatments. Numerous animal models have been used, including ovine and nonhuman pri-mate models (1). Rodents are among the most frequently studied model for fetal alcohol effects.

Artificial rearing has been used to examine the effects of alcohol exposure during the third trimester equivalent "brain growth spurt." The brain growth spurt is characterized by rapid brain development (2) and is a period during which the brain is particularly vulnerable to many insults, including alcohol exposure. In humans, the brain growth spurt begins at the end of the second trimester and con-tinues through the first 2 postnatal years. Rats, however, are altricial relative to humans, and so the brain growth spurt begins after birth. Therefore, to model the brain development that occurs during the third trimester in humans, the rat is exposed to alcohol during the early postnatal period, usually from postnatal days (PD) 4 through 9 (2).

Many experimental paradigms have been used to expose rat pups to alcohol during this period, including intragastric intubation (3) and vapor inhalation (4). Artificial rearing, commonly referred to as the "pup-in-a-cup" technique, was first used for examining the effects of early alcohol exposure by Diaz and Samson (5). It has been used by several groups to examine the effects of devel-opmental alcohol exposure on brain structure and function, as well as behavior (6–13). The decision to use artificial rearing as an administrative technique depends on several factors, including the level of alcohol exposure, the expo-sure paradigm (chronic vs binge), and outcome measures. Exposure to high levels of alcohol in rat pups can inhibit suckling, leading to potential nutritional confounds and changes in maternal-pup interactions. Artificial rearing allows the investigator to control for and manipulate nutritional variables to ensure adequate growth of the pups. In addition, dams may interact with the ethanol-treated pups differently from control pups. The artificial rearing procedure eliminates the impact of such differential interactions on pups' development. However, the major negative aspect of the artificial rearing procedure is the separation of the pup from the mother, which could potentially affect the pup's brain and behavioral development. Therefore, it is important to always include two control groups: an artificially reared control and a normally reared control group. These control groups will allow the investigator to determine the effects of alcohol as well as any potential effects of artificial rearing. Surprisingly, there are relatively few reported differences between artificially and normally reared subjects.

Before using this procedure, various materials, such as the artificial milk diet and surgical tools, must be prepared. For artificial rearing, rat pups are surgically implanted with an intragastric cannula to allow the artificial milk diet to be delivered via a timer-controlled infusion pump. Ethanol is mixed with the milk diet and administered to the rat pups directly via the intragastric cannula. The pups remain in a temperature- and humidity controlled artificial rearing environment throughout the ethanol treatment period and are then fostered to a lactating dam.

2   Materials

2.1   Milk Diet

2.1.1   Stock Milk Diet

The milk diet is intended to mimic maternal milk, providing the artificially reared pups with a nutritionally balanced diet, similar to that obtained by their suckle control siblings (*see* **Note 1**). To prepare approximately 2 liters of diet, the following ingredients are needed:

1. 1500 mL of Carnation® evaporated milk.
2. 450 mL of sterilized double-distilled H_2O.
3. 70 g of Purina™ Protein 710.
4. 120 g (~133 mL) of corn oil.
5. 2.0 g of methionine.
6. 1.0 g of tryptophan.
7. 10.0 g of ICN Nutritional Vitamin Diet Fortification Mix (refrigerated).
8. 11.0 g of calcium phosphate.
9. 0.2 g of deoxycholate acid.
10. To make the mineral mix, add the following: 0.24 g of zinc sulfate, 0.24 g of cupric sulfate, 0.24 g of ferrous sulfate, 4.0 g of magnesium chloride, 4.0 g of potassium chloride.

2.1.2   Treatment Diet

1. 95% ethanol.
2. Maltose-Dextrin.

2.2   Gastrostomy Surgery

1. Stainless steel rods (0.25 mm × 76 mm, A-M Systems, Inc.)
2. Silastic tubing (small; 0.30 mm ID; 0.64 mm OD, Dow Corning, Midland, MI).
3. Polyethylene tubing (small; PE 10, 0.28-mm ID; 0.61-mm OD, Clay Adams).
4. Polyethylene tubing (large; PE 50, 0.58-mm ID; 0.965-mm OD, Clay Adams).
5. Thin plastic sheet (e.g., polyethylene sandwich bag).
6. Thick plastic.

2.3   Artificial Rearing

1. Flexible plastic tubing for feeding lines (Tygon tubing; 0.60-mm ID, Tygon, Saint-Gobain PPL. Corp., Bridgewater, NJ).
2. Silastic tubing (large; 0.76-mm ID; 1.65-mm OD, Dow Corning, Midland, MI).
3. Syringes.
4. Soldering iron.

2.4   Artificial Rearing Components

1. Tank: Throughout the procedure, rat pups are housed in cups that float within a tank [made of stainless steel, Plexiglas, or glass (e.g., a fish tank or other specialty created tank will do)] filled with temperature-controlled water. The temperature of the water is calibrated so that the temperature inside the cups housing the pups is maintained close to 38°C. A cover above the tank may be used to help stabilize the water temperature (Fig. 1A,B).
2. Water circulator/heater: The water in the tank is maintained at a constant temperature and circulated, using an immersion water circulator/heater (e.g,. Fisher Scientific, Pittsburgh, PA; Model 71). This provides the cup environment with the desired temperature and offers additional vestibular stimulation to the isolated pups.
3. Pump: An infusion pump (similar to Model 2265 from Harvard Apparatus, South Natick, MA) able to hold several (8–30) syringes should be located in a slightly elevated position relative to the tank (Fig. 1A,B). The elevation helps reduce the possibility of air bubbles entering the feeding lines. The infusion pump should be calibrated with the same syringe size that will be used for the diet feedings and an infusion rate chart should be created. The infusion pump should be periodically calibrated to ensure accuracy.
4. Timer: The activation of the infusion pump is controlled by a programmable timer. Although some infusion pumps come equipped with an internal timer control, this component may need to be added separately (e.g. ChronTrol, Model XT; Fig. 1C).

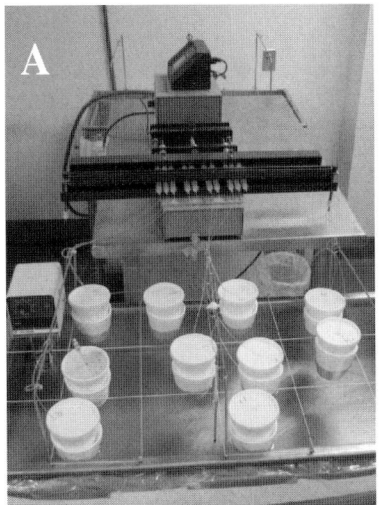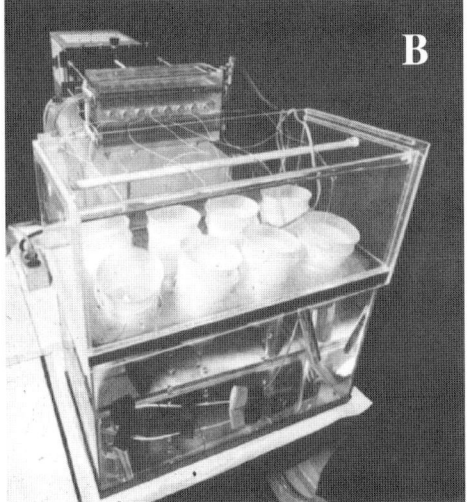

Fig. 1 Two examples of an artificial rearing environment (**A**) without a tank cover or (**B**) with a Plexiglas cover. The infusion pump may be controlled by an internal or external timer (**C**)

5. Graduated cylinder: A graduated cylinder (accurate to 0.01 mL) is placed next to the tank to measure the actual volume of milk diet delivered.

6. Cups: Pups are individually housed in cups (i.e., clean plastic or Styrofoam 16-oz containers), each covered with a lid containing punched air holes. To keep the cups afloat, each cup is placed within a ballast cup partially filled with sand or other substance to provide enough weight to maintain the cup in an upright position within the circulating water. Adding a grid of strings to reduce movement of the cups throughout the tank is suggested (Fig. 1A). Alternatively, the tank can be covered with a Plexiglas (or other material) top

Fig. 2 The cup to house the rat pup sits in a separate cup for ballast and contains bedding and artificial fur

with holes for each cup to limit cup movement (Fig. 1B). To reduce the effects of isolation, the cups are provided with elements that resemble the maternal niche, including cage bedding and a piece of artificial fur attached to the inside of the cup to provide tactile stimulation, simulating maternal contact (Fig. 2). Olfactory stimulation with familiar scents can be provided by adding a small amount of bedding from the maternal cage.

3 Methods

3.1 Milk Diet

3.1.1 Stock Milk Diet

The milk formula used to feed the pups should be prepared before the commencement of the artificial rearing. Typically, the diet is prepared days or even months before being used and kept frozen until needed.

1. To make the mineral mix, add ingredients 10–14 from Subheading 2.1.1 in 100 mL of sterilized double distilled water. Place on a stir plate at low heat until the minerals dissolve.

2. Add the mineral mix to ingredients 1–9 from Subheading 2.1.1 in a large container for mixing.
3. Blend for 1.5 h.
4. Allow the diet to settle for 0.5 h.
5. Siphon a known quantity of diet using sterile pipettes into sterilized bottles and seal the bottles.
6. Place the bottles in a water bath at 60–65°C for 30 min.
7. Place the bottles in the freezer until needed.

3.1.2 Treatment Diet

Ethanol is added to the stock milk diet at a dose determined by the investigator. To control for increased calories associated with ethanol, an isocaloric maltose-dextrin solution is added to the stock milk diet for artificially reared control subjects. That is, 1 mL of the maltose solution contains the same number of calories provided by 1 mL of 95% ethanol. To make isocaloric maltose-dextrin diet:

1. Add 27.415 g of maltose dextrin to 20 mL of double-distilled water.
2. Place the solution on a stir plate with a low level of heat until the maltose is completely dissolved.
3. The maltose solution should be stored in the freezer or refrigerator.

3.2 Gastrostomy Surgery

3.2.1 Gastrostomy Surgery Tools

The gastrostomy tools are used for implantation of the intragastric cannula. These tools are not commercially available, but can easily be constructed in the lab. Tools should be stored in an antiseptic substance (70% ethanol) in an airtight container and flushed with sterilized water before use (*see* **Note 2**).

3.2.1.1 Guide wire

A stainless steel rod is covered with the small silastic laboratory tubing (Fig. 3A). The silastic tubing should be approximately 5–8 mm longer than the wire to cover the sharp end and avoid perforation of the internal organs as the wire is introduced.

3.2.1.2 Intragastric Cannula

The intragastric cannula is ~15 cm of PE 10 tubing with a small plastic flange at one end (Fig. 3B). To make this tool, the end of the PE 10 tubing is placed close to

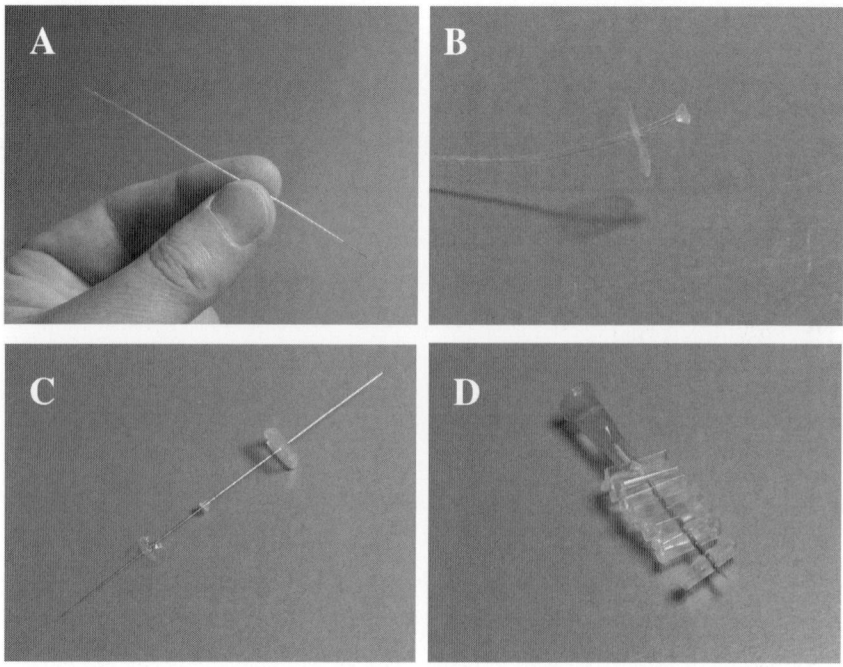

Fig. 3 Gastrostomy tools: guide wire (**A**), intragastric cannula (**B**), washers (**C**), and neck washers (**D**)

a heat source (e.g., soldering iron) until a small portion at the tip of the tubing softens. The melted plastic is immediately softly pressed against a clean cold metal surface, flattening the end without closing off the tube. A circular piece of soft and flexible plastic (~5 mm diameter; can be created with a hole punch) is needled onto the tubing and pressed onto the flat end of the intragastric cannula. The flange and washer prevent the intragastric cannula from coming out of the stomach.

3.2.1.3   Washers

To anchor the intragastric cannula, a washer is placed outside the abdominal wall, where the cannula exits the pup. The washer consists of a circular piece of soft and flexible plastic (5 mm diameter) similar to the one anchoring the cannula from the inside of the stomach. This washer sits next to the skin and is held in place by a short piece (3–4 mm) of PE 50 tubing that fits tight on the PE 10 intragastric cannula. One end of the PE 50 tubing is heated to create a flange similar to that on the intragastric cannula. In preparation for the surgery, the washer is needled onto a stainless steel rod, as indicated in Fig. 3C. One of the neck washers (see below) should be placed on the same rod for greater efficiency during the gastrostomy surgery. This tool is referred to as the "washers" tool.

3.2.1.4 Neck Washers

Two pieces of thick plastic (~5 mm) without sharp edges are needed for each pup. To prepare for surgeries, several washers can be pierced onto a 21-gauge needle to facilitate their use during the surgical procedure (*see* Fig. 3D; *see* **Note 3**).

3.2.2 Gastrostomy Surgery

All artificially reared subjects will undergo gastrostomy surgery, whereas suckle controls will be separated from the dam during the surgery time and then replaced with a foster dam (*see* **Note 4**). Litter size of the foster dam should be kept constant by including nonexperimental rat pups.

1. Subjects are anesthetized via halothane or isoflourane. Young pups are somewhat resistant to anesthesia and must be carefully monitored.
2. Once the pup is fully anesthetized, hold the pup loosely by the sides of the head, with the nose facing up, to create a clear and straight path through the esophagus to the stomach (Fig. 4A).
3. Dip the silastic-covered guide wire into corn oil for lubrication. Slowly place the silastic-covered guide wire down the esophagus, ensuring that the guidewire follows the dorsal side of the airway so that the wire enters the esophagus and not the trachea toward the lungs. If strong resistance is experienced, it is likely that the wire has been misplaced.
4. Once the guide wire is in the stomach (the stomach can be visible via the milk band), pressure is placed on the wire to puncture through the abdominal wall, so one end of the guide wire protrudes from the mouth and the other from the abdominal wall (Fig. 4B; *see* **Note 5**).
5. Attach the intragastric cannula to the end of the guide wire protruding from the mouth (this can also be done prior to the placement of the guidewire into the stomach). Ensure that the intragastric cannula overlaps with the silastic on the guide wire; otherwise, it may fall off of the guide wire during the procedure.
6. Slowly pull the guide wire and intragastric cannula from the abdominal wall (Fig. 4C,D). The washer at the flange end of the intragastric cannula will collapse; however, some slight resistance may be felt when the flange enters the esophagus and when the flange enters the stomach.
7. Once in the stomach, the outline of the washer on the intragastric cannula should be visible in the stomach (Fig. 4D).
8. Place antibiotic cream on the abdominal wound.
9. Remove the guide wire from the intragastric cannula and replace with the "washers" wire.
10. Slide the washers onto the intragastric cannula and slide to the outside of the abdominal wall (Fig. 4E). Ensure that the washers keep the intragastric cannula in place, but are not putting undue pressure on the abdomen.
12. Slide the neck washer onto the gastrostomy tube.
13. Wipe the back of the neck with antiseptic.

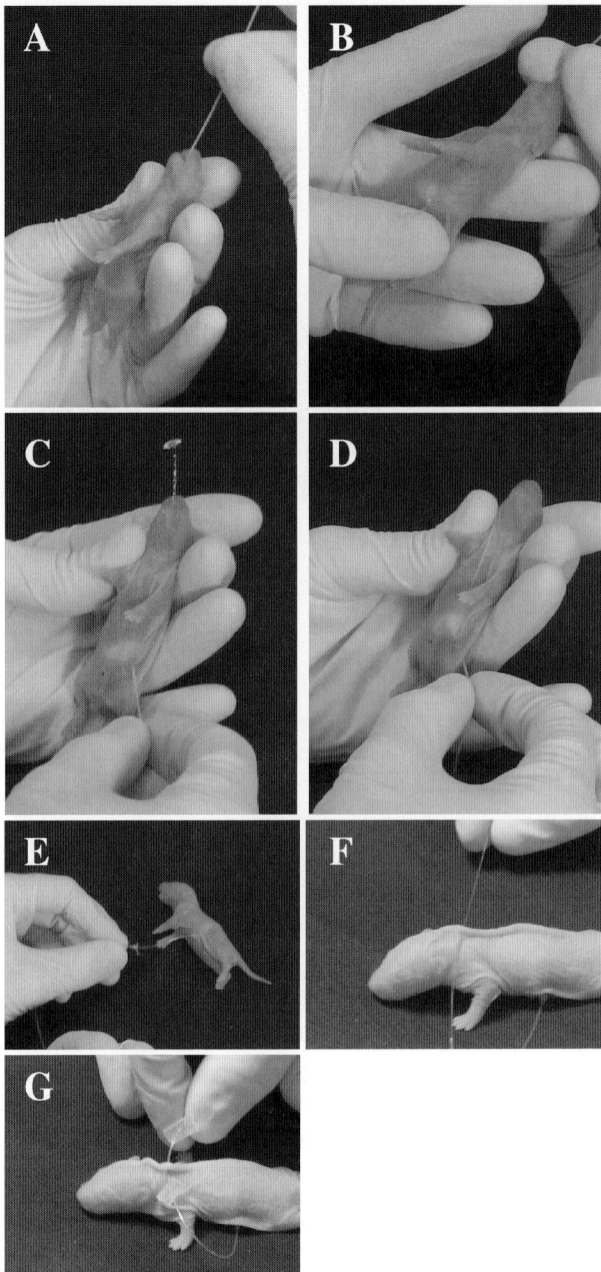

Fig. 4 The gastrostomy surgery. (**A**) The lubricated guide wire is introduced through the pup's mouth and guided through the esophagus to the stomach. (**B**) Gentle pressure on the guide wire against the abdomen perforates the abdominal wall and the guide wire emerges from the stomach. (**C**) Slowly pulling the guide wire from the abdomen, the intragastric cannula enters through the mouth. (**D**) As the intragastric cannula is pulled, the flange end anchors against the internal wall of the stomach. (**E**) The washers are positioned to anchor the intragastric cannula from the outside. (**F**) The skin from the back of the neck is pulled up and the washers' wire is used to puncture through the skin. (**G**) The final neck washer is placed on the intragastric cannula

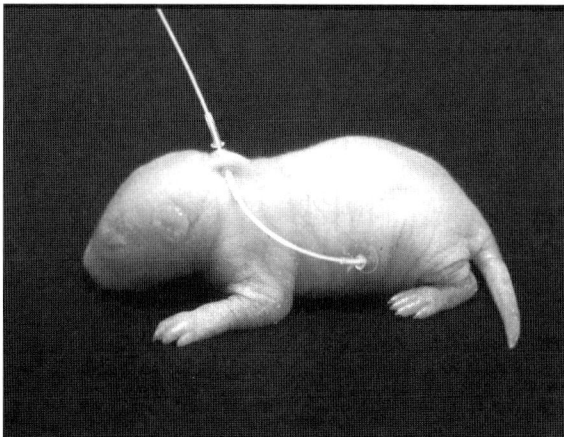

Fig. 5 A gastrostomized pup. Note that the neck washers are made of PE50 tubing rather than the thick plastic

14. The intragastric cannula will be placed through the skin on the neck to minimize the likelihood that the tube is pulled out of the stomach as the pup moves around. Pull the skin up from the back of the neck and use the "washers" wire to puncture through the skin (Fig. 4F).

15. Place the final neck washer onto the wire and onto the intragastric cannula. Ensure that the neck washers do not pinch the skin (Fig. 4G).

16. The distance between the washers at the abdominal wall and the neck needs to be long enough to allow full mobility of the rat pup, but not too long so that the tube can get easily caught and pulled. This distance will need to be adjusted as the rat pup grows (Fig. 5).

17. After gastrostomy surgery, subjects are allowed to recover in an appropriately heated environment (e.g., a heating pad) and are then placed in the artificial rearing environment. A piece of large Silastic tubing (~ 3 cm long) is placed on the end of the intragastric cannula to prevent twisting and kinking of the intragastric cannula as the subject moves around.

3.3 Artificial Rearing

The pups are maintained in the artificial rearing environment during the ethanol treatment and typically for 2 additional days to allow the pups to go through any withdrawal before fostered back to a lactating dam. The pups will be housed in the cups within the artificial rearing apparatus and fed automatically (*see* **Note 6**).

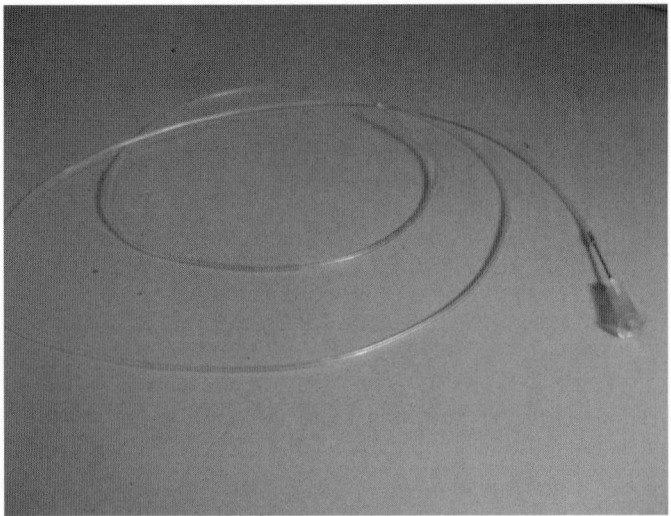

Fig. 6 The feeding line attaches the milk syringes to the intragastric cannula

3.3.1 Preparation

1. The artificial rearing environment should be prepared the night before use. The tank should be filled with water and warmed to the appropriate temperature so that the inside of the cup is 38°C.
2. Place the cups with fur in the tank so temperature is stabilized.
3. Feeding lines that will deliver the milk diet from the syringe on the pump to the intragastric cannulae are prepared. Each feeding line comprises a length (1–1.5 m) of flexible plastic tubing (Tygon tubing) connected to a needle (21-gauge that will be placed on the milk syringes (Fig. 6). Feeding lines should be stored in an antiseptic at least overnight (*see* **Note 2**) and flushed with sterilized water before use.

3.3.2 Determination of Diet Infusion Rate

1. Each morning, pups are weighed and the mean body weight of the pups in the artificial rearing environment is determined and recorded. Suckle control subjects should also be weighed each morning, but not included in the mean.
2. The amount of milk diet to be delivered within a 24-h period (12 feedings; a feeding every 2 h) is determined as follows: Mean body weight (g) × 0.33 = mL of diet/day. To determine the amount of milk diet to be delivered during each feeding, divide this number by 12. Refer to the pump calibration chart to determine the infusion rate (*see* **Notes 6–8**).

3.3.3 Delivery of Milk Diet

The milk diet is delivered into each intragastric cannula every 2 h via the timer-controlled infusion pump. Syringes filled with milk are typically changed twice a day, but this will depend on the alcohol treatment protocol.

1. Remove the milk diet from the refrigerator and warm to room temperature (*see* **Note 9**).
2. Record the amount of milk infused during the previous delivery period, as measured by the water in the graduated cylinder.
3. Disconnect pups from the feeding lines and remove the syringes from the infusion pump.
4. Rinse each feeding line with approximately 2 mL of sterilized water.
5. Fill syringes with the appropriate diet, color coding the syringes if more than one treatment diet is being delivered. Fill the syringe with at least 2 extra mL of milk to allow for priming.
6. Place the syringes in the pump and attach the feeding lines (attach the needles securely, but do not overtighten). Center the syringes evenly on the pump to ensure that pressure on the syringes is evenly distributed. Fill an additional syringe with water to measure the volume actually delivered.
7. Prime the feeding lines so they are filled with milk diet and ensure that there are no air bubbles present in the syringes or feeding lines.
8. Set the pump according to the appropriate infusion rate for the day.
9. Place the feeding line from the water syringe into the dry graduated cylinder to measure the actual volume of infusion.
10. Measure and record water temperature and the temperature inside the cups.
11. Re-connect the feeding lines to the intragastric cannulae. The feeding line should press-fit and stay attached to the intragastric cannula (*see* **Note 10**).

3.3.4 Care of Artificially Reared Pups

The investigator is responsible for the care of the pups; thus, it is critical that the pups be checked regularly to ensure their health and comfort. For example, the pups' ano-genital area should be gently stroked to facilitate urination and defecation (*see* **Notes 11–13**). In addition, every morning and afternoon, pups will be bathed and the intragastric cannulae flushed with sterilized water to keep them patent. Typically, bathing will co-occur with the changing of milk syringes.

1. Fill a cup with clean warm water (~38°C) for bathing the pups. Dip the lower half of each pup in the cup of water and wipe dry with a soft tissue. Gently stimulate the ano-genital region with a cotton ball to facilitate urination and defecation (*see* **Note 11**).
2. Flush each intragastric cannula with 0.1 mL of sterilized water. The water should flow easily. If the cannula gets clogged, insert a small wire into the cannula and then try flushing with water again.

3. Return the pup to the artificial rearing cup (*see* **Note 14**). Attach the feeding line to the intragastric cannula of each pup. The feeding line should press-fit and stay attached to the intragastric cannula (*see* **Note 10**). Check to ensure that each pup is getting the appropriate diet.
4. Ensure that the lid on the cup is securely attached (*see* **Note 15**).

3.3.5   Fostering of Pups to Lactating Dam

Once the artificial rearing period is complete, pups are removed from the artificial rearing environment and fostered to a lactating dam.

1. Intragastric cannulae are trimmed close to the abdominal wall and sealed with a soldering iron.
2. A small amount of milk diet is applied in the mouth via silastic tubing to stimulate suckling behavior.
3. To minimize possible maternal rejection, pups are covered in a slurry of the mother's feces before being placed in the dam's cage.

Acknowledgments Special thanks to Dr. Wei-Jung Chen and Dr. Susan Maier for their helpful comments and suggestions.

4   Notes

1. Various research groups have used slightly different diets for artificial rearing. A commonly used milk diet is that proposed originally by West et al. *(14)* as a modification from the one used in earlier studies *(15,16)*. This milk formula includes more water and protein than the original one, aiming to increase the body weight gain in artificially reared rat pups *(17)*. Wainwright et al. *(18)*, who have examined the effects of varying dietary components in artificial milk, use a slightly different milk stock solution and report no significant differences in body weight gain between artificially reared and normally reared pups. Thus, although not used so extensively in the literature, this milk diet should also be considered.
2. If the tools are left in the 70% ethanol for a period of months, the plastic may shrink or harden; thus, it is suggested that they be stored in the ethanol for no more than one month.
3. Alternatively, washers can be created with PE 50 tubing (Fig. 5).
4. Some investigators may decide to have the suckle controls undergo a sham surgery.
5. Occasionally, during the gastrostomy surgery, the guide wire pokes through the silastic covering when pressed out of the stomach. If this occurs, massaging the

area near the exit wound while gently pushing downward on the silastic at the top of the guide wire usually will cause the silastic to emerge. The silastic can not be simply taken off of the guide wire, as the press-fit of the silastic holds the intragastric cannula onto the guide wire. If this occurs, the silastic covering on the guide wire should be replaced before the next surgery, as it will likely re-occur.

6. Many investigators set the pumps to deliver the milk formula every 2 h for a 20-min delivery period, but this may vary according to the experimental design.

7. Some investigators will begin artificial rearing with a lower infusion rate *(18)*.

8. Occasionally, an individual pup may fail to gain weight or even lose weight. This typically occurs if the intragastric cannula becomes unhooked from the feeding line during the evening hours. The exclusion of these subjects from the study is a decision that needs to be made based on the particular conditions and history of the pups under consideration. If the subject remains in the study, the body weight should not be added to the mean as it will slow growth of all pups on the pump.

9. Some investigators have kept the milk syringes in a refrigerator during the milk infusion, allowing the milk to warm as it flows through the feeding lines *(19)*.

10. If, as the pup matures and becomes more active, the feeding line becomes unhooked, the connection may be taped.

11. While in the home cage, the dam licks the pups, stimulating their ano-genital area to induce urination and defecation. The pups in isolation will have difficulty excreting by themselves. Softly stimulating the ano-genital area with a cotton ball or a piece of soft tissue wet with warm water will stimulate excretion. This task needs to be performed with regularity (at least twice a day). Some investigators stimulate the pups three to four times a day, whereas others only once a day when the pups gets older.

12. Artificial rearing should be monitored periodically throughout the day. One should check that feeding lines do not become disconnected from the intragastric cannulae, that the pump is accurately set and operational, and that the temperature is stable. Pups should be checked to monitor health and to ensure that the cannulae and/or feeding lines do not become twisted.

13. Occasionally, pups will become bloated. Bloating may be related to ethanol-induced reductions in gastric motility or for other reasons. Ano-genital stimulation is critical for allowing the pups to urinate and defecate. In addition, bloating may occur if the temperature within the environment is reduced. If a subject becomes bloated, a warm bath may help stimulate defecation.

14. Some investigators apply cornstarch baby powder to the anogenital region of the pup after bathing to reduce irritation during urination.

15. As the rat pups get older, they become more active and it is important that the lid is on securely so that the pup can not exit the artificial rearing cup.

References

1. Riley, E. P., and Meyer, L. S. (1984) Considerations for the design, implementation, and interpretation of animal models of fetal alcohol effects. *Neurobehav. Toxicol. Teratol.* **6**, 97–101.
2. Dobbing, J., and Sands, J. (1979) Comparative aspects of the brain growth spurt. *Early Hum. Dev.* **3**, 79–83.
3. Goodlett, C. R., and Johnson, T. B. (1997) Neonatal binge ethanol exposure using intubation: timing and dose effects on place learning. *Neurotoxicol. Teratol.* **19**, 435–446.
4. Moore, D. B., Madorsky, I., Paiva, M., and Barrow Heaton, M. (2004) Ethanol exposure alters neurotrophin receptor expression in the rat central nervous system: Effects of neonatal exposure. *J. Neurobiol.* **60**, 114–126.
5. Diaz, J., and Samson, H. H. (1980) Impaired brain growth in neonatal rats exposed to ethanol. *Science* **208**, 751–753.
6. Pierce, D. R., and West, J. R. (1986) Alcohol-induced microencephaly during the third trimester equivalent: relationship to dose and blood alcohol concentration. *Alcohol* **3**, 185–191.
7. Klintsova, A. Y., Scamra, C., Hoffman, M., Napper, R. M., Goodlett, C. R., and Greenough, W. T. (2002) Therapeutic effects of complex motor training on motor performance deficits induced by neonatal binge-like alcohol exposure in rats: II. A quantitative stereological study of synaptic plasticity in female rat cerebellum. *Brain Res.* **937**, 83–93.
8. Kelly, S. J. (1996) Effects of alcohol exposure and artificial rearing during development on septal and hippocampal neurotransmitters in adult rats. *Alcohol Clin. Exp. Res.* **20**, 670–676.
9. Smith, A. M., Zeve, D. R., Grisel, J. J., and Chen, W. J. (2005) Neonatal alcohol exposure increases malondialdehyde (MDA) and glutathione (GSH) levels in the developing cerebellum. *Brain Res. Dev. Brain Res.* **160**, 231–238.
10. Clements, K. M., Girard, T. A., Ellard, C. G., and Wainwright, P. E. (2005) Short-term memory impairment and reduced hippocampal c-Fos expression in an animal model of fetal alcohol syndrome. *Alcohol Clin. Exp. Res.* **29**, 1049–1059.
11. Slawecki, C. J., Thomas, J. D., Riley, E. P., and Ehlers, C. L. (2004) Neurophysiologic consequences of neonatal ethanol exposure in the rat. *Alcohol* **34**, 187–196.
12. Allen, G. C., West, J. R., Chen, W. J., and Earnest, D. J. (2004) Developmental alcohol exposure disrupts circadian regulation of BDNF in the rat suprachiasmatic nucleus. *Neurotoxicol. Teratol.* **26**, 353–358.
13. Meyer, L. S., Kotch, L. E., and Riley, E. P. (1990) Alterations in gait following ethanol exposure during the brain growth spurt in rats. *Alcohol Clin. Exp. Res.* **14**, 23–27.
14. West, J. R., Hamre, K. M., and Pierce, D. R. (1984) Delay in brain growth induced by alcohol in artificially reared rat pups. *Alcohol* **1**, 213–222.
15. Messer, M., Thoman, E. B., Galofre, A., Dallman, T., and Dallman, P. R. (1969) Artificial feeding of infant rats by continuous gastric infusion. *J. Nutr.* **98**, 404–410.
16. Hall, W. G. (1975) Weaning and growth of artificially reared rats. *Science* **26**, 1313–1315.
17. Diaz, J., Moore, E., Patracca, F., Schacher, J., and Stamper, C. (1982) Artificial rearing of pups with a protein-enriched formula. *J. Nutr.* **112**, 841–847.
18. Ward, G. R., Huang, Y. S., Bobik, E., Zing, H. C., Mutsaers, L., Auestad, N., Montalto, M., and Wainwright, P. (1998) Long-chain polyunsaturated fatty acid levels in formulae influence deposition of docosahexaenoic acid and arachidonic acid in brain and red blood cells of artificially reared neonatal rats. *J. Nutr.* **128**, 2473–2487.
19. Moore, W. A., Goldberg, S. J., and Shall, M. S. (2007) Effects of artificial rearing on contractile properties of genioglossus muscle in Sprague-Dawley rat. *Arch. Oral Biol.* **52**, 133–141.

8
Intragastric Intubation of Alcohol During the Perinatal Period

Sandra J. Kelly and Charles R. Lawrence

Summary Animal models of fetal alcohol spectrum disorder (FASD) have been instrumental in isolating alcohol as a teratogen and demonstrating behavioral and neural effects. There are a number of different models for rodents with various strengths and weaknesses. A three-trimester model of FASD is described here; the model uses intragastric intubation of both pregnant dams and pups to mimic alcohol exposure across all three trimesters in humans. The model does not use expensive equipment and is relatively easy to accomplish. The model allows excellent control of alcohol dose and uses an oral route of administration. There are no undernutrition effects with the doses used here. A drawback of the model is the stress of the intubation procedures and ways in which to minimize this stress are discussed. In addition, a method to measure blood alcohol levels is described.

Keywords Ethanol; alcohol; fetal alcohol spectrum disorder; fetal alcohol syndrome; postnatal alcohol; prenatal alcohol.

L. E. Nagy (ed.), *Alcohol: Methods and Protocols*
© Humana Press 2008

1 Introduction

Fetal alcohol spectrum disorder (FASD) is the leading known cause of preventable mental retardation in the Western world *(1)*, and animal models have been instrumental in isolating alcohol as a teratogen and describing behavioral and neural deficits induced by alcohol *(2,3)*. Animal models of FASD are currently being used to conduct translational research directed towards delineating possible behavioral (e.g. *[4,5]*) and/or pharmacological treatments (e.g. *[6–8]*) of FASD and mechanisms of alcohol-induced damage (e.g. *[9,10]*). Alcohol administration during development in rodents (predominantly mice and rats but also including guinea pigs and ferrets) has been accomplished using a variety of different methods, and each of these methods has different strengths and weaknesses *(11)*. The most common models use rats or mice. The gestational period in rats and mice is only equivalent to the first two trimesters in the human with respect to brain growth *(12)* and so, to target a period equivalent to all three trimesters, alcohol administration in both the prenatal and postnatal period must occur.

An ideal method of alcohol administration in rodents would allow complete control over the actual alcohol exposure, entail no stress or handling of the animal, mimic the pharmacological time course of blood alcohol concentrations in the human fetus, be technically easy to do, and control for nutritional effects resulting from intoxication from alcohol. Such a method does not exist, but it is useful to go through the different methods and evaluate them along the different criteria (Table 1) *(13–22)*. The most commonly used methods include vapor inhalation, liquid diets, artificial rearing, and intragastric intubation. All of these methods can be considered stressful to the animal although the nature of the stressor varies across methods. Vapor inhalation and artificial rearing require the purchase of expensive equipment, whereas the other methods do not. In contrast to the other methods, the liquid diet procedure does not allow close experimental control of the dose given to the animal.

This chapter will focus on a three trimester model of FASD that uses intragastric intubation during the prenatal and postnatal period *(13,14,23)*. This procedure is relatively easy to do, gives excellent control over the dose, and allows administration across both the prenatal and postnatal period. The alcohol is administered orally so that the time course of blood alcohol levels is appropriate and there are no detectable nutritional effects using the doses described here (*see* **Note 1**) *(14)*. However, the administration procedure of intubation is stressful to the animals. As a result, it is important to include a control for the stress and allow a comparison with a nontreated control group. It is also very important to make efforts to minimize the stress; these efforts are emphasized in the Methods section. This alcohol administration method does not require expensive equipment and is not technically demanding. In addition, a method to evaluate blood alcohol concentrations is also included since this is typically done in many alcohol studies *(24)*. The procedure described here uses doses in dams and pups that, when given in a single bolus, results in equivalent peak blood alcohol concentrations between 300 and 400 mg/dL in both dams and pups.

Table 1 Evaluation of Alcohol Administration Methods for Rats During Development

Method[a]	Period of exposure	Experimenter control of dose	Nutritional control	Stress	Pharmacological time course	Technical difficulty and equipment
Vapor inhalation (15,16)	Prenatal	Yes	No, but no observed body weight differences	Yes, restricted to breathing ethanol fumes	Very swift increase phase compared with oral ingestion	Easy, requires inhalation chambers
Vapor inhalation (15,16)	Postnatal	Yes	No, but no observed body weight differences	Yes, restricted to breathing ethanol fumes	Very swift increase phase compared with oral ingestion	Easy, requires inhalation chambers
Liquid diet (17,18)	Prenatal	No	Yes, observed effects	Yes, diet restriction	Results from oral ingestion	Easy, requires graduated liquid containers
Artificial rearing (19,20)	Postnatal	Yes	Yes, observed effects	Yes, handling and maternal and sibling deprivation	Results from oral ingestion by the pup	Difficult, requires infusion pumps
Intragastric intubation (13,14, 21,22)	Prenatal	Yes	Yes, no observed body weight differences in dams	Yes, intragastric intubation and handling	Results from oral ingestion by dam	Easy, requires feeding tubes
Intragastric intubation (13,14, 21,22)	Postnatal	Yes	No, but no observed body weight differences in pups	Yes, intragastric intubation and handling	Results from oral ingestion by the pup	Moderate, requires PE tubing

[a]The references given for each method are representative and not exhaustive.

2   Materials

2.1   Alcohol Administration During the Prenatal Period

1. Feeding needles, curved, 18 gage (7.62 cm, 2.25-mm ball; VWR).
2. 10-mL syringes, slip tip connection.
3. Maltose-dextrin (Bio-Serv; 3.89 kcal/g) is dissolved in water to give a solution that is isocaloric with the ethanol solution. This solution must be heated and stirred to get the maltose-dextrin into solution.
4. Ethanol (Acros) (100%, 0.7893 g/mL or 95%, 0.7498 g/mL, 7 kcal/g) is dissolved in distilled water to give a solution of 0.225 g of ethanol per millilter. When injected in a volume of 20 mL/kg, this will give a dose of 4.5 g/kg.

2.2   Alcohol Administration During the Postnatal Period

1. *Intramedic* tubing, PE 10 Clay Adams and PE 50 Clay Adam (VWR).
2. 1-mL syringes and 23-gage needles.
3. A milk solution that is similar in composition to rat milk is made *(25)*. This solution is made under sterile conditions. A mineral mix is first made combining 1.2 g of $ZnSO_4$, 1.2 g of $CuSO_4$, 1.2 g of $FeSO_4$, 20 g of $MgCl_2$, and 20 g of KCl in 500 mL of sterile H_2O. It is a slurry mix that must be stirred before adding to the milk solution. A specially formulated vitamin mix is ordered from BioServ; it is listed as the custom mix for the University of Iowa. The milk solution is made by homogenizing 1500 mL of evaporated milk (purchased at a grocery store), 450 mL of sterile H_2O, 70 g of Supro 710 protein power (*see* **Note 2**), 130 mL of corn oil, 2 g of methionine, 1 g of tryptophan, 10 g of vitamin mix, 11 g of calcium phosphate dibasic, 0.2 g of deoxycholic acid, and 50 mL of the mineral mix. After homogenizing, the solution is put in bottles of a measured amount (usually 50 mL), and the bottle is sealed a rubber topper. The milk is then pasteurized by putting the bottles into an oven at 60 to 65°C for 30 min. The milk is then cooled rapidly and stored at −80°C, where it is stable indefinitely. When milk solutions are needed, a bottle of milk is thawed and used alone for the second injection for the pups. A separate bottle is used as the base of the ethanol intubation and is made such that 3.0 g/kg of ethanol is given when 0.0278 mL/g is given to the pup. The milk solutions that are being used are stored in the refrigerator (4°C) and will be stable for approximately 2 weeks.

2.3   Measurement of Blood Alcohol Concentrations (BACs)

1. Heparinized capillary tubes (10 μL, Drummond Scientific special ordered through VWR, vendor part 1-000-0100-H).

2. 0.53 *N* perchloric acid (Sigma).
3. 0.30 *M* potassium carbonate (Sigma).
4. Alcohol dehydrogenase (Sigma) is dissolved in distilled water to get a solution of 89.25 units/mL. The ADH solution is frozen at −4°C in 5-mL aliquots and thawed before the assay (*see* **Note 3**).
5. β-Nicotinamide adenine dinuculeotide (NAD) (Grade III, Sigma) is dissolved in 0.5 *M* TRIS buffer (Sigma 7-9) to give a solution of 1.875 m*M* NAD. This solution is kept refrigerated (4°C) until time of assay.
6. Ethanol standards are made such that there are solutions of 0, 50, 100, 200, 300, 400, 500, and 50 mg/dL of ethanol dissolved in water. These are kept refrigerated (4°C) until time of assay.

3 Methods

Subjects in our experiments are Long-Evans rats purchased from Harlan and housed in an animal colony with a 12-h light -dark cycle (7:00 AM lights on), temperature at 22°C, and humidity at 20%. Female animals are bred before alcohol administration in this model. The female rats are at least 90 d of age and are immediately put on breeder blocks (Purina, a diet specifically designed for breeding rats) and allowed a week of recovery after shipment from Harlan. They are housed in groups of two or three during this period. For breeding, they are put in groups of four or five overnight (from 5 PM on) with a proven male breeder. The next morning at 08:00 AM (*see* **Note 4**), vaginal smears are taken and examined for the presence of sperm. If sperm is detected, that day is designated gestational day (GD) 1 and the female rat is assigned to one of three experimental groups, which are ET (Ethanol), IC (intubated control), or NC (nontreated control; *see* **Note 5**). The experimental dams are singly housed in polypropylene cages on GD 1.

3.1 Alcohol Administration During the Prenatal Period

1. Each ET dam is allowed free access to rat chow (breeder blocks) and water; the amount of food intake is measured daily in order to provide data for pair-feeding of the IC dams.
2. From GD 1 through GD 22, the ET dam is weighed and then intubated with 4.5 g/kg of ethanol in a volume of 20 mL/kg via intragastric intubation. The ethanol solution is drawn up into a 10-mL syringe and then attached to an feeding needle. Intragastric intubations are given by first dipping the stainless steel feeding tube in corn oil to provide lubrication; the tube is then inserted down the esophagus of the rat (*see* **Note 6** and Fig. 1A). Handling time should be minimized. Intubation of ethanol is done in the late afternoon in order to minimize effects on circadian rhythms (*26*).

 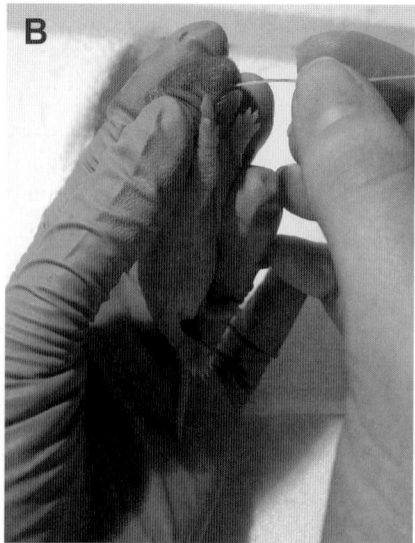

Fig. 1 (**A**) Intubation of a dam. Dams are held in a grip such that one of their forelimbs is over the thumb and the other is between the middle and ring fingers. The grip is firm such that the rat cannot slide down in the grip. This gavage tube is dipped in corn oil and inserted over the tongue. It should go down the esophagus without resistance and should feel as if it could be dropped into the rat's stomach. Note that the gavage tube is inserted all of the way down to the joint between the tube and syringe. (**B**) Intubation of a rat pup. Rat pups are held so that the esophagus is in a straight line. The PE 10 tubing is dipped in corn oil and inserted so that it slides over the tongue and follows the roof of the mouth to the esophagus. The tubing should slide with little resistance down the esophagus. The resistance that may sometimes be felt on rats on PD 2 is because the esophagus is still very narrow; the resistance is <u>not</u> that of the tubing feeling as if it should be forced into the rat

3. For IC dams, the treatment is similar except that food is restricted and the intubation is of isocaloric maltose-dextrin solution. On GD 1, the IC dam is matched
   by body weight to an ET dam of similar weight that has successfully given birth
   to a litter. The IC dam is then only given the amount of rat chow consumed by
   the matched ET dam on that particular gestational day. The IC dam is weighed
   daily and immediately intubated with a maltose-dextrin solution that is isocaloric
   with the ethanol solution in a volume of 20 mL/kg. The intubation process is the
   same as for the ET dams.
4. The NC dams are weighed on GD 1 and GD 22 to minimize handling during
   gestation. In earlier studies, NC dams were weighed on a daily basis in order to
   conclusively show that there were no dam weight differences, but it is felt that
   there are enough data showing this finding to warrant fully controlling handling
   to allow maximal detection of handling and stress effects.

3.2 Alcohol Administration During the Postnatal Period

1. The day of birth (typically GD 23) is designated postnatal day 1 (PD 1) and the dams and pups do not receive any intubations on this day (*see* **Note 7**). Litters are culled to 10 pups with as close to an even number of males and females as possible. On PD 2 through PD 10, pups from the ET and IC groups are removed from their litter, one at a time and weighed. They are then given their first intragastric intubation and then 2h later, they are given their second intragastric intubation. This is done on PD 2 through PD 10.

2. All intubations given to the pups are administered using PE10 *Intramedic* tubing attached to a 1-mL syringe via a short piece of PE 50 *Intramedic* tubing. A small amount of waterproof glue from a hot glue gun makes the connection between the PE 10 and PE 50 tubing tight. The PE 10 tubing is dipped in corn oil prior to the intubation in order to facilitate the procedure (*see* **Note 8** and Fig. 1B). ET pups receive a 3.0 g/kg dose of ethanol in a volume of 0.0278 mL/g milk solution (PD 2-10). Two hours after the first intubation, ET pups are intubated a second time with the milk solution only (0.0278 mL/g). This procedure does not result in deficits in body weight when this dose of alcohol of alcohol is used (*see* **Note 1**). The IC pups receive the same procedure (two intubations) as the ET pups except that solutions are not given (*see* **Note 9**). NC pups are weighed on PD 2 and PD 10 but not treated in any other way.

3. On PD 2, pups can be permanently paw-marked with India ink for identification purposes *(28)*. India ink is injected subcutaneously using a 1-mL syringe and 26-gage needle. The coding system resulting from this injection is such that the two forepaws represent the numbers 1 and 2 and the two hindpaws represent 4 and 8. With this numbering system, addition of the numbers represented by the different paws can number pups from 1 through 15. Alternatively, a rat tattooing system (Animal Identification and Marking systems, Inc., Hornell, NY) can be used to tattoo the paws of the rats.

3.3 BACs

1. On GD 20, 10 µL of blood is taken up using a heparinized capillary tube from a nick to the tail from the ET and IC dams 3h after the intubation procedure. On PD 10, 10 µL of blood is taken using a heparinized capillary tube from a nick to the tail from the ET and IC pups 2h after the alcohol intubation and just before the second intubation of milk only. The blood from the pups can be encourage to flow by holding the pup so that its tail is hanging and then slowly moving your fingers along the tail to encourage the blood drops to come out. It is important not to put too much pressure on the tail. The blood from the ET dams and pups is used to measure BACs via a colorimetric enzymatic assay as described below. The alcohol doses are chosen because they produce similar peak BACs prenatally

and postnatally and the time points are chosen to assay peak BACs *(29)*, which has been shown to be a critical determinant of the teratogenic effects of ethanol *(30, 31)*.

2. The blood alcohol concentrations are measured as described in Dudek and Abbott *(24)*. The 10 μL of blood of the experimental animal and 10 μL of distilled water are added to 190 μL of 0.53 N perchloric acid. Then, 200 μL of 0.30 M potassium carbonate is added. The solution is vortexed and then centrifuged in a refrigerated (4°C) centrifuge for 15 min at 12,000 rpm (14.4 g). On the same day, standards are made. Ethanol standards of 0, 50, 100, 200, 300, 400, 500, and 600 mg/dL (ethanol in distilled water) are made and stored at 4°C. Ten microliters of the standard and 10 μL of blood from a nontreated, nonexperimental animal are added to 190 μL of 0.53 N perchloric acid. The standards are then treated as the blood samples from the experimental animals. At this stage, the samples and standards can be frozen together at −80°C until the time of assay or the assay can be conducted immediately.

3. If the samples and standards are frozen, they should be thoroughly thawed and centrifuged for 15 min in a refrigerated (4°C) centrifuge at 12,000 rpm. The ADH solution should also be thoroughly thawed. Into a glass culture tube, 400 μL of the NAD-Tris solution and 50 μL of ADH should be combined. Then, 50 μL of the samples and standards should be added to each tube. The tubes should be vortexed and then incubated for one hour at room temperature. Each solution should then be read for absorbance of a light of 340 nm wavelength in a spectrophotometer. The standards should be used to construct a linear standard curve (with a correlation of greater than 0.95) and then the unknown samples are calculated from that curve.

Acknowledgments This work was supported by National Institute of Alcoholism and Alcohol Abuse grant RO1 11566 to S. J. K. A special thanks to Dr. Tuan D. Tran who was instrumental in the development of the three trimester model.

4   Notes

1. This procedure uses 4.5 g/kg and 3.0 g/kg of ethanol in dams and pups respectively, delivered in one bolus. Others have used higher doses and found that loss of body weight in the pups occurs. If higher doses or two ethanol intubations of a higher dose are used, careful pilot work and consideration of more than one supplemental milk injection should be considered. In addition, it is likely that the peak blood alcohol concentration will occur at a different point with two intubations.

2. The Supro 710 protein powder can be purchased from Purina only in bulk (40 kg). It is possible to get free samples from some of the sales representatives.

3. A common problem with this assay will appear as a flat standard curve. This occurs because the aliquots of ADH have been thawed and frozen too many times, resulting in a degradation of the enzyme and a failure of the reaction.

4. In general, the earlier the smears are taken, the better the detection of sperm.

5. Animals are assigned in cohorts such that all three groups are represented. Because survival of the litter tends to be less from ET dams because of neglect of the pups or spontaneous resorptions of the litters, more ET dams are designated than IC or NC dams.
6. Learning to do intragastric intubations in dams can be tricky, and it is highly recommended that extensive training take place prior to being allowed to intubate experimental animals. Training can be facilitated by first training with lightly anesthetized rats. Protective gloves should be thin enough to enable a good touch; we use gardening gloves. A good intubation should take less than 1 min.
7. Very rarely a litter is born either early or late. If this occurs, postnatal day is determined from the point of conception; postnatal day 2 then is always gestational day 24.
8. Intubating rat pups is the most demanding part of this technique and requires practice. Marking the PE 10 intubation tube to indicate the length that would reach the stomach helps gauge how deep the tube should be inserted. The tube should slide down the esophagus with little resistance although it is sometimes a tight fit in rats at PD 2. If resistance occurs, the tube should be removed and the procedure tried again. For training, starting with rat pups at PD 4 or 5 tends to result in more success.
9. Initially with this model, it was thought that the IC control animals should receive intubations of milk solution in attempt to match the ET groups. However, pups do not regulate their food intake well, and the intubation of milk resulted in IC control animals weighing considerably more that NC control animals; this has been shown in pilot studies in our laboratory and in a published study **(27)**.

References

1. May, P. A., and Gossage, J. P. (2001) Estimating the prevalence of fetal alcohol syndrome: A summary. *Alcohol Res. Health* **25,** 159–167.
2. Kelly, S. J., Day, N., and Streissguth, A. P. (2000) Effects of prenatal alcohol exposure on social behavior in humans and other species. *Neurotoxicol. Teratol.* **22,** 143–149.
3. Driscoll, C. D., Streissguth, A. P., and Riley, E. P. (1990) Prenatal alcohol exposure: comparability of effects in humans and animal models. *Neurotoxicol. Teratol.* **12,** 231–237.
4. Hannigan, J. H., and Berman, R. F. (2000) Amelioration of fetal alcohol-related neurodevelopmental disorders in rats: Exploring pharmacological and environmental treatments. *Neurotoxicol. Teratol.* **22,** 103–111.
5. Hannigan, J. H., O'Leary-Moore, S. K., and Berman, R. F. (2007) Postnatal environmental or experiential amelioration of neurobehavioral effects of perinatal alcohol exposure in rats. *Neurosci. Biobehav. Rev.* **31,** 202–211.
6. Thomas, J. D., Weinert, S. P., Sharif, S., and Riley, E. P. (1997) MK-801 administration during ethanol withdrawal in neonatal rat pups attenuates ethanol-induced behavioral deficits. *Alcoholism: Clin. Exp. Res.* **21,** 1218–1225.
7. Thomas, J. D., Garcia, G. G., Dominguez, H. D., and Riley, E. P. (2004) Administration of eliprodil during ethanol withdrawal in the neonatal rat attenuates ethanol-induced learning deficits. *Psychopharmacology (Berl)* **175,** 189–195.
8. Thomas, J. D., Garrison, M., and O'Neill, T. M. (2004) Perinatal choline supplementation attenuates behavioral alterations associated with neonatal alcohol exposure in rats. *Neurotoxicol. Teratol.* **26,** 35–45.

9. Heaton, M. B., Paiva, M., Madorsky, I., and Shaw, G. (2003) Ethanol effects on neonatal rat cortex: comparative analyses of neurotrophic factors, apoptosis-related proteins, and oxidative processes during vulnerable and resistant periods. *Dev. Brain Res.* **145,** 249–262.

10. Siler-Marsiglio, K. I., Shaw, G., and Heaton, M. B. (2004) Pycnogenol (R) and vitamin E inhibit ethanol-induced apoptosis in rat cerebellar granule cells. *J. Neurobiol.* **59,** 261–271.

11. Riley, E. P., and Meyer, L. S. (1984) Considerations for the design, implementation, and interpretation of animal models of fetal alcohol effects. *Neurobehav. Toxicol. Teratol.* **6,** 97–101.

12. Bayer, S. A, Altman, J., Russo, R. J., and Zhang, X. (1993) Timetables of neurogenesis in the human brain based on experimentally determined patterns in the rat. *Neurotoxicology* **14,** 83–144.

13. Cronise, K., Marino, M. D., Tran, T. D., and Kelly, S. J. (2001) Critical periods for the effects of alcohol exposure on learning in rats. *Behav. Neurosci.* **115,** 138–145.

14. Tran, T. D., Cronise, K., Marino, M. D., Jenkins, W. J., and Kelly, S. J. (2000) Critical periods for the effects of alcohol exposure on brain weight, body weight, activity, and investigation. *Behav. Brain Res.* **116,** 99–110.

15. Rogers, J., Wiener, S. G., and Bloom, F. E. (1979) Long-term ethanol administration methods for rats: advantages of inhalation over intubation or liquid diets. *Behav. Neural Biol.* **27,** 466–486.

16. Ryabinin, A. E., Cole, M., Bloom, F. E., and Wilson, M. C. (1995) Exposure of neonatal rats to alcohol by vapor inhalation demonstrates specificity of microcephaly and Purkinje cell loss but not astrogliosis. *Alcoholism: Clin. Exp. Res.* **19,** 784–791.

17. Weinberg, J. (1984) Nutritional issues in perinatal alcohol exposure. *Neurobehav. Toxicol. Teratol.* **6,** 261–269.

18. Lieber, C. S., and DeCarli, L. M. (1982) The feeding of alcohol in liquid diets: two decades of applications and 1982 update. *Alcoholism: Clin. Exp. Res.* **6,** 523–531.

19. Samson, H. H., and Diaz, J. (1982) Effects of neonatal ethanol exposure on brain development in rodents, in *Fetal alcohol syndrome* (Abel, E. L., ed.), CRC Press, Boca Raton, FL, pp. 131–150.

20. Kelly, S. J., Mahoney, J. C., Randich, A., and West, J. R. (1991) Indices of stress in rats: effects of sex, perinatal alcohol and artificial rearing. *Physiol. Behav.* **49,** 751–756.

21. Kelly, S. J., and Tran, T. D. (1997) Alcohol exposure during development alters social recognition and social communication in rats. *Neurotoxicol. Teratol.* **19,** 383–389.

22. Light, K. E., Kane, C. J., Pierce, D. R., Jenkins, D., Ge, Y, Brown, G., Yang, H., and Nyamweya, N. (1998) Intragastric intubation: Important aspects of the model for administration of ethanol to rat pups during the postnatal period. *Alcoholism: Clin. Exp. Res.* **22,** 1600–1606.

23. Tran, T. D., and Kelly, S. J. (2003) Critical periods for ethanol-induced cell loss in the hippocampal formation. *Neurotoxicol. Teratol.* **25,** 519–528.

24. Dudek, B. C., and Abbott, M. E. (1984) A biometrical genetic analysis of ethanol response in selectively bred long-sleep and short-sleep mice. *Behav. Genet.* **14,** 1–19.

25. West, J. R., Hamre, K. M., and Pierce, D. R. (1984) Delay in brain growth induced by alcohol in artificially reared rats. *Alcohol* **1,** 83–95.

26. Gallo, P. V., and Weinberg, J. (1981) Corticosterone rhythmicity in the rat: interactive effects of dietary restriction and schedule of feeding. *J. Nutr.* **111,** 208–218.

27. Goodlett, C. R., and Johnson, R. B (1997) Neonatal binge ethanol exposure using intubation: timing and dose effects on place learning. *Neurotoxicol. Teratol.* **19,** 435–446.

28. Geller, L. M., and Geller, E. H. (1966) A simple technique for the permanent marking of newborn rats. *Psychol.l Rep.* **18,** 221–222.

29. Marino, M. D., Cronise, K., Lugo Jr., J. N., and Kelly, S. J. (2002) Ultrasonic vocalizations and maternal-infant interactions in a rat model of fetal alcohol syndrome. *Devel. Psychobiol.* **41,** 341–351.

30. Pierce, D. R., and West, J. R. (1986) Blood alcohol concentration: a critical factor for producing fetal alcohol effects. Alcohol 3, 269–272.

31. Pierce, D. R., and West, J. R. (1986) Alcohol-induced microencephaly during the third trimester equivalent: relationship to dose and blood alcohol concentration. *Alcohol* **3,** 185–191.

Part III
Cell Culture Approaches to Studying Ethanol Exposure

9
Human Monocytes, Macrophages, and Dendritic Cells

Alcohol Treatment Methods

Gyongyi Szabo and Pranoti Mandrekar

Summary Both acute and chronic alcohol consumption have significant immunomodulatory effects of which alterations in innate immune functions contribute to impaired antimicrobial defense and inflammatory responses. Blood monocytes, macrophages, and dendritic cells play a central role in innate immune recognition as these cells recognize pathogens, respond with inflammatory cytokine production, and induce antigen-specific T-lymphocyte activation. All of these innate immune cell functions are affected in humans by alcohol intake. Here, we summarize the different effects of acute and chronic alcohol on monocyte, macrophage, and dendritic cell functions in humans and describe methods for separation and functional evaluation of these cell types.

Keywords Monocytes; macrophages; dendritic cells; NF-κB.

L. E. Nagy (ed.), *Alcohol: Methods and Protocols*
© Humana Press 2008

1 Introduction

Increased susceptibility to bacterial and viral infections has been described in individuals with chronic alcohol consumption *(1,2)*. Studies from humans, animal models, and in vitro systems demonstrate that a single dose of acute alcohol can also modulate host responses to infections or tissue injury induced by trauma or burn *(2-4)*. Many of these alcohol-related changes in host response can be linked to changes in innate immune cell functions. Circulating monocytes and tissue macrophages play a central role in recognition of pathogen- or host-derived danger signals through their expression of pathogen recognition receptors, including Toll-like receptors *(5)*. Tissue macrophages are mostly inflammatory cells whereas blood monocytes have plasticity in their recruitment to damaged/inflamed tissue sites where, depending on the signals from the tissue environment, monocytes can differentiate into macrophages or into dendritic cells *(6)*. Dendritic cells (DCs) represent the other end of the spectrum of the innate immune cells that are directed to accessory cell function. DCs also recognize pathogen- or host-derived signals and respond to those with maturation to acquire a strong antigen-presenting phenotype and T-cell stimulatory capacity. Antigen-primed DCs will induce antigen-specific T- cell activation and expansion. However, in the absence of appropriate co-stimulatory signals, DCs can induce anergy instead of activation in T cells, thereby regulating antigen-specific immune responses *(7,8)*.

Work from our laboratory has demonstrated that human monocyte-derived DCs have an impaired capacity to induce allogeneic or antigen-specific T-cell proliferation after exposure to either short- or long-term alcohol treatment *(9)*. We found that a single occasion of in vivo alcohol intake in normal volunteers results in abnormal DC capacity to induce T-cell proliferation *(9)*. Our studies in humans and of others in animal models demonstrated that acute alcohol intake also modulates monocyte production of pro- and anti-inflammatory cytokines *(10–14)*. For example, acute alcohol inhibits monocyte and macrophage production of tumor necrosis factor (TNF)-α, interleukin (IL)-1, IL-8, and cytokines regulated by nuclear regulatory factor kappa B (NF-κB) *(15,16)*. Consistent with this finding, we found that acute alcohol inhibits stimulation-induced activation of NF-κB in monocytes both in vitro and in vivo *(10–12)*. The anti-inflammatory effects of acute alcohol also extend to induction of anti-inflammatory mediators such as IL-10, PGE$_2$ and transforming growth factor-β in monocytes *(15)*. Importantly, all of these anti-inflammatory mediators exert immunomodulatory function by inhibiting antigen presentation and T-cell proliferation *(9)*.

Investigation of inflammatory mediators in chronic alcohol models revealed an induction pattern opposite from the acute alcohol effects. Animal models of chronic alcohol feeding, studies in human alcoholics, as well as prolonged in vitro treatment of cells with alcohol, revealed increased inflammatory cell activation, suggesting that chronic alcohol has different effects on inflammation *(17–19)*. Human monocytes produce increased levels of TNF-α and show elevated NF-κB activation after prolonged alcohol treatment in vitro *(20)*. Furthermore, alcoholic hepatitis and pancreatitis, two major clinical complications of chronic alcohol use, are associated with increased circulating levels of proinflammatory cytokines that predict poor clinical outcomes *(18,21)*. Therefore, investigation of monocyte-macrophage and dendritic functions in acute and chronic alcohol-induced modulation of host responses is a key component in further studies understanding alcohol-related pathology.

2   Materials

2.1   Isolation of Monocytes and Cell Culture of Macrophages and DCs

1. RPMI 1640 (Gibco/BRL, Bethesda, MD).
2. RPMI 1640 with 15% fetal bovine serum (FBS, Hyclone, Ogden, UT) for monocyte isolation.
3. Hank Balanced Salt Solution (HBSS, Gibco/BRL, Bethesda, MD) with 3% fetal bovine serum.
4. Ficoll-Hypaque (Amersham Biosciences, Piscataway, NJ).
5. Monocyte culture medium: Iscove's Modified Dulbecco's Medium (IMDM, Gibco/BRL, Bethesda, MD) with 10% fetal bovine serum.
6. Dynabeads MyPure Monocyte kit (Dynal Biotech/Invitrogen Co., Carlsbad, CA).
7. IL-4 and granulocyte macrophage colony-stimulating factor (GM-CSF) (Peprotech, Rocky Hill, NJ).
8. DC culture medium: RPMI with 10% fetal bovine serum + β-mercaptoethanol at a final concentration of 5×10^{-5} (EMD/VWR International, West Chester, PA).
9. RPMI 1640 with %10 FBS plus 10 mM ethylenediamine tetraacetic acid (EDTA, Sigma-Aldrich, St. Louis, MO).
10. Macrophage culture medium: RPMI with 18% autologous serum.
11. Absolute alcohol (VWR International, West Chester, PA).
12. Incubator Culture Chamber (C.B.S., Del Mar, CA).

2.2   Cytokine Measurements

ELISA kits: Total IL-12 (Endogen/Pierce Biotechnology, Rockford, IL), IL-10 (eBioscience, San Diego, CA), TNF-α, IL-1β, and IL-6 (Pharmingen/BD Bioscience, San Jose, CA).

2.3   Electrophoretic Mobility Shift Assay for Assaying NF-κB

1. Dulbecco's Phosphate-Buffered Saline (DPBS, Gibco/BRL, Bethesda, MD).
2. Buffer A: 10 mM HEPES, pH 7.9, 10 mM potassium chloride (KCl), 1 mM EDTA, 1 mM ethyleneglycol tetraacetic acid (EGTA), 1 mM dithiothreitol,

1 mM phenylmethylsulfonyl fluoride, 2 mM sodium orthavanadate, 1 µg/mL sodium fluoride, 10 µg/mL leupeptin, 10 µg/mL aprotinin, and 10 µg/mL antipain. Balance with nuclease-free water.

3. Buffer B: 20% glycerol, 20 mM HEPES, pH 7.9, 400 mM potassium chloride (KCl), 1 mM EDTA, 1 mM EGTA, 1 mM dithiothreitol, 1 mM phenylmethylsulfonyl fluoride, 20 mM sodium orthavanidate, 10 µg/mL sodium fluoride, 10 µg/mL leupeptin, 10 µg/mL aprotinin, and 10 µg/mL antipain. Balance with nuclease-free water.

4. 6% PAGE gel: Bis/acrylamide **29**:1 (Boston Bioproducts, Worcester, MA) Tris-borate-EDTA (TBE, Boston Bioproducts, Worcester, MA), 1.6% ammonia persulfate (EMD/VWR International, West Chester, PA), and, *N*.,*N*,*N*,*N*''-tetramethyl-ethylenediamine (TEMED, EMD/VWR International, West Chester, PA).

5. T4 Polynucleotide kinase (Promega Corp., Madison, WI).

6. NF-κB Oligonucleotide (5′-AGTTGAGGGGACTTTCGC-3′; Promega Corp., Madison, WI).

7. BioSpin 6 columns (BioRad Laboratories, Hercules, CA).

8. 0.25% Tris-borate-EDTA (TBE) Running buffer

9. Gel loading buffer: .05% bromophenol blue (BPB, Pharmacia/Amersham Biosciences, Piscataway, NJ), 30% gylcerol (EMD/VWR International, West Chester, PA), Tris-HCl, pH7.5 (Sigma-Aldrich, St. Louis, MO).

10. Bio-Max film (VWR International, West Chester, PA).

2.4   Mixed Lymphocyte Reaction

1. T-cell negative isolation kit (Dynal Biotech/Invitrogen Corp., Carlsbad, CA).

2. Anti-CD8 and anti-CD45RO antibodies (Pharmingen/BD Bioscience, San Jose, CA).

3. Anti-rabbit IgG beads (Dynal Biotech/Invitrogen Corp., Carlsbad, CA).

4. Tritiated thymidine (PerkinElmer, Wellesley, MA).

3   Methods

3.1   Isolation of Peripheral Blood Monocytes

1. Heparin (10 USP units/mL) anticoagulated peripheral blood (*see* **Note 1**) is diluted with warm (37°C) tissue culture medium (RPMI) at a **1**:2 ratio and 30-mL portions are carefully layered over warm (37°C) Ficoll-Hypaque solution (10 mL/per 50 mL plastic screw-top tube).

2. After centrifugation for 20 min at 450 g at 37°C, cells at the interface from each tube are carefully collected into new 50-mL tubes and washed with 45 mL of RPMI by centrifugation at 600 g for 10 min in two repeated cycles.

3. After the second wash, cells (pellets) are pooled into one tube and cell count is obtained. This cell population represents the mononuclear cells from the peripheral blood including monocytes, circulating DCs, T cells, B cells, NK, and NK-T cells. Although the cell yield varies between individual blood donors depending on their circulating blood counts, the average range obtained from 100 mL of blood is approx $200–350 \times 10^6$ peripheral blood mononuclear cells (PBMCs).

4. For isolation of monocytes, the traditional, less expensive monocyte isolation method relies on the properties of monocytes as adherent cells. For this, 5×10^6 PBMCs are plated in 30-mm diameter wells on 6-well plates for 2–24 h to allow adherence. Nonadherent cells are then removed by gentle washing with warm tissue culture medium (twice with a 3-mL volume; *see* **Note 2**). Routinely, the purity of this population is 95–98% monocytes as we determined by CD14 expression with some contaminating T cells, B cells, and few NK cells. Isolated monocytes are then stimulated with stock lipopolysaccharide (*Escherichia coli* 0111: B4, SIGMA-Aldrich Co.) to induce cytokine production.

5. Alternatively, for positive isolation anti-CD14-coated beads (Dynabeads, CD14) are used routinely according to the recommendations of the manufacturer (Dynal Biotech Co.; *see* **Note 3**). This method provides essentially 100% purity.

6. For negative selection of monocytes, the recommended method is to use a cocktail of magnetic beads covered with CD3, CD7, CD16 (a and b), CD19, CD56, CDw123, and CD235a (Glycophorin A) provided in the Dynabeads MyPure Monocyte kit for untouched monocytes (Dynal Biotech Co.). Additional advantage of the positive and negative bead isolation techniques for monocytes over adherence isolation is that this provides cells in solution that could be used for further studies.

3.2 Generation of Human Monocyte-Derived Macrophages

1. Plastic tissue culture flasks are coated for 2 h with 2% sterile, endotoxin-free gelatin and dried overnight in a 37°C 5% CO_2 incubator.

2. On the next day, the gelatin-coated flasks are coated with platelet-free serum isolated from the same monocyte-donor at 37°C for 1 h and rinsed with Hank's Balanced Buffer Solution.

3. PBMCs are isolated as described previoulsy and placed on gelatin-coated flasks *(22,23)*.

4. After a 1-h incubation at 37°C, 5% CO_2, adherent monocytes are washed three times with RPMI and harvested from the flasks with 10% RPMI and 10 mM EDTA (**1**: 1 dilution) treatment for 15 min at 37°C.

5. Isolated monocytes are then washed three times with RPMI, resuspended, and plated at a concentration of 5×10^5/mL in 18% autologous serum and RPMI for 8 d and macrophages phenotyped for CD16 and CD68 *(22,23)* were then used in functional assays. Autologous serum is separated by centrifugation from clotted, nonheparinized blood.

3.3 DC Isolation and Generation of Human Monocyte-Derived DCs

Human DCs have various phenotypes depending on their lineage and maturation stage. In peripheral blood, there are circulating myeloid dendritic cells (mDC), plasmacytoid dendritic cells (pDC), and monocytes that can be precursors for DCs. Generation of monocyte-derived DCs has been described earlier by our and other laboratories *(6–9)*.

1. DCs are generated from adherent blood monocytes in DC media (RPMI 1640 with 10% FBS, 2 m*M* L-glutamine, 50 μm ß mercaptoethanol) with IL-4 (0.1 μg/mL) and GM-CSF (0.01 μg/mL) for 7 d.
2. On days 3 and 5, 1 mL of DC media is replaced with fresh DC media. DC viability on day 7 is >95% (trypan blue exclusion) and the cell culture contains <5% contaminating lymphocytes (CD3, CD56, CD19). Microscopic examination of 7-d immature DCs reveals distinct dendrite-like projections in comparison with no projections in monocytes (Fig. 1).
3. DC maturation can be achieved by adding stock lipopolysaccharide 0.1 μg/mL on day 5 for 48 h.
4. On day 7, DC phenotypic markers can be evaluated by labeling with FITC- or PE-labeled monoclonal antibodies (anti-human-CD1a, -CD3, -CD11c, -CD14, -CD19, -CD56, -CD80, -CD81, -CD83, -CD86, -HLA-ABC, -HLA-DR, or relevant isotype controls) according to the manufacturer's directions.
5. Briefly, cells (1×10^5) are incubated with 20 μL of antibody in a total volume of 100 μl (2% FBS-PBS) for 60 min at 4°C in the dark.
6. Stained cells are then extensively washed and resuspended in 500 μL of 1% paraformaldehyde in PBS.
7. Using a FACSCalibur (BD Bioscience, Immunocytometry system, San Diego, CA) flow-cytometer, cells are gated according to their size (FSC) and granularity (SSC) and surface marker expression of gated large and granular cells is analyzed using the FlowJo (Tree Star Inc, San Carlos, CA), and CellQuest (BD Bioscience) programs. Immature DCs will express CD80, CD86, CD40 HLA-DR but little CD14 (Table 1) *(24)*. In contrast, monocytes express CD14 as well as HLA Class II. DC maturation will be indicated by increased expression of CD80/86 and CD83 (Table 1).
8. The microscopic appearance of the cells changes as they differentiate and mature from monoctyes to DCs (Fig. 1). The round adherent monocytes form loose clusters of more spread-out cells during DC differentiation, and mature DCs have long extended dendrites.

3.4 Isolation of Plasmacytoid DCs (PDCs)

PDCs can be isolated from PBMCs with positive selection using anti-BDCA2 antibodies, according to the manufacturer's recommendations (Miltenyi Biotech, Auburn, CA) as described *(25)*.

Monocytes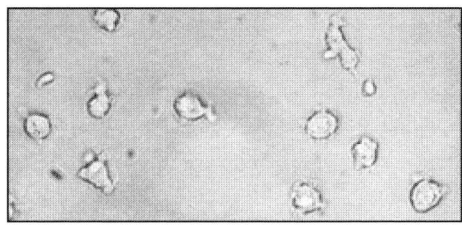

Dendritic cells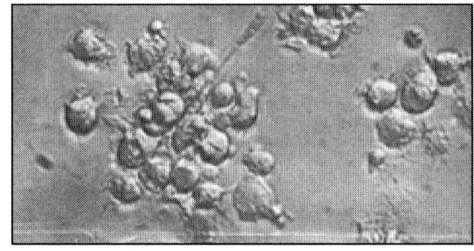

Fig. 1 Microscopic evaluation of the isolated monocytes and 7-d-old immature DCs. The microscopic appearance of the cells changes from round adherent monocytes to more spread-out cells in loose clusters as they differentiate. Mature DCs have long extended dendrites

Table 1 Phenotypic Markers of Human Monocytes and Myeloid DCs

Markers	Monocytes	Immature DCs	Mature DCs
CD 14	+++	+	−
CD1A	+	+++	+
HLA-DR	++	+++	+++
CD80	++	+++	+++
CD86	++	+++	+++
CD40	++	+++	+++
CD83	−	+	+++

1. PBMCs (10^8 cells/100 µL of separation buffer (ice-cold PBS, 0.5% BSA, 2 mM EDTA) are incubated with FcR blocking reagent (50 µL/10^8 cells) and anti-BDCA-2-Biotin antibody (100 µL/10^8 cells) for 10 min at +4°C.
2. This is followed by addition of 400 µl separation buffer/10^8 cells, FcR blocking reagent (150 µl/10^8 cells) and anti-Biotin MicroBeads (200 µl/10^8 cells) for 15 minutes at +4°C.
3. After incubation, PBMCs are washed with 20 volumes of cold separation buffer and filtered through a magnetic column. To increase the purity of BDCA2$^+$ PDCs, the magnetic separation can be repeated using a fresh column. The purity of these isolated PDCs, determined by flow cytometry analysis after staining with Streptavidin-PE and anti-CD123APC antibodies, is greater than 95%.

3.5   Acute Alcohol Treatment of Cells In Vitro

In vitro acute alcohol treatment of monocytes/macrophages/DCs can be achieved under regular tissue culture conditions by adding 100 proof alcohol directly to the tissue culture medium. Alcohol concentrations in the range of 10–500 mM have been reported in various studies *(2,9,14,16)*; however, the higher end of this concentration range is clearly nonphysiological. The 25 mM in vitro concentration is used in many studies considering that this in vitro concentration is close to 0.1 g/dL blood alcohol level (BAL) reached in normal nonalcoholic individuals after 4–5 drink equivalents. The 0.1 g/dL BAL is above the legal limit (0.08 g/dL) allowed for driving in most of the United States. Therefore, 0.1 g/dL represents a "moderately drunk" equivalent. It is important to note that BALs can reach much greater concentrations in chronic alcoholic individuals because of accelerated alcohol metabolism by increased alcohol dehydrogenase.

The length of acute alcohol treatment can vary depending on the experimental question, but the alcohol effects are considered as "acute" for up to 24 h. During a 24-h culture alcohol concentrations decrease in the tissue culture medium as a result of evaporation in 6-, 24-, or 96-well tissue culture plates, flasks, or Petri dishes placed in incubators. The initial alcohol concentration of 25 mM decreases over time and by 24h is in the 12–15 mM range. One of the potential concerns with in vitro alcohol administration is induction of cell death or apoptosis. We reported that in human monocytes or DCs, alcohol does not increase the rate of apoptotic cells in the 10–100 mM concentration range.

3.6   Chronic Alcohol Treatment of Cells In Vitro

1. For long-term "chronic" ethanol experiments monocytes/macrophages/DCs are incubated in 25 mM ethanol for 7–12 d in 6-well tissue culture plates kept in modular incubator chambers (Incubation culture chambers) that are sealed and filled with standard gas mixture at 37°C.
2. Alcohol concentration is maintained by placing a 2-fold greater concentration of alcohol in a Petri dish (open) at the bottom of the modular chamber. For example, treated cells contain 25 mM alcohol, but the alcohol concentration in the Petri dish is 50 mM.
3. One milliliter of culture media is replaced with fresh medium containing one-third of the final alcohol concentration per well on days 3, 5, and 8. This method maintains the initial alcohol concentration ± 15 % *(20)*. Alcohol concentration in the tissue culture supernatants can be checked daily using an Analox alcohol instrument (Lenox, MA) (Table 2).

3.7   Cell Stimulation With Pathogen-Derived Agents

1. Monocytes, macrophages, and DCs have distinct Toll-like receptor (TLR) expression profile, as summarized in Table 3. In general, stimulation with

Table 2 Maintenance of Alcohol Concentrations in 6-Well Plates 25 mM Over the Course of 7 Days

	Day 1	Day 2	Day 3	Day 4	Day 5	Day 6	Day 7
Etoh Conc. (mM)	24 ± 1	21.5 ± 0.5	25 ± 1	22 ± 2	24 ± 0.7	22 ± 1	19.5 ± 0.5

Table 3 Stimulation with TLR Ligands

Receptor	Ligand	Concentrations used	Cell expressing TLR
TLR2/TLR1	PAM_3CSK_4	0.1–1 µg/mL	Monocyte, DC Macrophage
TLR2/TLR6	$PAM_2 CSK_4$	10–20 µg/mL	Monocyte, DC Macrophage
	PGN	1 µg/mL	
TLR4	Phenol-purified LPS Stock LPS[1]	0.1–1 µg/mL	Monocyte, DC Macrophage
TLR5	Flagellin	0.1–1 µg/mL	Monocyte, DC Macrophage
TLR7	R848	0.5 µM	DC
TLR8	R848	0.5 µM	DC
TLR9	CpG DNA	25 µM	Monocyte (?), DC

Most stock lipopolysaccharide stimulates both TLR4 (endotoxin) and TLR2 (lipopepetide contamination).

TLR2, TLR4, and TLR5 ligands induces production of pro-inflammatory cytokines (TNF-α, IL-6, IL-12) in all of these cells with measurable cytokine levels at 8–24 h after stimulation. IL-10 is induced with a slower kinetics with maximal levels at 24–48 h after TLR stimulation. TLR7/8 and TLR9 stimulation primarily induce Type I interferon production *(5)*.

3.8   Cytokine Measurements

1. Levels of cytokines in cell free supernatants can be analyzed using commercially available ELISA kits (BD Pharmingen, San Diego, CA) for the respective cytokines.
2. ELISA kits for TNF-α, IL-1β, and IL-6 in most of our studies were obtained from BD Pharmingen, whereas kits to determine total IL-12 were obtained from Endogen and IL-10 from eBiosciences Co. (San Diego, CA).

3.9   Isolation of Nuclear Proteins

1. Cells are washed with ice-cold PBS once by centrifugation at 600 g for 10 min.
2. Cell pellets are then harvested and resuspended in 400 µL of cold hypotonic buffer A (*see* Subheading 3) for 20 min on ice.
3. Cells are then lysed in 25 µL of 10% NP-40, by vortexing for 10-20 s.

4. Nuclei separated from cytosol by centrifugation at $12,000\,g$ for 30 s. Supernatants recovered and stored at $-80°C$ as cytoplasmic extracts.
5. The nuclei are washed by resuspending in Buffer A ($300\,\mu L$) followed by centrifugation at $12,000\,g$ for 30 s three times.
6. Nuclear proteins are then extracted by resuspending the pellets in $50\,\mu L$ of Buffer B (see materials) and rocking at $4°C$ for 30 min.
7. Nuclear extracts are then obtained by centrifugation at $20,000\,g$ for 10 min.
8. Total protein concentration is estimated in the nuclear extracts by the Biorad Protein Assay kit (Biorad Co.).

3.10   Electrophoretic Mobility Shift Assay for NF-κB

1. Prepare a 1.5-mm thick 6% gel by mixing 7.5 mL of 40% acrylamide/bisacrylamide (**29,** 1), 2.5 mL of 10X TBE, 2.5 mL of 1.6% APS, 37.5 mL of sterile water, and $25\,\mu L$ of TEMED. Pour the gel between two glass plates separated with 1.5 mm spacers and polymerize for approx 30 min.
2. Wash the wells with water and warm up the gel by running in running buffer (1X TBE) empty for 20–30 min at 180 volts.
3. In a microcentrifuge tube mix $5\,\mu g$ of nuclear extract (volume varies for each sample), $4\,\mu L$ of 5X Gel Buffer (see Subheading 2), $4\,\mu L$ of dI-dC, 50,000 cpm of $\gamma^{32}P$-labeled NF-κB oligonucleotide, and nuclease-free water to bring up the reaction to $20\,\mu L$ of total volume. The contents of the tube are centrifuged to ensure mixing of all components.
4. The reaction is incubated at room temperature for 30 mins and, the at the end of the incubation, samples are put on ice.
5. Add 1X loading buffer to one well in the gel and run 10 more minutes to track the course of the gel (no loading buffer added to the samples).
6. Turn off the gel and load samples into wells and run the gel until marker reaches bottom (approx 30–45 min running time).
7. Remove gel and dry between cellophane support sheets.
8. For Supershift assays- Follow all previous steps as above. In addition, add $2\,\mu L$ of anti-p65 or anti-p50 or anti-c-rel antibodies in a separate gel shift reaction tube with the desired sample.
9. After addition of the antibodies incubate for an additional 30 min at room temperature. Keep samples on ice and load the gel.

3.11   Mixed Lymphocyte Reaction

Allostimulatory capacity of day 7 irradiated DCs (3000 Rads; stimulator cells) can be tested in a one-way mixed lymphocyte reaction.

1. Allogeneic T lymphocytes are isolated from normal mononuclear cells via roset-ting with neuraminidase-treated sheep red blood cells *(16)*. This T-cell popula-tion showed 98% CD3 staining by FACS analysis.
2. Normal, allogeneic T lymphocytes (2×10^5 cells/well; responder cells) are mixed with moncoytes or DCs (stimulator cells) at various stimulator/responder cell ratios (**1**, 20-**1**, 80) in triplicates in 96-well plates (Corning Glassware, Corning, NY).
3. T-cell proliferation as a read-out of monocyte or DC allostimulatory capacity is measured by methyl ^{3}H-thymidine (DuPont-NEN, Boston, MA) incorporation evaluated during the final 16 hours of the 5-day assay. Results are expressed as mean counts per minute ± standard error.

4 Notes

1. Monocyte can also be separated from blood anticoagulated with citrate or EDTA.
2. Monocytes isolated by adherence can be detached as early as 2 h after adherence for functional studies by using EDTA for detachment.
3. While nonspecific monocyte activation due to phagocytosis of the magnetic beads has been a concern with the first-generation of beads using positive isola-tion, recent formulations provide negligible background activation.

References

1. Cook, R. T. (1998) Alcohol abuse, alcoholism, and damage to the immune system —a review. *Alcoholism: Clin. Exp. Res.* **22,** 1927–1942.
2. Szabo, G. (1999) Consequences of alcohol consumption on host defense. *Alcohol Alcoholism* **43,** 830–841.
3. Nelson, S., Kolls, J. K. (2002) Alcohol, host defence and society. *Nat. Rev. Immunol.* **2,** 205–209.
4. Szabo, G., Mandrekar, P., Verma, B., Isaac, A., and Catalano, D. (1994) Acute ethanol consumption synergizes with trauma to increase monocyte TNFα production late postinjury. *J. Clin. Immunol.* **14,** 340–352.
5. Akira, S., Uematsu, S., and Takeuchi, O. (2006) Pathogen recognition and innate immunity. *Cell* **124,** 783–801.
6. Palucka, K., Taquet, N., Sanchez-Chauis, F., and Gluckman, J. C. (1998) Dendritic cells as the terminal stage of monocyte differentiation. *J. Immunol.* **160,** 4587–4595.
7. Thery, C., and Amigorena, S. (2001) The cell biology of antigen presentation in dendritic cells. *Curr. Opin. Immunol.* **13,** 45–51.
8. Steinman, R. M., Hawiger, D., and Nussenzweig, M. (2003) Tolerogenic dendritic cells. *Annu. Rev. Immunol.* **21,** 685–711.
9. Mandrekar, P., Catalano, C., Dolganiuc, A., Kodys, K., Szabo, G. (2004) Inhibition of mye-loid dendritic cell accessory cell function by alcohol correlates with reduced CD80/CD86 expression and decreased IL-12 production. *J. Immunol.* **173,** 3398–3407.

10. Mandrekar, P., Catalano, D., Szabo, G. (1997) Alcohol-induced regulation of nuclear regulatory factor-κB in human monocytes. *Alcoholism: Clin. Exp. Res.* **21,** 988–994.
11. Mandrekar, P., Catalano, D., Szabo, G. (1999) Inhibition of LPS-mediated NF-κB activation by ethanol in human monocytes. *Int. Immunol.* **11,** 1781–1790.
12. Mandrekar, P., Catalano, D., White, B., and Szabo, G. (2006) Moderate alcohol intake in humans attenuate moncoytes inflammatory responses: Inhibition of NF-κB and induction of IL-10. *Alcoholism: Clin. Exp. Res.* **30,** 135–139.
13. Mandrekar, P., Catalano, D., and Szabo, G. (1994) Macrophage IL-10 induction and its regulatory effects after acute ethanol exposure. *J. Leuk. Biol.* Suppl, 36.
14. Nelson, S., Bagby, G., and Summer, W. R. (1989) Alcohol suppresses lipopolysaccharide-induced tumor necrosis factor activity in serum and lung. *Life Sci.* **44,** 673–676.
15. Szabo, G., Fogarasi, M., Catalano, D. (1991) Alteration of monocyte TNFα, IL-6, PGE2, and TGFβ responses by ethanol. *Alcoholism Clin. Exp. Res.* **15,** 361.
16. Szabo, G., Mandrekar, P., and Catalano, D. (1995) Inhibition of superantigen-induced T cell proliferation and monocyte IL-1β, TNFa, and IL-6 production by acute ethanol treatment. *J. Leuk. Biol.* **58,** 342–351.
17. Kishore, R., McMullen, M., and Nagy, L. E. (2001) Stabilization of TNFα mRNA by chronic ethanol. *J. Biol. Chem.* **276,** 41930–41937.
18. McClain, C., Barve s, Deaciuc, I., Kugelmas, M., and Hill, D. (1999) Cytokines in alcoholic liver disease. *Semin. Liver Dis.* **19,** 205–219.
19. Arteel, G., Marsano, L., Mendez, C., Bentley, F., and McClain, C. (2003) Advances in alcohol liver disease. *Best. Pract. Res. Clin. Gastroenterol.* **17,** 625–647.
20. Zhang, Z., Bagby, G., Stoltz, D., Oliver, P., Schwarzenberger, P. O., and Kolls, J. K. (2001) Prolonged ethanol treatment enhances LPS/PMA-induced TNFα production in human monocytic cells. *Alcoholism: Clin. Exp. Res.* **25,** 444–449.
21. Pandol, S., Gukovsky, I., Satoh, A., Lugea, A., and Gukovskaya, A. (2003) Emerging concepts for the mechanisms of alcoholic pancreatitis *J. Gastroentrol.* **38,** 623–628.
22. Peotte, V., Rebeiro Gomes, F., Arnholdt, A., and Nagao, P. (2001) Human monocytes and monocyte-derived macrophage phagocytosis of serotype III group B streptococci strains. *Curr. Microbiol.* **43,** 64–68.
23. Wang, J., Kurt-Jones, E., Shin, O., Manchak, M., Levin, M., and Finberg, R. (2005) Varicella-Zoster virus activates inflammatory cytokines in human monocytes and macrophages via toll-like receptor 2. *J. Virol.* **79,** 12658–12666.
24. Steinman, R., and Inaba, K. (1999) Myeloid dendritic cells. *J. Leuk. Biol.* **66,** 205–208.
25. Dolganiuc, A., Chang, S., Kodys, K., Mandrekar, P., Bakis, G., Cormier, M., and Szabo, G. (2006) Hepatitis C virus (HCV) core protein induced, monocyte-mediated mechanisms of reduced IFN-alpha and placmacytoid dendritic cell loss in chronic HCV infection. *J. Immunol.* **177,** 6758–6768.

10

Generation and Use of Primary Rat Cultures for Studies of the Effects of Ethanol

Amanda Lindke, Barbara Tremper-Wells, and Michael W. Miller

Abstract   In vivo studies are ideal for identifying the phenomenology of ethanol toxicity and teratology. They are limited in being able to explore cellular and molecular mechanisms of action. Two types of culture models have proven to be very instructive: monolayer primary cultures of dissociated cells and organotypic slice cultures. Dissociated cell preparations have the advantage of being enriched populations of cells, whereas the organotypic cultures have the advantage of providing normal cell associations. Details for the methods used to generate these preparations are described. As ethanol is a volatile liquid, the success of a culture model depends upon stabilizing the ethanol content in the culture medium. A method to maintain the ethanol concentration is described.

Keywords   Primary neuronal cultures; ethanol; fetal alcohol syndrome; organotypic slice cultures.

L. E. Nagy (ed.), *Alcohol: Methods and Protocols*
© Humana Press 2008

1   Introduction

Studies relying on animal models demonstrate that the development and organization of the brain are profoundly effected by ethanol. In vivo models identify targets of ethanol action and provide context to the effects, however, they are limited in their utility as a substrate for exploring molecular, intracellular, and intercellular mechanisms. In vitro systems lend themselves to such analyses because the composition of the culture, the environment of the cells, and the manipulation of the environment can be carefully controlled. This is particularly important in examinations of growth factor actions, intracellular signal transduction, and genetic manipulations such as electroporation and blocking RNAs.

2   Two Culture Models

Two types of cultures have been particularly useful in alcohol studies. These are 1) primary cultures in which samples are derived directly from animals and 2) cultures of cell lines (e.g., cancer cells and genetically engineered/selected lines). The present chapter discusses the primary cultures. Various types of primary cultures have been used in alcohol studies: dissociated cell cultures and organotypic explants. Each has its uses and limitations.

2.1   Dissociated Cell Cultures

Primary cell cultures are composed of neurons, glia, or stem/progenitor cells. The constituents are relatively homogeneous, nontransformed cells. Common cell types used in these cultures are cortical neurons, cerebellar granule neurons, and astrocytes. These cultures are enriched by various procedures so that the cell of choice comprises 90–95% of the cultured cells. The enrichment method varies with the type of cell to be examined, and the present chapter describes the method for obtaining primary cultures of cortical neurons. For example, procedures used for generating cultures of astrocytes (1–4) and cerebellar granule neurons (5–7) can be found elsewhere.

Dissociated cultures can be criticized for their nonphysiological arrangement, i.e., removing cells from their normal microenvironments. Neurons are intimately dependent on interactions with other cells, thus, it becomes particularly important to preserve in vivo cellular organization before extrapolating the effects of exogenous substances on neuronal development. This need drove investigators to develop systems in which quasi-normal relationships are maintained organotypic (slice) cultures (8–11).

2.2   Organotypic Slice Cultures

Many studies have focused on the developing cerebral cortex. Neurons in the cerebral cortex are derived from two proliferative zones, the ventricular zone (VZ) and the subventricular zone (SZ) *(12,13)*. Intrinsic activities of cells in the prolif-erative zones proceed normally in slice preparations. This includes interkinetic nuclear migration of VZ cells and the migration of postmitotic cells *(14–17)*. Indeed, these metrics describing these activities are similar in slices and in vivo preparations *(18,19)*. Neurogenesis in the neocortical VZ is influenced by both intrinsic intracellular factors and extrinsic intercellular factors *(15,20)*. Using such explants has provided unique insight into the mechanism of ethanol toxicity.

Chief criticisms of the cortical explant model are that not all relationships in the developing cortex are maintained *(1)*. The formation of the slice is likely to disrupt paths of migration or axonal trajectories. Long-distance projections such as the ascending monoaminergic and thalamic afferents are often severed in the produc-tion of the slice *(21)*. Likewise, under normal circumstances, many cortical local circuit neurons are generated in another site, the ganglionic eminence (GE). In generating the slices, either the GE is eliminated from the preparation or the path-ways connecting the GE to the developing cortex may be eliminated. Thus, the contribution from the young local circuit neurons could be eliminated from the developing cortex of the slice *(2)*. A consequence of producing the slice is that cells on the surface of the slice are damaged. The response is the death of many of these cells and in the formation of a glial coat. These can lead to an altered microenviron-ment of cells near the surface of the slice and possibly a change in the microenvi-ronment of other strata of the slice. Conceivably, the increased amount of glia are elaborating growth factors that are expressed in higher amount than would be expressed in vivo.

3   Materials

3.1   Animals

Timed pregnant Sprague-Dawley rats: Depending upon the issue to be addressed, rats may be anywhere from 13 to 19 days pregnant. Commonly, fetal rats are har-vested on gestational (G) day 16. The first day a sperm-positive plug is detected is designated as G1 (*see* **Note 1**).

3.2   Primary Cultures of Cortical Neurons

3.2.1   Surgical Equipment

1. Operating scissors.
2. Microdissecting scissors.
3. Microdissecting forceps.
4. Microdissecting tweezers #5 ($\times$2).

5. Spatula.
6. 60-mm or 10-mm round tissue culture dishes.
7. Dissecting microscope.
8. 1.5-mL microcentrifuge tubes.
9. Glass aspirating pipets.
10. Alcohol burner.

3.2.2 Reagents

1. Hank's balanced salt solution (HBSS, Gibco, Carlsbad, CA).
2. 70% ethanol in water.
3. Poly-D-lysine molecular (PDL) weight 70–100 kDa (Sigma, St. Louis, MO) is dissolved in sterile water at 10 mg/mL to make a 100× stock solution that is aliquoted and frozen at −20°C. Working solutions of PDL are prepared by dilution in phosphate-buffered saline (PBS, Gibco), to a final concentration of 100 µg/mL.
4. Eagle's Minimal Essential Medium (MEM) containing glutamine (Gibco) and supplemented with 10% fetal bovine serum (FBS, Gibco), 33 mM dextrose, and penicillin/streptomycin (Gibco).

3.2.3 Culturing Methods

3.2.3.1 Preparation of the Culture Dishes

1. Coat the experimental tissue culture plates with PDL sufficient to cover the plating surface for a minimum of one hour at room temperature.
2. Chill two 100-mm tissue culture plates containing 10 mL each of HBSS on ice.
3. Chill a 1.5-mL centrifuge tube containing 1.0 mL of HBSS on ice.
4. Sterilize surgical instruments in 70% ethanol.

3.2.3.2 Dissection of the Rat Cortices

1. Anesthetize the timed pregnant G16 dams with a cocktail of ketamine (1.0 mg/kg) and xylazine (1.0 mg/kg). Open the abdominal cavity to expose the uterine horns and extract the pups, one at a time.
2. Hold the pup between the thumb and index finger and cut a shallow incision from the upper spine to the base of the brain, and then laterally around the head from the base of the brain to the eye. Use the forceps to peel the skin and skull from left to right up over the head and expose the brain. Scoop out the entire brain with a spatula and place in one of the culture dishes with chilled HBSS to rinse blood away. Repeat this procedure for all pups. Transfer the brains to a fresh culture dish with chilled HBSS.
3. Under the dissecting scope, use the sharp #5 tweezers to remove the meninges and vasculature from the cortex as fully as possible (*see* **Note 2**). After removal

of the meninges, use the tweezers to pinch off the cortical wall by removal of the olfactory bulb, the striatum, and the hippocampus.
4. Place all cortices in a 1.5-mL centrifuge tube with HBSS. Use sterile glass pipets to gently triturate the cortices and dissociate the cells (see **Note 3**).
5. After the initial cell dissociation, allow heavier fragments (meninges, tissue clumps) to settle for a few minutes to the bottom of the conical tube. Carefully transfer the cell suspension to a 15 mL conical tube leaving the fragments in the bottom of the centrifuge tube. For optimal cell yield, add 1 mL of HBSS to the fragments and repeat the trituration, breaking up as many tissue chucks as possible, while leaving any meninges present following dissection to settle out of the cell suspension. Transfer the cell suspension to the 15 mL conical tube.
6. Centrifuge the cells at $600 \times g$ for 3 min to pellet cells. Add 2.0 mL of 10% serum containing MEM and resuspend cells. Dilute the cell suspension to a concentration suitable for counting.

3.2.3.3   Yield and Plating of the Cells

The total number of primary neuronal cells recovered after the dissociation of G16 cortices is highly dependent on a variety of factors. These factors include the number of pups in utero, the extent to which areas other than the cortical wall are included for culture, and the quality of dissociation. Two pregnant dams with 12 pups each generate approximately 100 million cells with considerable flux dependent on the variables stated. This cell yield would result in approximately four to six 100-mm tissue culture plates, if treatment with growth factors and ethanol with intent to generate a protein lysate, is planned. Plating 2×10^7 cells per 100 mm tissue culture dish yields more than 1.0 mg of total protein and is approximately 70% confluent.

3.3   An Alternative Method for Obtaining Primary Cultures of Cortical Neurons

3.3.1   Reagents

1. HEBSS: 25 mM HEPES, 13.8 mM NaCl, 5.0 mM KCl, 4.2 mM NaHCO$_3$, 1.0 mM NaH$_2$PO$_4$, 0.01% phenol red.
2. MgSO$_4$ stock solution (3.82%): 1.14 g MgSO$_4$ in 30 mL of sterile water.
3. Solution A: 50 mL of HEBSS containing 2.5 mg/ml dextrose, 0.30% bovine serum albumin fraction V (BSA), and 0.50 mL of MgSO$_4$ stock solution.
4. Solution B: 12.5 mL of HEBSS containing 4.0% BSA and 0.10 mL of MgSO$_4$ stock solution.
5. Solution C: 10 mL of Solution A containing 2.5 mg trypsin (1000–1500 U/mg; Sigma).

6. Solution D: 10 mL of Solution A containing 0.10 ml $MgSO_4$ stock solution, 7.5 mg trypsin inhibitor (Sigma), 130 kU/ml DNase (Sigma).
7. Solution E: Add 8.4 mL of Solution A to 1.6 ml of Solution D.

3.3.2   Methods

1. Prepare all solutions described in Subheading 2.3 (A through E) before performing dissection. Filter Solutions A to E through a syringe with a 0.22-μm filter attached.
2. Dissect out the cerebral cortices of all pups with a spatula and place the cortical tissue in a 1.5-mL microcentrifuge tube containing 1.0 mL of solution A.3. In a sterile hood, carefully pour the contents of the tube onto the frosted side of a sterile frosted glass plate. Use a sterile razor blade to homogenize tissue.
4. Pipet the homogenized tissue into a 15-mL conical tube. Add Solution C to the conical tube and place in a hot water bath at 37°C for 15 min. Swirl occasionally to mix.
5. Add Solution E to the conical tube and swirl gently. Centrifuge at 2000 rpm for 5 s to pellet cells.
6. Decant the supernatant and add Solution D. Triturate 10–15 times with a pipet for maximum cell yield. Let the solution sit for 5 min to allow debris and clumps to settle. Remove the supernatant and place in a fresh 15-mL conical tube.
7. Add Solution B to the bottom of the conical tube. There will be a visible separation between Solution B and the cell-rich supernatant. Centrifuge at 2,000 rpm for 5 min to pellet the cells.
8. Decant the supernatant and add MEM medium containing 10% serum. Mix gently and triturate to resuspend the cells for counting.
9. Plate cells as described in Subheading 3.2.3.3.

3.4   Organotypic Slice Cultures

3.4.1   Equipment

1. MacIlwain Tissue Chopper (Mickle Lab Engineering, Gomshell, UK).
2. Razor blade.
3. Plastic disc for tissue chopper (Brinkman Instruments, Westbury, NY).
4. 70% ethanol.
5. Tube of commercial Krazy Glue.
6. Glass Pasteur pipet.
7. 4-inch microdissecting forceps with 0.8-mm tip width.
8. Fine-tip paintbrush with trimmed bristles.
9. Spatula.

10. Plastic transfer pipet.
11. Ice.
12. 60-mm tissue culture dishes (Falcon, Lincoln Park, NY).
13. Millicell filter inserts with $0.40\,\mu M$ pores.
14. Plastic container with lid.

3.4.2 Reagents

1. Neurobasal medium (Gibco, Carlsbad, CA) supplemented with B27 (Gibco), $200\,\text{m}M$ L-glutamine (Gibco), $100\,\mu M$ penicillin/streptomycin (Gibco), and $100\,\text{m}M$ dextrose.
2. Krebs buffer ($10\times$) containing NaCl ($73.6\,\text{g/L}$), KCl ($1.87\,\text{g/L}$), $NaH_2PO_4 \cdot H_2O$ ($1.66\,\text{g/L}$), $MgCl_2$ ($2.44\,\text{g/L}$), $CaCl_2$ ($3.68\,\text{g/L}$). This can be stored at $4°C$ for 2–3 months.
3. Krebs buffer with additives containing $10\times$ Krebs buffer (final concentration): $126\,\text{m}M$ NaCl, $2.5\,\text{m}M$ KCl, $1.2\,\text{m}M$ NaH_2PO_4, $1.2\,\text{m}M$ $MgCl_2$, $2.5\,\text{m}M$ $CaCl_2$), $11\,\text{m}M$ dextrose, $25\,\text{m}M$ $NaHCO_3$, $10\,\text{m}M$ HEPES, double-distilled H_2O.
4. 4.0% (w/v) paraformaldehyde in PBS (pH 7.4) at room temperature, freshly prepared.

3.4.3 Methods

3.4.3.1 Dissection

Pregnant C57/B6 mice are anesthetized on G15 or G17 with a cocktail of ketamine ($1.0\,\text{mg/kg}$) and xylazine ($1.0\,\text{mg/kg}$). The fetuses are collected and their brains are removed. The forebrains are dissected out with forceps and a spatula and placed into 60-mm tissue culture dishes containing Krebs buffer with additives on ice.

3.4.3.2 Preparation and Culture of Mouse Cortical Slices

A brain is removed from a tissue culture dish and prepared for cutting. Krazy glue is used to affix the rear of the brain to the plastic disc to prevent movement while slicing. A reservoir of Krebs buffer with additives is placed around the perimeter of the brain to allow slices to float away from the uncut brain during the slicing process. A trimmed paintbrush, which allows more control for gentle manipulation, is used to gently push down on and flatten out each hemisphere to ensure a uniform thickness of slices. The brain is cut into 300- to 400-μm coronal slices with the MacIlwain Tissue Chopper. The slices are gently retrieved with a plastic transfer pipet containing Krebs with additives and transferred to the original Krebs buffer containing 60 mm tissue culture dish. Approximately six coronal slices are obtained per fetal brain.

Slices from each brain are transferred to 3.0 mL of neurobasal medium supplemented with B27, L-glutamine, penicillin/streptomycin, and dextrose, as indicated previously, in a 60-mm tissue culture dish containing a 0.40 μM Millicell filter insert. Slices are carefully arranged flat with forceps in the center of the insert. The residual Krebs buffer remaining in the insert, following the transfer of the slices, is slowly removed and replaced with neurobasal medium from the surrounding dish. The level of medium is then lowered to form a meniscus just above the slices. Slices are cultured in an incubator for 2 h at 37°C with 95% O_2 / 5.0% CO_2 to allow for recovery before experimental treatments.

3.4.3.3   Preparation and Culturing of Rat Cortical Slices

There are slight variations in the preparation and culturing of rat cortical slices that are worthy of mention. Rat cortical slices do not tolerate Krebs buffer well. Therefore, to preserve the integrity of the cortical slices, artificial cerebrospinal fluid (ACSF) is substituted for Krebs. Brains are dissected out of fetuses and placed in ACSF containing 124 mM NaCl, 5.0 mM KCl, 1.23 mM NaH_2PO_4~H_2O, 18.5 mM $MgSO_4$, 2.0 mM $CaCl_2$, 10 mM dextrose, and 26 mM $NaCO_3$ in sterile water. Slices are cultured in neurobasal medium containing L-glutamine, penicillin/streptomycin and 20% FBS. Neurobasal medium supplemented with B27 or N2, 5.0 mM L-glutamine, and penicillin/streptomycin without serum. After 2 h, the serum-free medium is replaced with neurobasal medium supplemented with B27 or N2, 5.0 mM L-glutamine and penicillin/streptomycin without serum. For shorter ethanol treatments (less than 90 min), slices are cultured in a serum-free medium for the duration of the experiment.

3.5   Controlled Ethanol Exposure

A potential confounding factor for studies of ethanol exposure is the volatility of ethanol. This problem is overcome by incubating primary cultures within a sealed chamber containing ethanol *(3,21–23)*. These chambers maintain a consistent vapor pressure established by ethanol evaporation, prevent loss of ethanol from the culture medium, and can stabilize the concentration of ethanol for more than 72 h.

3.5.1   Equipment

1. 1.6-liter Rubbermaid "Flex and Seal" food storage containers (Rubbermaid, Fairlawn OH).
2. "Picnic table" plastic 24-slot microcentrifuge tube rack with the handles snapped off to make a flat surface for incubation of tissue culture dishes and plates.
3. Syringe (60 mL) with 6-inch tubing attached for CO_2 injection.
4. GM7 Micro-Stat Analyzer (Analox Instruments, Hammersmith, London UK).

3.5.2   Reagents

1. Medium with MEM supplemented with $33\,mM$ dextrose, penicillin/streptomycin, and serum as needed.
2. 95% ethanol.
3. Compressed carbon dioxide gas (CO_2).

3.5.3   Methods

1. Set up the chambers prior to initiating ethanol treatment. Each experimental concentration of ethanol requires a separate chamber. For example, to treat neuronal cultures with ethanol at 0, 200, and 400 mg/dl, a minimum of three chambers is required. Add 200 mL of water to each chamber and set a rack inside the chamber. The 1.60-liter chamber system described here holds up to two 100-mm culture dishes per rack.
2. The amount of ethanol added to the treatment medium (in μL/mL) is determined by the following equation: (molar concentration of ethanol treatment) × (46.07 g/M)/ (ethanol concentration = 0.95) × (specific gravity of ethanol = 0.7750 g/mL). For exposure to 400 mg/dL ($86.6\,mM$) of ethanol, $5.43\,\mu L/mL$ of 95% ethanol is added to both the culture medium and the water in the chamber (*see* **Note 4**). Cell cultures not subject to ethanol exposure are incubated in identical chambers, but without ethanol addition.
3. Place the culture dishes securely on the rack in the ethanol concentration appropriate chamber. Make sure the dishes are level and that medium adequately covers the cells. Position the lid on the chamber and secure all but one corner. Fill a syringe with 60 mL of CO_2 by using a short tubing attachment to connect to the valve on the CO_2 tank. Pinch off the tubing with a hemostat and, removing the tubing from the valve, insert the tubing into the unsecured corner of the chamber. Release the constriction on the tubing and plunge the CO_2 from the syringe into the closed chamber. Quickly close the open corner to tightly seal off the chamber and place the chamber in the incubator. These instructions assume the use of a 37°C incubator with 6.0% CO_2 and saturating humidity.

4   Application of Exogenous Substances and Harvesting the Cells/Tissue

4.1   Fresh Samples

Primary neuronal cell cultures or organotypic slice cultures are easily manipulated for a variety of biochemical assays. Growth factors, biological inhibitors, or other modulatory substances may be added to the medium alone, or in combination with ethanol, using the chambered treatment system. Conditioned medium from the cultures may be saved for detection of cell released cytokines by enzyme-linked immunosorbant assay (ELISA).

Brief exposure of cell cultures to 0.25% trypsin/EDTA (Gibco) generates a dissociated cell suspension suitable to process for cell sorting by flow cytometry. Protein expression profiles from slice cultures or dissociated primary cultures may be determined by Western immunoblot or ELISA after lysis in an appropriate lysis buffer.

4.2   Fixed Samples

Cortical slices are best prepared for immunohistochemistry by fixation. Slices are fixed in freshly made 4% paraformaldehyde for 35 min at room temperature after experimental treatments. Fixative is gently removed from the filter and culture dish and slices are washed once with $0.10 M$ PBS. Slices are transferred to a 12-well plate (each well labeled with the original treatment condition) containing 10% sucrose and incubated for 24 h. We have found it useful to use a spatula with the tip bent to a 90° angle to remove slices from the culture dish and transfer them to the 12-well plate. After 24 h in 10% sucrose, slices are incubated in 15% sucrose for at least 3 h before freezing in preparation for cutting.

In preparation for cryosectioning, slices are removed from the sucrose and transferred to a precleaned Superfrost®/Plus microscope slide (Fisher Scientific, Pittsburgh, PA). Typically three slices or groups of slices are placed on each slide. Sucrose is aspirated off as much as possible. Slices are covered with just enough Optimal Cutting Temperature (OCT; Sakura Finetechnical, Tokyo, Japan) embedding medium to allow for complete coverage of the slices. The slide is pushed down firmly into a flat block of dry ice until completely frozen. Slides are wrapped in aluminum foil and placed in a −20°C freezer until ready for cryosectioning in 12 μm increments. We have found that slice integrity and antigenicity diminish if left in the freezer for longer than a week.

Before immunohistochemistry, sections are incubated in $0.010 M$ citric acid in sterile ddH$_2$O (pH 6.0) for 20 min followed by two rinses in PBS. Standard protocols are then used for the remaining immunolabeling process.

Acknowledgments   The investigators thank a number of others who have worked with us over the years and have helped perfect the culturing methods described in this chapter, including Steve Alcott, Julie Jacobs, Jia Luo, Sandra Mooney, and Julie Siegenthaler. Our work has been funded by the National Institute of Alcohol Abuse and Alcoholism and the Department of Veterans Affairs.

5   Notes

1. Samples of neurons from pups and older animals are less likely to be viable, though allowances (mostly in the composition of the medium) can be made.
2. This takes practice since the cortex is thin and will tend to peel away with the meninges. One method that seems to work well is to start at the olfactory bulbs or the thicker striatal areas.

3. Overly vigorous trituration will result in low cell recovery. Heat the glass pipette tip over an alcohol burner to serially narrow the pipet tip for complete dissociation of cells (cool the tip before triturating).
4. It is recommended that the concentration of ethanol in the medium be measured with an Analox GM7 Micro-Stat Analyzer (or another similar instrument) before and after treatment. Ethanol concentrations will vary with chamber selection and culture conditions and may need to be optimized experimentally. The calculation given for ethanol addition is intended as a starting point to achieve the desired ethanol concentration. In practice, the mathematically determined ethanol concentration generally results in a lower measured ethanol concentration for the chambers described. Increasing the ethanol added to the chamber water and the treatment medium by 10–20% attains the desired ethanol concentration. Samples collected for ethanol measurement may be frozen at −20°C for several weeks without detectable loss of ethanol.

References

1. Gill, T. H., Young, O. M., and Tower, D. B. (1974) The uptake of 36C1 into astrocytes in tissue culture by a potassium-dependent, saturable process. *J. Neurochem.* **23**, 1011–1018.
2. Kennedy, L. A., and Mukerji, S. (1986) Ethanol neurotoxicity. 1. Direct effects on replicating astrocytes. *Neurobehav. Toxicol. Teratol.* **8**, 11–15.
3. Luo, J., and Miller, M. W. (1999) Platelet-derived growth factor-mediated signal transduction underlying astrocyte proliferation: site of ethanol action. *J. Neurosci.* **19**, 10014–10025.
4. Guerri, C., Rupert, G., and Pasqual, M. (2006) Glial targets of developmental exposure to ethanol, in *Brain Development. Normal Processes and the Effects of Alcohol and Nicotine.* (Miller, M. W., ed.). Oxford University Press, New York, pp. 295–312.
5. Seil, F. J., and Herndon, R. M. (1970) Cerebellar granule cells in vitro. A light and electron microscope study. *J. Cell Biol.* **45**, 212–220.
6. Levi, G., Aloisi, F., Ciotti, M., Thangnipon, W., Kingsbury, A., and Balazs, R., (1989). Preparation of 98% pure granule cell cultures, in *Dissection and Tissue Culture Manual of the Nervous System* (Shahar, A., deVellis, J., Vernadakis, A., Haber, B., eds.), Alan R. Liss, New York, pp. 211–214.
7. Pantazis, N. J., Dohrman, D. P., Goodlett, C. R., Cook, R. T., and West, J. R. (1993) Vulnerability of cerebellar granule cells to alcohol-induced cell death diminishes with time in culture. *Alcohol. Clin. Exp. Res.* **17**, 1014–1021.
8. Raine, C. S., and Bornstein, M. B. (1974) Unusual profiles in organotypic cultures of central nervous tissue. *J. Neurocytol.* **3**, 313–325.
9. Gahwiler, B. H. (1981) Organotypic monolayer cultures of nervous tissue. *J. Neurosci. Meth.* **4**, 329–342.
10. Bolz, J., Novak, N., Gotz, M., and Bonhoeffer, T. (1990) Formation of target-specific neuronal projections in organotypic slice cultures from rat visual cortex. *Nature* **346**, 359–362.
11. Siegenthaler, J. A., and Miller, M. W. (2004) Transforming growth factor ?1 regulates cell migration in rat cortex: effects of ethanol. *Cereb. Cortex* **14**, 602–613.
12. Popolo, M., and McCarthy, D. M. (2004) Influence of dopamine on precursor cell proliferation and differentiation in the embryonic mouse telencephalon. *Dev. Neurosci.* **26**, 229–244.
13. Miller, M. W. (2006) Effect of prenatal exposure to ethanol on glutamate and GABA immunoreactivity in macaque somatosensory and motor cortices: critical timing of exposure. *Neuroscience* **138**, 97–107.

14. Haydar, T. F., Wang, F., Schwartz, M. L., and Rakic, P. (2000) Differential modulation of proliferation in the neocortical ventricular and subventricular zones. *J. Neurosci.* **20**, 5764–5774.
15. Siegenthaler, J. A., and Miller, M. W. (2005) Transforming growth factor β1 promotes cell cycle exit through the cyclin-dependent kinase inhibitor p21 in the developing cerebral cortex. *J. Neurosci.* **21**, 8627–8636.
16. Siegenthaler, J. A., and Miller, M. W. (2005) Ethanol disrupts cell cycle regulation in developing rat cortex: interaction with transforming growth factor β1. *J. Neurochem.* **95**, 902–912.
17. Gal, J. S., Morozov, Y. M., Ayoub, A. E., Chatterjee, M., Rakic, P., and Haydar, T. F. (2006) Molecular and morphological heterogeneity of neural precursors in the mouse neocortical proliferative zones. *J. Neurosci.* **26**, 1045–1056.
18. Miller, M. W., and Nowakowski, R. S. (1991) Effect of prenatal exposure to ethanol on the cell cycle kinetics and growth fraction in the proliferative zones of fetal rat cerebral cortex. *Alcohol Clin. Exp. Res.* **14**, 229–232.
19. Miller, N. W. (1993) Migration of cortical neurons is altered by gestational exposure to ethanol. *Alcohol Clin. Exp. Res.* **17**, 304–314.
20. Haydar, T. F., Bambrick, L. L., Krueger, B. K., and Rakic, P. (1999) Organotypic slice cultures for analysis of proliferation, cell death, and migration in the embryonic neocortex. *Brain Res. Brain Res. Protoc.* **4**, 425–437.
21. Agmon, A., and Connors, B. W. (1991) Thalamocortical responses of mouse somatosensory (barrel) cortex in vitro. *Neuroscience.* **41**, 365–379.

11
Development and Properties of HepG2 Cells That Constitutively Express CYP2E1

Defeng Wu and Arthur I. Cederbaum

Summary CYP2E1, a member of the cytochrome P450 family, is induced by ethanol. CYP2E1 activates many hepatotoxins to their reactive toxic intermediate form, and generates reactive oxygen species (ROS) during its catalytic cycle. Induction of CYP2E1 plays an important role in ethanol-induced oxidant stress and ethanol toxicity. To study the biochemical and toxicological properties of CYP2E1, our laboratory developed a HepG2 cell line which constitutively expresses the human CYP2E1 form. These cells displayed elevated oxidative stress, loss of mitochondrial function and loss of viability when challenged with prooxidants such as ethanol, polyunsaturated fatty acids (PUFA) such as arachidonic acid, iron, or when depleted of the critical antioxidant glutathione, as compared with control HepG2 cells which do not express CYP2E1. In the sections below, protocols are described for use of these cell lines to assay for CYP2E1-dependent oxidant stress and toxicity. Methods are described as to how the cell lines were established and maintained, how CYP2E1 is assayed, how cellular viability, mitochondrial function and generation of oxidant stress are determined.

Keywords CYP2E1; oxidant stress; ethanol-inducible; lipid peroxidation; mitochondrial membrane potential, HepG2 cells; cell viability; glutathione.

1   Introduction

CYP2E1, an ethanol-inducible form of P450, metabolizes and activates many toxicologically important substrates, including ethanol, carbon tetrachloride, acetaminophen, and N-nitrosodimethylamine to more toxic products (1–3). Induction of cytochrome P450 2E1 by ethanol appears to be one of the central pathways by which ethanol generates a state of oxidative stress (4–6). In the intragastric model of ethanol feeding, significant alcohol injury occurs. In these models, large increases in lipid peroxidation have been observed, and the ethanol-induced liver pathology has been shown to correlate with CYP2E1 levels and elevated lipid peroxidation, and to be blocked by inhibitors of CYP2E1 (7–10). Although there is disagreement of the role of CYP2E1 in alcohol liver injury (11–13), convincing evidence has been presented that CYP2E1 in the hepatocyte is a key source of reactive oxygen production and in DNA damage (14–18).

   An approach that our laboratory has used to try to understand basic effects and actions of CYP2E1 is to establish cell lines that constitutively express human CYP2E1. HepG2 cell lines, which overexpress CYP2E1, were established either by retroviral infection methods (MV2E1–9 cells, or E9 cells) or by plasmid transfection methods (E47 cells) (19,20). Hepatotoxins such as CCl4 or acetaminophen were more toxic in E9 cells than control HepG2 cells validating the utility of the model to study CYP2E1-dependent toxicity (21,22). We have characterized the toxicity of ethanol, polyunsaturated fatty acids (PUFA) such as arachidonic acid (AA) and iron in the E9 and E47 cell lines (23–27). Concentrations of ethanol or AA or iron that were toxic to the CYP2E1-expressing cells had no effect on control HepG2 cells not expressing CYP2E1 or to HepG2 cells expressing a different P450, CYP3A4 (3A4 cells). Toxicity to CYP2E1-expressing cells was found when glutathione (GSH) was depleted by treatment with 1-buthionine sulfoximine (27). Inhibitors of CYP2E1 prevented the toxicity by the aforementioned treatments. Antioxidants such as vitamin E, trolox, and catalase also prevented toxicity. The aforementioned treatments of CYP2E1-expressing cells with ethanol, AA, iron, or BSO resulted in an increase in oxidative stress to the cells. In other experiments, damage to mitochondria, e.g., decline in mitochondrial membrane potential, was found in CYP2E1-expressing cells treated with AA or iron or BSO. Please refer to Caro and others *(28–30)* for recent reviews of this work and for CYP2E1 biochemistry and toxicology.

This chapter describes protocols used with the HepG2 cells developed to express CYP2E1 to study CYP2E1-dependent toxicity, mitochondrial damage, and oxidant stress. Assays for CYP2E1 expression, for loss of cell viability, for apoptosis, for decline in mitochondrial membrane potential, for lipid peroxidation and generation of reactive oxygen species (ROS), and for activation of mitogen activated protein kinases (MAP kinases) when the CYP2E1-expressing HepG2 cells are challenged with pro-oxidants will be described herein.

2   Materials

Unless otherwise mentioned, solutions are prepared in distilled, deionized water and, except for buffers, prepared fresh.

2.1   Cell Line Maintenance and Storage

1. Dulbecco's Modified Eagle's Medium (DMEM, Sigma, St Louis, MO), supplemented with 10% fetal bovine serum (FBS, Sigma, St Louis, MO) and 1% penicillin-streptomycin-glutamine (GIBCO, Grand Island, NY).
2. Trypsin (0.25%) and ethylenediamine tetraacetic acid (EDTA; 1 mM) from GIBCO.
3. G418 antibiotic (GIBCO, Grand Island, NY).
4. Cell freezing medium, DMEM containing 20% FBS, 10% dimethyl sulfoxide (DMSO, Sigma), 1% penicillin-streptomycin-glutamine mixture.

2.2   CYP2E1 Oxidation Activity

1. 1 M KH$_2$PO$_4$/K$_2$HPO$_4$ buffer, pH 7.4.
2. 10 mM *p*-nitrophenol (FLUKA) in water.
3. 10 mM NADPH (Sigma) in water.
4. 20 % (W/V) Trichloroactic acid (Sigma).
5. 10 N NaOH.

2.3   Cell Toxicity

1. MTT Test Solution, Thiazolyl Blue Tetrazolium (Sigma), 3.3 mg/mL in PBS, filter before use. N-propanol (Sigma).

2. LDH test solution: A. 50 mM KH$_2$PO$_4$/K$_2$HPO$_4$, pH. 7.3; B. NADH, 4.1 mg/mL in solution A; C. pyruvic acid, 2.1 mg/mL in solution A. In 100 mL LDH Test Solution: Solution A 94 mL, Solution B 3 mL, Solution C 3 mL.
3. Trypan Blue Exclusion Test, 0.4 % Trypan Blue Solution (Sigma).

2.4 Measurement of Malondialdehyde and ROS

1. Stock TCA-TBA-HCl reagent: 15% w/v trichloroacetic acid, 0.375 % w/v thiobarbituric acid, 0.25 N hydrochloric acid. 0.1 mM butylated hydroxytoluene (BHT). This solution may be mildly heated to assist in the dissolution of the thiobarbituric acid.
2. Dichlorofluorescin diacetate: 5 mM stock solution in DMSO (DCF-DA).

2.5 Determination of GSH

1. Buffer A, 0.1 M sodium phosphate, pH 7.4, 5 μM EDTA, 0.6 mM 5,5'-dithio-bis (2- Nitrobenzoic acid) (DTNB, Sigma), 0.2 mM NADPH (Sigma), pH 7.5.
2. Buffer B, 10 U/mL glutathione reductase (Roche) in 0.1 M sodium phosphate buffer, pH 7.5.
3. 10% trichloroacetic acid (TCA).
4. Reduced-form GSH (Sigma).

2.6 Apoptosis Analysis

1. Cell lysis buffer: 10 mM Tris-HCl, pH 7.5, 10 mM NaCl, 10 mM EDTA, 100 μg/ mL proteinase K, and 0.5 % sodium dodecyl sulfate.
2. Chloroform/isoamyl alcohol (24:1) (Sigma), 0.1 μg/mL RNase (Sigma), 1 μg/ mL ethidium bromide (Sigma).
3. Rhodamine 123 (5 μg/mL final), propidium iodide (25 μg/mL final) (Sigma).

2.7 Caspase Activity

1. Lysis buffer: 50 mM Tris HCl, pH 7.4, 50 mM β-glycerophosphate, 15 mM MgCl$_2$, 15 mM EDTA, 10 μM phenylmethy sulfonyl fluoride; 1 mM dithiothretol; 150 μg/mL digitonin.
2. Caspase 3—substrate II, AC-DEVD-AMC.
3. Caspase 8 substrate II, fluorogenic Z-IETD-AFC.

4. Caspase 9 substrate 1, fluorogenic AC-LEHD-AFC (CalBiochem).
5. Prepare 50 mM stock solution in DMSO.
6. Reaction buffer: 100 mM HEPES (pH. 7.5), 10% sucrose, 10 mM dithiothreitol 0.5 mM EDTA (Sigma).

2.8   Mitogen-Activated Protein Kinase (MAPK) Phosphorylation

1. pp38/p38, p-JNK/JNK, p-ERK/ERK antibodies (Santa Cruz Biotechnology, Inc., Santa Cruz, CA).
2. p38 inhibitor SB203580, JNK inhibitor SP600125, ERK inhibitor U0126 (CalBiochem). Stock solutions of 5 mM dissolved in DMSO. Horseradish conjugated anti-IgG antibody (Sigma)
3. Enhanced chemiluminescence immune blot detection reagent (ECL, Amersham Biosciences Inc.).

3   Methods

The HepG2 E47 cell line which constitutively expresses human CYP2E1 was established as follows (20). Human CYP2E1 cDNA, excised from a plasmid p91023(B)-2E1 (kindly provided by Dr. F. J. Gonzalez, National Cancer Institute, Bethesda, MD) was inserted into the EcoRI restriction site of pCI-neo expression vector (Promega, Madison, WI) in the sense and antisense orientation to form the plasmids pCI-2E1 and pCI-as-2E1. The empty pCI plasmid was used to prepare the control, non-CYP2E1-expressing HepG2 C34 cells. Transfections of HepG2 cells were carried out with the use of LipofecAMINE reagent (Life Technologies, Grand Island, NY). Cells were then selected with 0.8 mg/mL geneticin in the medium, colonies formed from the survivor cells were grown to large scale, and subjected to Western blot analysis and PNP oxidation activity assay to determine CYP2E1 activity. Positive clones were subjected to another two rounds of limited dilution screening to create stable cell lines.

3.1   Cell Culture and Storage

1. HepG2 E47 or HepG2 C34 cells are maintained in DMEM containing 10 % FBS and 1 % penicillin-streptomycin-glutamine mixture and 0.1 mg/mL G418. Cells are passaged when approaching confluence by splitting with trypsin/EDTA in 1:5 ratio to provide new maintenance cultures (*see* **Notes 1 and 2**).
2. An aliquot of cells is centrifuged at 600 g for 5 min, the pellet is resuspended in the cell freezing medium at a concentration of 1 × 10^6 /mL and aliquoted into

cryogenic vials in 1 mL each, and the vials are stored at −20°C for 2 h and then at −80°C for 1 h and finally the vials are stored in a rack in liquid nitrogen.
3. When experiments are to be initiated, one or two frozen vials of cells are taken out from the liquid nitrogen and immediately put in a 37°C water bath for 5 min. Cells are centrifuged at 1000 rpm for 5 min, the supernatant is discarded and the cells are resuspended in DMEM in two 75-mL culture flasks. Cells are cultured for about 1 wk, and are now ready to set up new experiments (*see* **Note 3**).

3.2   *CYP2E1 Activity*

1. When the culture of HepG2 E47 or C34 cells grown in the 75-mL flask approaches confluence, cells are harvested with a cell policeman and centrifuged 600 g for 5 min. The cell pellet is resuspended in 5 mL of 0.15 M KCl and soni-cated for 10 s in an ice bath with a W-375 sonicator (50% duty cycle, output at 4 W) to obtain the cellular lysate.
2. Cell lysate is centrifuged at 9000 g for 15 min, and the supernatant is further centrifuged at 105,000 g for 1 h. Afterward, the microsomal pellet is collected.
3. The microsomal pellet is resuspended in 200 µL of 0.15 M KCl with a miniho-mogenizer, and the protein concentration is determined by a protein assay kit from BioRad (Hercules, CA).
4. 100 µg of microsomal protein is incubated in 100-µL reaction buffer containing 0.1 M KH_2PO_4/K_2HPO_4, pH 7.4, 0.4 mM p-nitrophenol, 1 mM NADPH in a 37°C shaking water bath for 1 h.
5. At the end of this period, 30 µL of 20% TCA is added to stop the reaction and the mixture is centrifuged at 10,000 g for 10 min. The supernatant is collected and mixed with 10 µL of 10 N NaOH. The absorbance of the pink-yellow color supernatant is determined in a spectrophotometer at 510 nm, and the CYP2E1 activity is calculated using an extinction coefficient of 9.53×10^5 M^{-1} CM^{-1}. The activity is expressed as pmoles/mg/min.
6. HepG2 E47 cell CYP2E1 protein content can be determined using a regular Western blot method with a CYP2E1 antibody.

3.3   *MTT Assay to Determine Cytotoxicity*

1. HepG2 E47 and control C34 cells are seeded onto 24-well plates at concentra-tions of 1×10^4/1 mL DMEM / well (*see* **Note 4**). Cells are cultured for 24 h and the medium is replaced with fresh DMEM.
2. Arachidonic acid (30–60 µM, diluted with serum to prepare 5 mM stock solution, aliquot and keep in −20°C), BSO (0.1 mM), (prepare 0.1 M stock solution with 1X PBS), Ethanol (50–100 mM) or iron (25 µM, prepared as Fe/nitrilotriacetate = 1:3 couple at concentration of 25 mM stock solution) are added to the cells for 24 h, or varying time points (*see* **Note 5**).

3. Two hours before ending the treatment, 75 μL of MTT solution (tetrazolium salts) is added onto each well and continue to incubate at 37°C for 2 h. The medium is then discarded followed by adding 1 mL of *n*-propanol to stop the metabolism, place in room temperature for 1 h in the dark.

4. The absorbance at 570 and 630 nm are recorded. The net A_{570}-A_{630} is taken as the index of cell viability. The net absorbance from the wells of cells cultured with control medium is taken as the 100% viability value. The percent viability of treated cells is calculated by the formula $(A_{570} - A_{630})_{sample} / (A_{570} - A_{630})_{control} \times 100\%$.

3.4 LDH Assay to Determine Cytotoxicity

1. HepG2 E47 or HepG2 C34 cells are seeded onto six-well plates at a concentration of $5 \times 10^5/2$ mL DMEM/well for 24 h. The medium is replaced with fresh medium and treated as described in Subheading 3.3., step 2.

2. At the end of treatment, the culture medium and the cells are collected separately.

3. The cells are resuspended in 0.5 mL of PBS and sonicated for 5 s in an ice bath (duty cycle 50 %, output control 4 W).

4. Mix 50 μL of medium or cell lysate with 950 μL of LDH reaction solution and measure LDH activity (Δ Absorbance per min) due to NADH formation by scanning in a spectrophotometer at 340 nm, 35°C for 1 min. The cytotoxicity index is expressed as the ratio of $LDH_{medium} / LDH_{lysate + medium} \times 100\%$.

3.5 Trypan Blue Exclusion Assay to Determine Cytotoxicity

1. HepG2 E47 or C34 cells are seeded onto six-well plates at a concentration of $5 \times 10^5/ 2$ mL DMEM per well for 24 h. The medium is replaced with fresh medium.

2. Treat the cells as mentioned previously in Subheading 3.3., step 2.

3. At the end of treatment, collect the medium and cells (trypsin/EDTA harvesting).

4. Collect cells by trypsinization and resuspend the cells in 1 mL of PBS and treat the cells with 0.1 mL of 0.05 % trypan blue for 5 min.

5. Count the excluding non-staining or staining cells in a hemocytometer under a light microscope. The cell toxicity is expressed as: $Staining_{total}/Staining_{total}+ Excluding_{total} \times 100\%$.

3.6 Lipid Peroxidation Analysis

1. HepG2 E47 or C34 cells are seeded onto six-well plates at a concentration of 5×10^5 /well and cultured for 24 hours.Fresh medium is added and the cells are then treated with different toxins as mentioned in Subheading 3.3., step 2.

2. Cells are harvested by trypsinization and resuspended in 0.5 mL of PBS containing 0.1 mM butylated hydroxytoluene (BHT) and sonicated for 10 s in an ice bath using an Ultrasonic Cell Disruptor (50% duty cycle, output at 6) (*see* **Note 6**). Protein concentration of the lysate is determined as mentioned previously.

3. Mix 0.5 mg of lysate protein with 0.5 mL of trichloroacetic acid-thiobarbituric acid-HCl solution (containing 0.1 mM BHT) and heat in a 100°C water bath for 1 h.

4. Centrifuge the reaction solution at 14,000 rpm in a microcentrifuge for 10 minutes and collect the supernatant.

5. The formation of TBA-reactive substances (TBARS) is determined by measuring absorbance at 535 nm and using an extinction coefficient of $1.56 \times 10^5 \, M$/cm to calculate malondialdehyde equivalents (nmoles/mg protein).

3.7   Measurement of ROS Production

1. HepG2 E47 or C34 cells are seeded onto six-well plates at a concentration of 5×10^5 / 2 mL DMEM per well and cultured for 24 h. Replace with fresh medium and treat with different toxins as mentioned in Subheading 3.3., step 2.

2. One-half hour before ending treatment, 2',7'–dichlorofluorescin diacetate is added at a final concentration of 2 µg/mL for 30 min.

3. Cells are harvested by trypsinization and resuspended in 3 mL 1× PBS (avoid direct exposure to strong light).

4. Fluorescence is immediately determined in a Perkin-Elmer 650–105 fluorescence spectrometer using wavelengths of 490 nm for excitation and 525 nm for emission. Results are expressed as arbitrary units of fluorescence per milligram of protein.

3.8   Measurement of GSH Levels

1. HepG2 E47 or C34 cells are seeded onto six-well plates at a concentration of 5×10^5 / 2 mL DMEM per well for 24 h and treated with different toxins as mentioned in Subheading 3.3., step 2.

2. Cells are scraped and harvested in 0.5 mL 1× PBS and mixed with 0.5 mL 10% TCA solution and vortexed briefly.

3. Centrifuge the cell mixture solution at 10,000 rpm for 5 min and collect the supernatant.

4. In a 1-mL cuvette, add 900 µL of Buffer A, 100 µL of Buffer B and 50 µL of sample and determine absorbance at 412 nm for 2 min.

5. Prepare a standard curve with four different concentrations of reduced form of glutathione from 0 to 100 µM in 5% TCA and determine the absorbance at 412 nm for 2 min as step 4 and evaluate the GSH levels (nmoles/mg protein) by comparing with the standard curve.

3.9   Mitochondrial Membrane Potential Analysis

1. Mitochondrial membrane potential is analyzed from the accumulation of rhodamine 123. Cells are seeded onto six-well plates at a concentration of 5×10^5/well for 24 h and treated as detailed previously.
2. One hour before ending treatment, rhodamine 123 is added at a final concentration of 5 µg/mL for 1 h.
3. Cells are harvested by trypsinization and resuspended in 1 mL of PBS containing 25 µg/mL PI and the intensity of the fluorescence from rhodamine 123 and PI is determined by Fas-Scan Flow Cytometry. Cells with high mitochondrial potential and viable are mainly in the lower right quadrants of the PI vs rhodamine 123 curves (*see* **Note 7**).

3.10   Apoptosis Determination by Flow Cytometry

1. E47 or C34 cells are grown in 6 well plates and treated with different toxins as mentioned previously.
2. Harvest the cells by trypsinization and fix with 70% ethanol for 15 min at room temperature.
3. Cells are washed with PBS and incubated with 50 µg/mL RNase A for 2 h at room temperature.
4. Stain the cells with 50 µg/mL PI for 30 min and the fluorescence (excitation, 488 nm; emission, 575 nm) is measured with a flow cytometer. The cells undergoing apoptosis are calculated from the sub-G hypodiploid area programmed by the analyzer (*see* **Note 8**).

3.11   Apoptosis Analysis by Determination of DNA Fragmentation (DNA Ladder)

1. E47 or C34 cells at concentration of 5×10^5/2 mL DMEM per well are grown on six-well plates and treated as detailed previously.
2. The cells are harvested and resuspended in 1 mL of lysis buffer and incubated for 1 h at 50°C.
3. The lysis samples are then extracted with 2 mL of phenol (neutralized with TE buffer, pH 7.5), followed by extraction with 1 mL of chloroform/isoamyl alcohol (24:1).
4. The aqueous supernatants are precipitated with 2.5 volumes of ice-cold ethanol plus 10 % volume of 3 M sodium acetate, pH 5.2, at −20°C overnight.
5. Centrifuge the samples at 13,000 g for 10 min and the pellets are air-dried and resuspended with 50 µL of TE buffer, pH 7.5, supplemented with 0.1 µg/mL RNase A.
6. Samples are electrophoretically separated on a 1.5% agarose gel in 0.5 × TBE buffer containing 1 µg/mL ethidium bromide at 100 V for 2 h.
7. Pictures of the separated DNA on the gel are taken under UV transillumination (*see* **Note 8**).

3.12 Immunohistochemistry Analysis to Determine Nuclear DNA Fragmentation

1. Nuclear DNA fragmentation is determined by a DNA Fragmentation Detection Kit from CalBiochem. HepG2 E47 cells or C34 cells are grown on a cover slide in six-well plates and treated with different toxins as mentioned previously.
2. After treatment, the slides are fixed with 4% para-formaldehyde at room temperature for 15 min and are dehydrated by using ethanol (concentrations are from 70% to 95%).
3. Permeabilize the cells with proteinase K. Inactivate endogenous peroxidase by incubating the cells on the coverslide with 3 % H_2O_2.
4. The fragmented DNA is labeled with labeling solution of the kit and the labeled DNA is mixed with a conjugate solution provided by the kit.
5. Incubate the cells on the slide with 3, 3'-diaminobenzidine (DAB)/H_2O_2/urea solution (provide by the kit).
6. Nuclear DNA fragmentation is detected by stained dark-brown cell nuclei and the number of stained cells is counted under a light microscope.

3.13 Caspase Activity

1. Cells are grown on six-well plates at the concentration of 5×10^5 / 2 mL DMEM per well and treated with different toxins. Cells are scrapped and harvested.
2. The cell pellet is suspended in 300 μL of lysis medium and incubated for 30 min at room temperature.
3. The cell suspension is centrifuged at $200\,g$ for 5 mins and the supernatant is collected.
4. 100 μg of supernatant protein is added into 0.5 mL of the assaying solution containing 50 μM caspase substrates (caspase 3- substrate II, AC-DEVD-AMC; casepase 8 substrate II or Granzyme B substrate II, Z-IETD-AFC; caspase 9 substrate I, fluorogenic, AC-LEHD-AFC respectively, CalBiochem) and incubated for 12 h in a 30°C shaking water bath overnight.
5. Caspase activities are determined by measuring the fluorescence associated with released AMC (7-amino-4-methyl-coumarin, excitation at 380 nm, emission at 460 nm) or AFC (excitation at 400 nm, emission at 505 nm) in a spectrofluorimeter. Results are expressed as arbitrary units per 100 μg cell lysate protein.

3.14 Activation of p38 MAPK

1. E47 or C34 cells are grown on six-well plates for 24 h, and fresh medium is added.
2. The cells are treated with either 60 μM AA or 0.1 mM BSO in the absence or presence of 5 μM of the p38 MAPK inhibitor SB203580 for 0.5, 1.0, 2.0, 4.0, 8.0, or 12 h, respectively (*see* **Note 9**).

3. Collect the cell pellets, resuspend in 200 μL of PBS and sonicate for 10 s in an ice bath (50% duty cycle, output at 5–7 W).

4. Prepare a 8% SDS-polyacrylamide gel, loading 20 μg of each lysate protein onto the gel and carry out the gel electrophoresis for 2 h at 50 V.

5. The blotted membranes are incubated with pp38 MAPK antibody (Santa Cruz, 1:200 diluted) for 4 h at room temperature (or overnight in the cold room).

6. Wash the blotted membranes with PBS containing 0.05 % Tween 20 and incubate with anti-IgG antibody conjugated with horseradish peroxidase (Sigma, 1:10,000 diluted) for 1 h.

7. Wash the membrane and develop the fluorescence using the enhanced chemiluminescence immune blot detection agent. The arbitrary units of density of each lane are determined after scanning with computer software ImageJ (National Institutes of Health).

8. The blotted membranes are washed in a Re-Blot Plus Strong Solution for 15 minutes to wash out the previous bound pp38 MAPK antibody and re-incubate the membrane with p38 MAPK antibody, and repeat step 6 and 7 to determine the total p38 MAPK. The final pp38 MAPK activation levels are expressed as the ratio of pp38 MAPK/p38 MAPK.

9. Similar procedures can be carried out to determine the activation of JNK or ERK or Akt by use of specific antibodies recognizing these kinases.

4   Notes

1. HepG2 E47 cells are a human CYP2E1 cDNA transfected cell line; because of different growth conditions at different times and altered cell density, medium nutrition, and other conditions, the CYP2E1 expression levels may vary considerably. To insure the cells are expressing high CYP2E1 before initiating the experiment, the CYP2E1 PNP oxidation activity should be determined regularly (at least weekly). Because the preparation of microsomes from the cells usually is time consuming, in such case, cell lysate can be used for quick determination of CYP2E1 activity but the lysate protein used for this measurement should be three to five times more than the microsomal protein (100 μg per 100 μL) level. To validate that the oxidation of *p*-nitrophenol is via CYP2E1, one can evaluate whether an inhibitor of CYP2E1 such as 1 mM diallyldisulfide or 0.1 mM diethyldithiocarbamate or anti-CYP2E1 antibody blocks this oxidation.

2. The medium should contain high concentrations of G418 (0.1–0.4 mg /mL) to exclude those cells which do not express CYP2E1. Because G418 is a very expensive antibiotic, for long-term use of these cell lines according our experience, G418 can be used for 1 wk per month in the culture medium to maintain this cell line.

3. We generally find that 2 wk after thawing out fresh cells is an appropriate time when CYP2E1 levels are high to carry out experiments. We typically discard cultures 4 to 6 wk after thawing. Thus, it may be helpful to thaw out new aliquots and start new cultures about a week before discarding the old culture to keep experiments moving ahead.

4. To use E47 cells for toxicology studies, the initial seeding amount of the cells should be controlled carefully. Too many cells increase the density which will enhance resistance to the added toxin which will lower the toxicity index. In many cases, if the treatment period is not longer than 24 h, low serum medium can be used (2 %). This will enhance sensitivity of the cells to added toxins. In this case, control groups should be carefully set up.

5. To validate a role for oxidant stress or caspase or p38 MAPK activation in the cytotoxicity produced by the addition of ethanol or AA or iron or BSO to the E47 cells, the decline in viability as reflected by the decrease in MTT reduction or the increase in the LDH out/LDH total ratio or the increase in trypan blue uptake should be blocked by, e.g., trolox and α-tocopherol, or N-acetylcysteine and GSH ethyl ester or by SB203580 or Ac-DEVD- FMK and Z-VAD-FMK. Inhibitors of other MAPK, e.g., PD98059 for ERK 1/2 or SP600125 for JNK can be studied to assess any specificity for p38 MAPK in the overall toxicity.

6. For the determination of lipid peroxidation, 0.1 mM BHT should be added in the PBS used to sonicate cells to prepare cell lysate, and in the incubation buffer when heating the reaction in boiling water in order to prevent artifactual lipid peroxidation. Besides the cells, lipid peroxidation in the culture medium can be determined because peroxidized lipid products, e.g., lipid hydroperoxides and malondialdehyde, can be secreted from the cells into the medium. We have found approx 15% of total lipid peroxidation products are present in the medium when E47 cells were treated with 60 μM AA for 24 h. To validate that the TBARS assay is reflecting lipid peroxidation, antioxidants such as 0.1 mM trolox, or α-tocopherol can be added to inhibit the absorbance at 535 nm. To validate the 2',7'–dichlorofluorescin fluorescence reflects ROS production; antioxidants such as 1 to 5 mM N-acetylcysteine or GSH ethylester can be added to inhibit the arbitrary units of fluorescence.

7. Mitochondrial membrane potential is analyzed by double staining with rhodamine 123 and PI and fluorescence of the two dyes is detected by flow cytometry. When preparing the sample for determination by flow cytometry, make sure to prepare four additional samples that are nonstained, stained by rhodamine alone, stained by PI alone, and stained by both rhodamine and PI as reference control to adjust the flow cytometer for the following measurements.

8. To validate that the DNA fragmentation or the PI fluorescence reflects apoptosis, inhibitors of specific caspases, e. g., 0.05 mM Ac-DEVD-FMK (caspase inhibitor) or a pan caspase inhibitor such 0.05 mM Z-VAD-FMK can be added. These inhibitors should also prevent any increases in caspase 3 (or other caspase) activities. Moreover, these inhibitors would provide evidence that any apoptosis is caspase-dependent, since caspase-independent apoptosis may also occur.

9. MAPK in E47 cells can be activated at very early time periods (as early as 15 min) depending on different toxins. P38 MAPK can be activated by treatment with $60 \mu M$ AA in 0.5 h, but treating the cells with 0.1 mM BSO requires 2 h or longer to activate p38 MAPK. To study the activation of MAPK in E47 cells by different toxins, a time and dose course study should be conducted to select the correct conditions that may active MAPK in E47 or C34 cells.

References

1. Guengerich, F. P., Kim, D. H., and Iwasaki, M. (1990) Role of human cytochrome P450 IIE1 in the oxidation of many low molecular weight cancer suspects. *Chem. Res. Toxicol.* **14,** 168–179.
2. Yang, C. S., Yoo, J. S.H., Ishizaki, H., and Hong, J. (1990) Cytochrome P450 IIE1: roles in nitrosamine metabolism and mechanism of regulation. *Drug Metab. Rev.* **22,** 147–159.
3. Koop, D. R. (1992) Oxidative and reductive metabolism by cytochrome P4502E1. *FASEB J.* **6,** 724–730.
4. Esktrom, G., and Ingelman-Sundberg, M. (1989) Rat liver microsomal NADPH-supported oxidase activity and lipid peroxidation dependent on ethanol-inducible cytochrome P450. *Biochem. Pharmacol.* **38,** 1313–1319.
5. Tsukamoto, H., and Lu, S. C. (2001) Current concepts in the pathogenesis of alcoholic liver injury. *FASEB J.* **15,** 1335–1349.
6. Cederbaum, A. I. (2001) Introduction: serial review: alcohol, oxidative stress and cell injury. *Free Rad. Biol. Med.* **31,** 1524–1526.
7. Castillo, T., Koop, D. R., Kamimura, S., Triadafilopoulos, G., and Tsukamoto, H. (1992) Role of cytochrome P4502E1 in ethanol-, carbon tetrachloride-, and iron-dependent microsomal lipid peroxidation. *Hepatology* **16,** 992–996.
8. Morimoto, M., Zern, M. A., Hagbjork, A. L., Ingelman-Sundberg, M., and French, S. W. (1994) Fish oil, alcohol, and liver pathology: role of cytochrome P4502E1. *Proc. Soc. Exp. Biol. Med.* **207,** 197–205.
9. Nanji, A. A., Zhao, S., Sadrzadeh, S. M.H., Dannenberg, A. J., Tahan, S. T., and Waxman, D. J. (1994) Markedly enhanced cytochrome P4502E1 induction and lipid peroxidation is associated with severe liver injury in fish oil-ethanol-fed rats. *Alcohol. Clin. Exp. Res.* **18,** 1280–1285.
10. Gouillon, Z., Lucas, D., Li, J., Hagbjork, A. L., French, B. A., Fu, P., Fang, C., Ingelman-Sundberg, M., Donohue, T. M. Jr., and French, S. W. (2000) Inhibition of ethanol-induced liver disease in the intragastric feeding rat model by chlormethiazole. *Proc. Soc. Exp. Biol. Med.* **224**; 302–308.
11. Koop, D. R., Klopfenstein, B., Iimuro, Y., and Thurman, R. G. (1997) Gadolinium chloride blocks alcohol-dependent liver toxicity in rats treated chronically with intragastric alcohol despite the induction of CYP2E1. *Mol. Pharmacol.* **51,** 944–950.
12. Kono, H., Bradford, B. U., Yin, M., Sulik, K. K., Koop, D. R., Peters, J. M., Gonzalez, F. J., McDonald, T., Dikalova, A., Kadiiska, M. B., et al. (1999) CYP2E1 is not involved in early alcohol-induced liver injury. *Am. J. Physiol.* **277,** G1259–G1267.
13. Isayama, F., Froh, M., Bradford, B. U. Mckim, S. E., Kadiiska, M. B., Connor, H. D., Mason, R. P., Koop, D. R., Wheeler, M. D., and Arteel, G. E. (2003) The CYP inhibitor 1-aminobenzotriazole does not prevent oxidative stress associated with alcohol-induced liver injury in rats and mice. *Free Rad. Biol. Med.* **35,** 1568–1581.
14. Morgan, K., French, S. W., and Morgan, T. R. (2002) Production of a cytochrome P450 2E1 transgenic mouse and initial evaluation of alcoholic liver damage. *Hepatology.* **36,** 122–134.

15. Bardag-Gorce, F., Yuan, Q. X., Li, J., French, B. A., Fang, C., Ingelman-Sundberg, M., and French, S. W. (2000) The effect of ethanol-induced cytochrome P450 2E1 on the inhibition of proteasome activity by alcohol. *Biochem. Biophys. Res. Commun.* **279**, 23–29.
16. Bradford, B. U., Kono, H., Isayama, F., Kosyk, O. Wheeler, M. D., Akiyama, T. E., et al (2005) Cytochrome P4502E1 but not NADPH oxidase is required for ethanol-induced oxidative DNA damage in rodent liver. *Hepatology* **41**, 336–344.
17. Tanaka, E., Terada, M., and Misawa, S. (2000) Cytochrome P450 2E1: its clinical and toxicological role. *J. Clin. Pharm. Ther.* **25**, 165–175.
18. Gonzalez, F. J. (2005) Role of cytochromes P450 in chemical toxicology and oxidative stress I studies with CYP2E1. *Mutat. Res.* **569**, 101–110.
19. Dai, Y., Rashba-Step, J., and Cederbaum, A. I. (1993) Stable transfection of human cytochrome P4502E1 in HepG2 cells: characterization of catalytic activities and production of reactive oxygen intermediates. *Biochemistry* **32**, 6928–6937.
20. Chen, Q., and Cederbaum, A. I. (1998) Cytotoxicity and apoptosis produced by cytochrome P450 2E1 in HepG2 cells. *Mol. Pharmacol.* **53**, 638–648.
21. Dai, Y., and Cederbaum, A. I. (1995) Cytotoxicity of acetaminophen in human cytochrome P4502E1-transfected HepG2 cells. *J. Pharmacol. Exp. Ther.* **273**, 1497–1505.
22. Dai, Y., and Cederbaum, A. I. (1995) Inactivation and degradation of human cytochrome P4502E1 by CCl4 in a transfected HepG2 cell line. *J. Pharmacol. Exp. Ther.* **275**, 1614–1622.
23. Wu, D., and Cederbaum, A. I. (1996) Ethanol toxicity to transfected HepG2 cells expressing human cytochrome P4502E1. *J. Biol. Chem.* **271**, 23914–23919.
24. Chen, Q., Galleano, M., and Cederbaum, A. I. (1997) Cytotoxicity and apoptosis produced by arachidonic acid in HepG2 cells overexpressing human cytochrome P4502E1. *J. Biol. Chem.* **272**, 14532–14541.
25. Wu, D., and Cederbaum, A. I. (1999) Ethanol-induced apoptosis to hepatic cell lines expressing human cytochrome P4502E1. *Alcohol. Clin. Exp. Res.* **23**, 67–76.
26. Sakurai, K., and Cederbaum, A. I. (1998) Oxidative stress and cytotoxicity induced by ferric-nitrilotriacetate in HepG2 cells expressing CYP2E1. *Mol. Pharmacol.* **54**, 1024–1035.
27. Wu, D., and Cederbaum, A. I. (2001) Removal of glutathione produces apoptosis and necrosis in HepG2 cells overexpressing CYP2E1. *Alcohol. Clin. Exp. Res.* **25**, 619–628.
28. Caro, A. A., and Cederbaum, A. I. (2004) Oxidative stress, toxicology and Pharmacology of CYP2E1. *Annu. Rev. Pharmacol. Toxicol.* **44**, 27–42.
29. Kessova, I. and Cederbaum, A. I. (2003) CYP2E1: Biochemistry, toxicology and regulation. *Curr. Mol. Med.* **3**, 509–518.
30. Jimenez-Lopez, J. M., and Cederbaum, A. I. (2005) CYP2E1-dependent oxidative stress and toxicity: role in ethanol-induced liver injury. *Expert Opinion Drug Metab. Toxicol.* **1**, 671–685.

12

Modeling the Impact of Alcohol on Cortical Development in a Dish

Strategies from Mapping Neural Stem Cell Fate

Rajesh C. Miranda, Daniel R. Santillano, Cynthia Camarillo, and Douglas Dohrman

Summary During the second trimester period, neuroepithelial stem cells give birth to millions of new neuroblasts, which migrate away from their germinal zones to populate the developing brain and terminally differentiate into neurons. During this period, large numbers of cells are also eliminated by programmed cell death. Therefore, the second trimester constitutes an important critical period for neuronal proliferation, migration, differentiation and apoptosis. Substantial evidence indicates that teratogens like ethanol can interfere with neuronal maturation. However, there is a paucity of good model systems to study early, second trimester events. In vivo models are inherently interpretatively complex because cell proliferation, migration, differentiation, and death mechanisms occur concurrently in regions like the cerebral cortex. This temporal overlap of multiple developmental critical periods makes it difficult to evaluate the relative vulnerability of any individual critical period. Our laboratory has elected to utilize fetal rodent cerebral cortical-derived neurosphere cultures as an experimental model of the second-trimester ventricular neuroepithelium. This model has enabled us to use flow cytometric approaches to identify neuroepithelial stem cell and progenitor sub-populations and to show that ethanol accelerates the maturation of neural stem cells. We have also developed a simplified mitogen-withdrawal/matrix-adhesion paradigm to model the exit of neuroepithelial cells from the ventricular zone towards the subventricular zone and cortical plate, and their maturation into multipolar neurons. We can treat neurosphere cultures with ethanol to mimic exposure during the period of neuroepithelial proliferation and by using the step-wise maturation model, ask questions about the impact of prior ethanol exposure on the subsequent maturation of neurons as they migrate and undergo terminal differentiation. The combination of neurosphere culture and stepwise maturation models will enable us to dissect out the contributions of specific developmental critical periods to the overall teratology of a drug of abuse like ethanol.

Keywords Neural stem cells; CD133; Sca-1; ABCG2; c-kit; neurosphere culture; Mitogen; FGF; EGF; LIF.

1   Introduction

1.1   *The Fetal Alcohol Syndrome*

Heavy ethanol consumption during pregnancy can persistently alter fetal development and lead to a constellation of craniofacial, brain, and cardiovascular defects that are collectively termed the fetal alcohol syndrome or, F.A.S *(1)*. The constellation of brain defects includes microencephaly, malformations of gyri, diminution or loss of interhemispheric communicating fiber tracts like the corpus callosum (reviewed in *(2,3)*), and the presence of "brain warts" or heterotopias containing displaced neurons *(3,4)*. Because of an increasing recognition that lower levels of ethanol consumption during pregnancy also can lead to neurological, behavioral, and cognitive deficits, the range of defects associated with *in utero* alcohol exposure have collectively been termed fetal alcohol spectrum disorders or, F.A.S.D *(5,6)*.

In *utero* alcohol exposure is the most important nongenetic cause of mental retardation. Genetic susceptibility factors do increase vulnerability to *in utero* ethanol exposure. However, the identified risk factors include genes like alcohol dehydrogenase (e.g., the, A. D.H1B*2 allele *(7)*), which control the metabolism of ethanol and, consequently, titrate either maternal or fetal blood alcohol levels. It follows, therefore, that a dose of ethanol that results in permanent alteration of the nervous system must produce its effects by disrupting events underlying critical periods of neural development, rather than recruiting genetic susceptibility factors per se. In other words, the effect of ethanol during a particular developmental period is determined by the specific biological events that occur during that period. Therefore, to study the effects of ethanol on brain development, we need to pay close attention to appropriately modeling relevant underlying biological events. It is critically important that models of neural development closely mimic pertinent aspects of *in utero* developmental biology.

1.2 Effects of Alcohol During Brain Growth and Development

The brain growth spurt period (the second and third trimester equivalent of human gestation, comparable with the latter half of gestation and the early postnatal period of rodent development *(8)*), is characterized by rapid neuronogenesis (*(9)*, the initial period of neuron generation), compensatory apoptosis *(10,11)*, neuroblast migration out of the ventricular zone (VZ), and early neuronal maturation. This developmental phase constitutes a period of particular vulnerability to alcohol. Most studies on the effects of alcohol on brain development have focused on the third-trimester model. For example, early studies from West and his colleagues used a third-trimester model in rats to demonstrate neuronal cell loss in the cerebellum, olfactory bulb, and hippocampus after alcohol exposure *(12-17)*. Rodent cell culture models and tumor-derived cell lines also have been used extensively to understand the effects of ethanol on neuronal survival and differentiation, i.e., events that occur during the third trimester. Using these models, we and others, have shown that, in differentiated neural tissue, part of alcohol's neurotoxicity may be caused by the induction of death mechanisms *(18-21)*, the loss of growth and survival factors *(22-28)*, alterations in neuronal migration *(29)*, and in neurotransmitters systems *(30)*, among others. Further, in vitro studies found that the state of differentiation of the cells determined the degree of sensitivity to alcohol insult *(31,32)*, confirming in vivo work showing windows of vulnerability that reflected maturation of the neurons *(16,33,34)*.

In contrast to the extensive research that has focused on the third-trimester effects of ethanol, the second trimester represents a poorly understood and less-studied period of vulnerability to ethanol. The second trimester is an important time frame for study of alcohol's effects because during this period, millions of new neurons are born, migrate away from neuroepithelial germinal zones (primarily the ventricular and subventricular zones), and populate various brain regions *(35,36)*, laying down a cellular framework for the rest of brain development. The maturation of neural stem cells plays a crucial role in the process of neuronogenesis. An early study by Barnes and Walker *(37)* reported a loss of hippocampal neurons from a second-trimester equivalent alcohol exposure, and significant work from Miller's lab showed alcohol-induced changes in cortical neuronogenesis *(38-40)*. Others have shown that rats exposure to ethanol over a 2-d window from, G. D.14-15 exhibit an immediate enlargement of the, S.V.Z, suggesting, N.S.C/NPC maturation *(41)*, disorganized cortical architecture at the end of the neuronogenic period *(42)*, and a persistent thinning of lamina V of the rodent cerebral cortex *(43)*, suggesting that the effects of second-trimester ethanol exposure are persistent.

Until recently, we have lacked good culture models to examine the cell and molecular biological underpinnings of ethanol's actions. In this chapter, we will discuss cell and tissue culture models that we use in our laboratory to model pertinent events that occur during the second trimester of human gestation, with specific reference to the formation of the cerebral cortex.

1.3 Neurodevelopment During the Second-Trimester Period

During the second trimester-equivalent prenatal period of neurogenesis, the number of neuroepithelial cells expands rapidly to generate most of the neurons of the adult brain *(44)* requiring, as with other tissues (Fig. 1. *(45)*), the conversion of uncommitted stem cells to more fate-restricted progenitors, then to blast-cells, and ultimately neurons (Fig. 2). The fetal murine cerebral cortical ventricular zone initiates proliferation by gestational day (GD) 11, and a neural stem cell (NSC) that starts proliferating on, G. D.11 will undergo ~11 integer cell cycles during the period required to generate cortical plate (CP) neurons *(36)*. During the peak period of neuronogenesis, the rodent cerebral cortex is estimated to add ~2400 new neurons to the cerebral CP per minute *(46)*. Neuroepithelial cells are coupled to each other by gap junctions during S and G2 phases of cell cycle *(47)* and are vulnerable to apoptosis during these cell cycle phases *(11)*. Consequently, the fate of a single neuroepithelial cell is likely to be tied to the fate of its neighbors, and to its cell cycle stage, and the collective neuroepithelial response to a teratogen may be more critical than the response of individual neuroepithelial cells. A teratogen that alters the rate of proliferation and death in neuroepithelial cells is likely to significantly alter the structure of the mature brain.

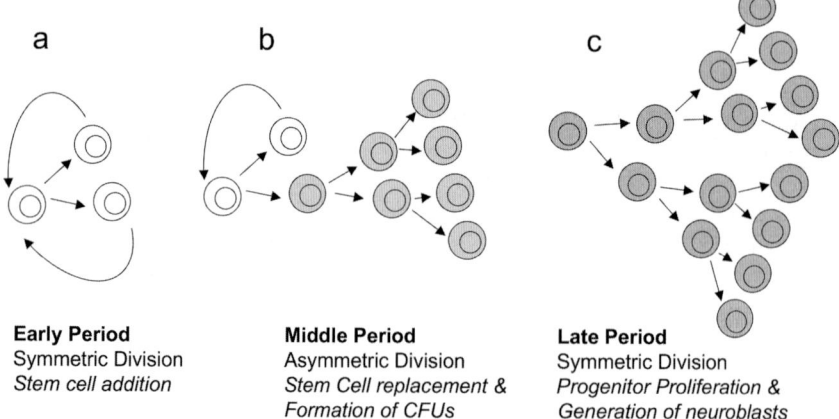

a	b	c
Early Period	**Middle Period**	**Late Period**
Symmetric Division	Asymmetric Division	Symmetric Division
Stem cell addition	*Stem Cell replacement &*	*Progenitor Proliferation &*
	Formation of CFUs	*Generation of neuroblasts*

Fig. 1 Schematic for modes of cell division observed in the stem cell beds. Symmetrical division **(a)** results in the generation of two daughter stem cells and permits the early expansion of the stem cell pool. Asymmetric division **(b)** in contrast, results in the generation of one stem and one more mature, progenitor daughter cell (e.g., in bone marrow, this results in the generation of a common lymphoid or myeloid progenitor, and a replacement stem cell). Asymmetric cell division therefore results in the maintenance of the stem cell pool. Subsequent cell divisions result in the clonal expansion of the progenitor pool (e.g., the transformation of a common myeloid progenitor to erythroid or myeloid CFUs). During the late period of maturation **(c)** symmetric cell division results in two daughter cells that are more mature, blast-type cells, resulting in a depletion of the parent progenitor pool (e.g., transformation of an erythroid CFU to a proerythroblast, resulting finally in the formation of an erythrocyte)

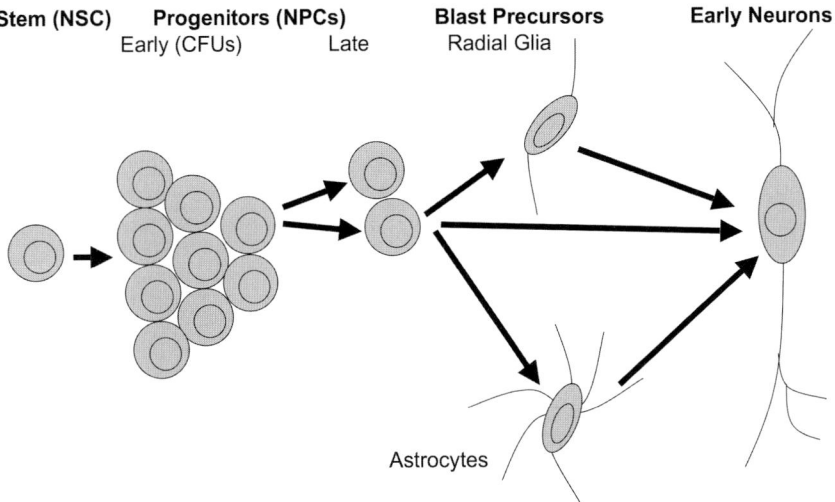

Fig. 2 Presumptive maturation program of neural stem cells. Neural stem cells give rise to progenitors, which give rise to blast-type cells (radial glia and astrocyte-type cells), which can give rise to neurons

1.4 The Neurosphere Culture Model Recapitulates the Second-Trimester Neuroepithelium

The nonuniform distribution of cell proliferation cycles throughout the neuronogenic period suggests that classes of rapidly proliferating neuroepithelial progenitors that are present early in the neuronogenic period are either eliminated or otherwise suppressed as the ventricular zone matures. We and others have modeled the fetal neuroepithelium using neurosphere cultures derived from the fetal mouse and rat ventricular zones. We typically isolate cortical tissue from GD 12.5 mouse fetuses corresponding to the initial period of CP neuronogenesis (although we have also isolated tissue from rat fetuses and, at later gestational ages, corresponding to the peak of CP neurogenesis). Isolated cells are dispersed into defined culture medium and may be further fractionated by immunomagnetic separation (MACS, Miltenyi Biotech), or by fluorescence-assisted cell sorting (FACS). In the presence of mitogenic factors, the individual neuroepithelial cells form floating clonal colonies, or spheroid bodies. In the absence of extracellular matrix molecules, cells preferentially adhere to each other and are often referred to as "neurospheres." An individual neurosphere is quite heterogenous, with respect to maturation state. Cells expressing the nestin, glial fibrillary acidic protein (GFAP), and the early neuronal maturation marker, microtubule-associated protein-2 (MAP-2), can all be identified within neuronal cultures. However, other markers such as NeuN, which identify mature neurons, are not present in neurosphere cultures. Neurospheres grown from single cells can assume varying sizes during a 48- to 72-h period, suggesting that their parental neuroepithelial cells are intrinsically variable with respect to cell cycle kinetics, perhaps reflecting an emerging heterogeneity of fate within the neuroepithelium.

1.5   Mapping the Diversity of Fetal Neuroepithelial Cells

Increasing evidence suggests that cortical neuron heterogeneity results from early diversification of the neuroepithelium (48-52), well before the advent of external influences in the form of cortico-fugal projections from other brain nuclei. For example, pyramidal neurons of the mature cerebral cortex are generated within the cortical ventricular zone, wherea interneurons are generated within the ventral ganglionic eminences and migrate tangentially into the CP (53-55). Neuronal heterogeneity in terms of gene expression patterns and lamina preference emerges within each neuroepithelial zone as well. Transcription factors like Lhx-1&2, TBr-1, and Emx-1 identify distinct subpopulations of neuronal precursors (53,56). Because of these and other data, it has been suggested that the cortical neuroepithelium contains a "protomap" (56) of the mature adult cortex and that the principal determinants of cortical structure arise from diversification of stem and progenitor cells within the cortical neuroepithelium itself.

Transcription factor identity, however, is not useful for isolating live neuroepithelial subpopulations. Such cell-types are more easily isolated based on their differential expression of cell-surface antigens. For example, we and others have shown that cell-surface molecules like the Fas/Apo-1/CD95 suicide receptor (10,11), β2-microglobulin, a, M. H.C class I antigen (50), and receptor tyrosine kinase, EphA (56), all mark subsets of neuroepithelial precursors within the, V. Z. Interestingly, these proteins (57,58), or their family members (59), also mark cells in various stages of hematopoiesis, supporting the notion that stem and progenitor cells in dissimilar tissues nevertheless express a common repertoire of surface antigens.

Hematopoietic-derived cells express different surface antigens at various developmental transitions (60) and, thus, surface antigen expression reflects the status of development along a competency continuum. In our research, we have elected to use cell-surface markers derived from the hematopoietic system, like, C. D.133/prominin-1, Sca-1 (Ly6A/E), CD117/c-kit, and, A. B.CG2 (ATP-binding cassette, sub-family G (WHITE), member 2), to define early cortical neuroepithelial subpopulations. These cell surface markers have been used successfully to monitor stem cell heterogeneity in a variety of tissues (61-64). ABCG2 in particular is most likely to mark stem cells uniquely. Stem cells in the hematopoietic and nervous system (61,65,66) display a unique ability to rapidly induce efflux of Hoechst dye (#33342), thereby generating a characteristic staining-pattern (termed the side population or 'SP'), which can be assessed by flow cytometry (67,68). The, S. P.-population in hematopoietic tissues represents less than 1% of the total population and is a rare event, as is expected for a stem cell group. The protein, ABCG2 (69), confers a, S. P.-population phenotype to stem cells, and is downregulated in more differentiated cells *(70-72)*. In our research, we have therefore used, A. B.CG2 expression to mark neural stem cells, and to monitor this population after ethanol exposure. Stem cells can exhibit two alternate modes of cell division *(73)*; the first, symmetric division (Fig. 1a), results in the formation of two daughter stem cells, whereas the second, i.e., asymmetric division, results in the formation of one daughter stem cell, and a second, more mature, daughter progenitor cell (Fig. 1b). Symmetric division boosts the pool of stem cells while asymmetric division maintains the stem cell pool.

In the mouse, a major burst of cell proliferation occurs between, G.D.11 and, G.D.14, encompassing 63% of the integer cell cycles that span the neuronogenic period *(36)*. Within the hematopoietic system, this burst pattern is consistent with the appearance of colony-forming units (CFUs or early progenitors), which proliferate rapidly by symmetric division (Fig. 1b) to regenerate lineage-specific colonies, and are present in large numbers. Recent evidence indicates that such symmetric division of neural, C. F.U-equivalents occurs at a high rate in the apical portion of the fetal neuroepithelium, and is an important source of neurons *(73)*. Our data show that the antibodies to Sca-1, CD133, and, C.D.117 label ~30%, 12%, and 20% of neuroepithelial cells, respectively, by flow cytometry *(74)*. Because these antigens mark a significantly larger number of cells than, A. B.CG1, it is likely that these antigens mark more mature (CFU and later) stages of neuronal lineage. Later symmetric division may result in the continued formation of lineage committed, blast precursors (Fig. 1c).

1.6 Modeling Early Cortical Neuronal Differentiation With Culture Models

Cells leave the, V.Z. to directly form CP neurons *(75)*, or to form radial glial and astrocytic intermediate blast precursors to additional neurons (Fig. 2 *(76,77)*). Our laboratory has been interested in developing models to study this transition between cell proliferation and differentiation in the cerebral cortex. In our initial model of cortical maturation, we infected fetal rodent-derived cortical cells with an adenovirus expressing a temperature sensitive, S.V.40-large T antigen (*ts*TA) *(78)*. At the permissive temperature for T-antigen expression (+*ts*TA) fetal cortical cells proliferate rapidly, assume an epithelioid morphology, and express the intermediate filament protein, nestin. However, at the non-permissive temperature for T-antigen expression (-*ts*TA), neuroepithelial cells exit cell cycle, and large numbers of cells undergo apoptosis by p53-dependent mechanisms *(11,78)*. Surviving cells, however, undergo morphological transformation and initiate the growth of neurites. Treatment with retinoic acid results in further differentiation along a neuronal lineage *(78)*. In this model, SV40-*ts*TA served as a molecular switch between states of cell proliferation on one hand, and differentiation and apoptosis on the other. Although this model did partially segregate critical periods, it was subject to several limitations. First, it was not clear whether specific subpopulations of the neuroepithelium are selectively eliminated by apoptosis in the –*ts*TA condition. Second, the, S.V.40-T antigen maintains cells in cell cycle by suppressing cell cycle arrest factors like p53 and p21/Waf-1, thereby permitting the accumulation of gene mutations and ultimately, escape from cell cycle controls.

1.6.1   Naturalistic Models of Neuronal Differentiation

In comparison with, S.V.40-T antigen-transformed cells, neurosphere cultures can be differentiated under well-defined, naturalistic conditions, to model the exit of neuroblasts out of the, V.Z., and their maturation into multipolar neurons. In the model outlined herein, we empirically defined conditions that include sequential withdrawal of mitogenic factors and addition of extra-cellular matrix (to activate integrin signaling *(79))*, resulting in the sequential appearance of two unique morphological phenotypes of migratory bipolar, or multipolar neurons. We classify the migratory, bipolar cells as belonging to the subventricular zone (SVZ) or "early-neuronal differentiation" phenotype. The multipolar morphology is characteristic of both, S.V.Z *(75)* and CP neurons. We refer to this phenotype as the, C. P. or "late neuronal differentiation" phenotype. This model provides for several advantages. We have shown that the transformation between proliferation and differentiation phenotypes is not accompanied by a significant alteration in apoptosis *(80)*, unlike our previous, S. V.40-T antigen-transformed model. Second, this model recapitulates the major second-trimester events, neuroepithelial proliferation, migration, and neuronal maturation, which are necessary for the development of brain regions like the cerebral cortex. Finally, and most importantly, this model permits us to expose cells to a presumptive teratogen during one stage of differentiation, and to examine the immediate (activational) effects on cells within that stage, as well as the persistent (organizational) effects of that teratogen on subsequent neuronal differentiation. This model therefore enables us to capture the key feature of a teratogen, which is its ability to persistently and permanently alter the development of tissues, even though it is no longer present within that tissue's environmental milieu.

2   Materials

1. Dulbecco's phosphate-buffered saline (DPBS, cat. no. 14040-133).
2. Dulbecco's minimal essential medium (DMEM, cat. no. 10313-021).
3. Hanks's balanced salt solution (HBSS, cat. no. 14175-095).
4. Trypsin/EDTA (cat. no. 25300-112).
5. Bovine serum albumin, fraction-V, (BSA-V, cat. no. 1526037).
6. Laminin (cat. no. 23017-015).
7. Ethylene diamine tetraacetic acid (EDTA), DMEM/F12 (cat. no. 11330-032).
8. Basic fibroblast growth factor (bFGF; cat. no. 13256-029).
9. Human recombinant epidermal growth factor (EGF; cat. no. 53003-018).
10. ITS-X supplement (Insulin, Transferrin, Selenium, Invitrogen cat. no. 51500-056).
11. Heparin (Invitrogen cat. no. 15077-019).
12. Recombinant human Leukemia Inhibitory Factor (LIF; cat. no. L200) (Alomone Labs. Progesterone, cat. no. P6149).
13. 95% Ethanol is obtained from Sigma.

3   Methods

3.1   Isolation of Embryonic Neural Precursors

Timed-pregnant C57/Bl6 mice were generated by timed-mating for 1 h, and GD
−0.5 was defined as the day when female mice exhibited postcoital sperm plugs.
Mice were maintained in the animal housing facility at Texas A&M University
System Health Sciences Center, College of Medicine, on a **12,** 12-h light-dark
schedule. At, G. D. 12.5, mice were anesthetized with a mixture of ketamine
(0.09 mg/g body weight) and xylazine (0.106 mg/g body weight) by intramuscular
injection. The abdomen of the anesthetized pregnant mouse was swabbed first with
80% ethanol (vol/vol), and then with Betadine (10% povidone-iodine; *see* **Note 1**).
A laparotomy was then performed with fresh sets of dissection instruments to make
sequential incisions into the skin and underlying peritoneum. The gravid uterus was
dissected, rinsed in chilled, D. P.BS (*see* **Note 2**), the fetuses dissected, and fetal
brains removed and placed in chilled, H. B.SS supplemented with glucose and
magnesium chloride. Meningeal tissue was removed (*see* **Note 3**), regions of the
mouse fetal brain corresponding to the structural precursor of the neocortex were
isolated, and care was taken to exclude the structural precursors to the striatum and
hippocampus. Individual cortical fragments (*see* **Note 4**) are collected in sterile 15-
mL conical tubes and gently triturated in trypsin/EDTA. Trypsin is inactivated with,
D. M.EM containing 10% fetal bovine serum. The cell suspension is centrifuged
for 5 min at 18°C, 1000 rpm (300 g). Cell pellets are resuspended in chilled, D. P.
BS containing 0.5% BSA, Fraction-V, and 2.0 mM EDTA. Total cell counts are
determined using a hemocytometer. Dispersed neuroepithelial precursors are established in culture at an initial density of 10^6 cells in T-25 flasks containing serum-
free mitogenic media (DMEM/F12, 20 ng/mL bFGF, 20 ng/mL EGF, 0.15 ng/mL
(*see* **Note 5**) LIF, ITS-X supplement, 0.85 Units/mL heparin, and 20 nM progesterone). Cultures are incubated at 37°C, 5% CO_2 in a humidified environment to generate neurospheres (Fig. 3). Cells are allowed to proliferate as neurospheres until
cultures achieved a density of 2×10^6 cells per T25 flask (*see* **Note 6**), and after
approximately 6-8 passages (at ~3 d per passage), used for experiments.

3.2   Ethanol Treatment

Neurosphere cultures in T25 flasks are randomly assigned to control or ethanol
treatment groups. In our laboratory, we have treated neurosphere cultures with a
wide range of ethanol doses, ranging from 60 mg/dL (~13 mM) to 920 mg/dL
(200 mM). We monitor culture media ethanol concentrations with gas chromatography. Ethanol containing medium is prepared freshly before use, from 95% ethanol. Each flask is defined as a single sample. Culture medium is changed every 2 d.
Control and ethanol-treated flasks are capped tightly with phenolic caps, and sealed
with parafilm to limit the loss of ethanol. These measured doses range in equivalence

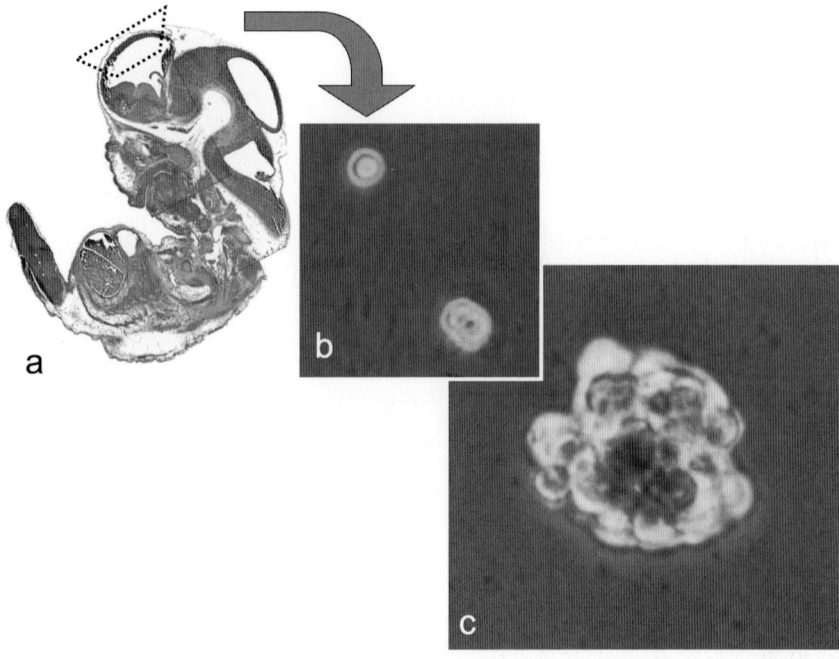

Fig. 3 (**a**) Boxed area indicates the region from which neuroepithelial cells are isolated. (**b** and **c**) Neuroepithelial cells dispersed by trituration, and cultured in defined medium, generate nonadherent aggregates of cells referred to as neurospheres. The cellular composition of neurospheres is heterogeneous, and includes cells that express stem and progenitor markers (ABCG2, Sca-1, c-kit/CD117, CD133), nonselective markers of immature cells (nestin), markers for blast cells (Glial Fibrillary Acidic Protein, GFAP), as well as markers for early neuronal maturation (MAP-2). However, neurosphere cells do not express the neuron-specific marker, NeuN

to consumption levels that can be attained by social drinkers (60 mg/dL) to those attained by chronic alcoholics (320–620 mg/dL, 70–135 mM (**81,82**)), to levels above those typically attainable in chronic alcoholics (920 mg/dL). Our analysis indicates that there is no significant change in ethanol content within culture dishes over the exposure time period (Fig. 4).

3.3   Differentiation of Neurosphere Cultures

We characterize our neurosphere cultures as a fetal ventricular-zone "neuroepithelial proliferation" or, V. Z. model. To initiate the differentiation program, neurosphere cultures are transferred to fresh T-25 flasks, Petri-dishes, or microwell plates coated with laminin (at 50 µg/mL in, D.M.EM/F12 for 1 h). The presence of laminin by itself is enough to cause neurospheres to become adherent and permits neuroepithelial cells to migrate away from the parent neurosphere (**80**), indicating that laminin-mediated activation of integrins is a strong migratory stimulus. However, migrating neuroepithelial cells acquire a squamous epithelioid appearance and do not exhibit neurites and

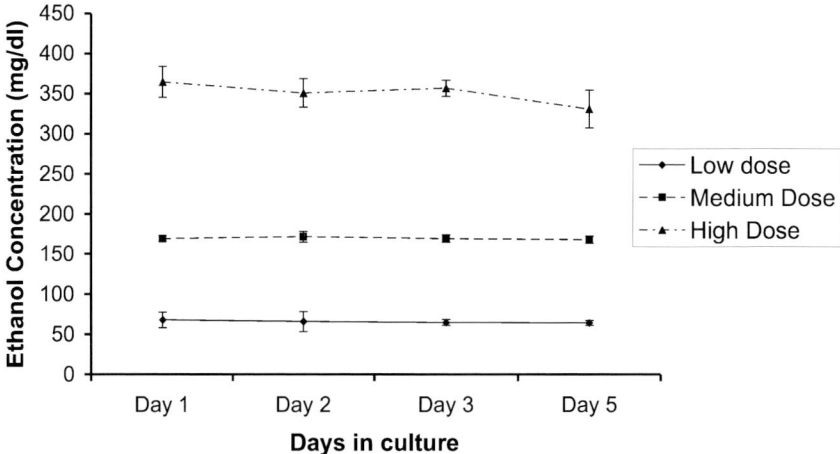

Fig. 4 Gas chromatographic analysis of ethanol content in culture medium over a 5-d exposure period. Culture medium is replaced on day 3. The data show that ethanol concentrations remain stable over the period of the experiment

growth cones that are typical of migratory neurons. Within 24 h after concurrent removal of EGF and LIF from the culture medium (i.e., the +Laminin/+FGF/-EGF/-LIF condition) these migratory cells lose their epithelioid morphology and assume a bipolar appearance (Fig. 5). Unlike cells of the neurosphere cultures, these migrating cells express the neuronal nuclear antigen, NeuN, within their nuclei and consequently can be defined as early, migratory neurons. We refer to the +Laminin/+FGF/-EGF/-LIF condition as the 'Early Neuronal Differentiation' or SVZ model condition.

Neuroepithelial cells cultured on Laminin, without any mitogenic factors (i.e., in the +Laminin/-FGF/-EGF/-LIF condition), continue to express NeuN in their nuclei, and additionally, assume a multipolar morphology, characteristic of more mature neurons. We refer to this condition as the "late neuronal differentiation condition" or the CP model. Neurosphere cultures can be directly transferred to either the SVZ or CP conditions to produce the early or late neuronal phenotypes or, alternatively, differentiated sequentially through the, S. V.Z and, C. P. condition. We typically culture cells in these early or late differentiation conditions for between 24 to 72 h. Flow cytometric analysis has shown that, in contrast to our previous +/–*ts*TA model of conditionally immortalized cortical cells *(78)*, neuroepithelial cells differentiated by mitogen-withdrawal and extracellular matrix addition do not undergo a significant change in apoptosis as a function of differentiation state *(80)*. Neuroepithelial cells undergoing differentiation in this model system also exhibit significant changes in their profile of secreted cytokines. Vascular endothelial growth factor (VEGF)-A and monocyte chemotactic protein (MCP)-1 are significantly decreased as neuroepithelial cells differentiate, whereas levels of granulocyte-macrophage colony stimulating factor (GM-CSF) increase *(80)*. Furthermore, we have been able to use this model to show that ethanol exposure during the proliferation phase has a persistent (organizational) effect on the secreted cytokine profile during the neuronal differentiation period. Therefore, this model facilitates the temporal separation

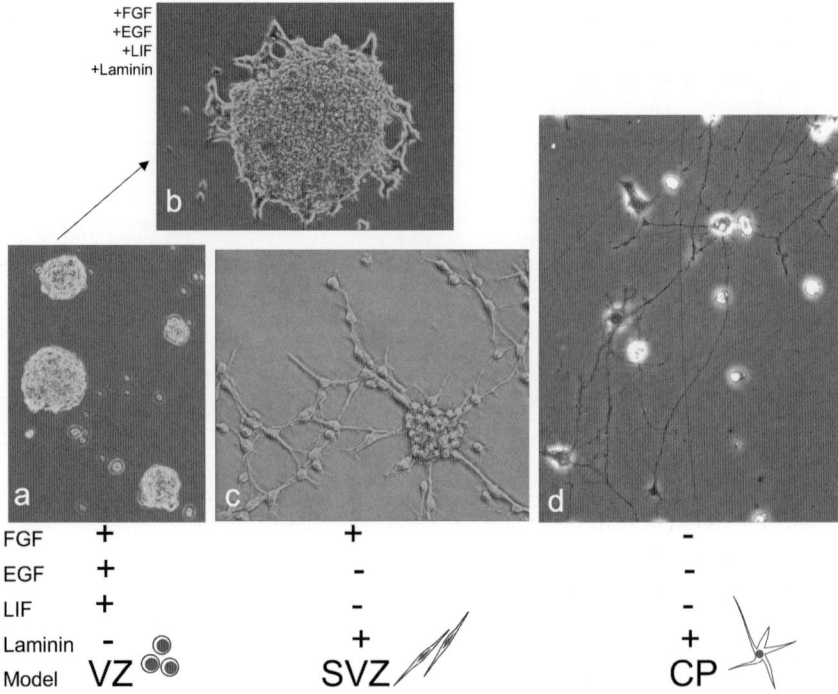

Fig. 5 Schematic of the Mitogen-withdrawal, extra-cellular matrix addition paradigm. **(a)** The mitogenic condition results in an expansion of the nonadherent neurosphere population. Exposure to laminin in addition to mitogenic medium **(b)** results in neurospheres becoming adherent. However, migratory neuroepithelial cells retain an immature squamous epithelioid appearance. Withdrawal of EGF and LIF, and provision of laminin as an adhesion matrix **(c)**, results in neurospheres becoming adherent to the culture dish, and the appearance of bipolar migratory cells. These cells express nuclear NeuN, but do not exhibit nestin immunoreactivity, showing that these cells have transformed into migratory neurons. The additional removal of FGF (d) results in the appearance of multipolar neurons between 24 and 72 h

of major second-trimester developmental programs, that is, neuronognesis, migration and neuronal differentiation, which normally occur together in the developing brain. The advantage of this simplified model is that we can now expose neuroepithelial cells to a teratogen during the progression of one developmental program, and study the impact of that teratogen on subsequent developmental programs, thereby teasing apart the activational and organizational effects of teratogens.

Acknowledgments   Supported by a grant from the NIH/NIAAA (#AA13440) to RCM.

4   Notes

1. Betadine (10% povidone-iodine) can be purchased as an over-the-counter antimicrobial agent from any local pharmacy or supermarket. It is easy to apply if you first drench the mouse abdomen with 80% ethanol, to wet the fur. Apply Betadine

liberally, ensuring coverage of the entire lower abdomen, including the ano-genital region and the proximal tail. Wait 5 min to let the Betadine dry, then use a separate pairs of scissors and forceps to open up the skin and underlying peritoneum so to limit the carryover of contamination from one layer to the next.

2. While dissecting out the uterus, be careful to prevent contact with the retracted skin and peritoneal flap. Avoid perforating the gastrointestinal system while dissecting away the uterine horns, to prevent bacterial contamination. This dissection may be performed under aseptic conditions. Place the dissected uterus into a Petri dish containing sterile DPBS and transfer the Petri dish to a sterile laminar-flow hood or workbench. Rinse out the uterine horns by three transfers to new PBS-filled Petri dishes. Serial rinsing serves to eliminate any bacteria or fungal spores that become adherent to the uterine horns during the initial dissection. The uterine horns can then be sliced open with a fresh pair of scissors, and the fetal amniotic sacs should then be extracted and transferred to a new PBS-filled Petri dish. Using a fresh set of Dumont-style forceps, dissect the individual fetuses away from their amniotic sacs and placenta, and transfer them to a new HBSS-filled Petri dish. Further microdissections are performed on a chilled stage of a dissecting microscope, within the confines of a sterile, laminar-flow workbench. The stage is chilled to 4°C by circulating a chilled mixture of polyethylene glycol and water through channels imbedded into the stage. Using a fresh set of microscissors and forceps, the fetal cranial skin is removed. At this developmental stage, the calvarial bones are extremely thin and can be dissected away with scissors and microforceps. Cuts are made at the level of the olfactory bulb and the brainstem, and the brain lifted away. The key to successful sterile dissection is to serially dilute bacterial and fungal inoculums by repeated changes of sterile PBS. The use of fresh, sterile sets of instruments at each step of the dissection also limits carry-over of contaminants.

3. It is easier to start peeling away meningeal tissue from the ventral portion of the brain first. Using a pair of fine, Dumont-style forceps, flip the brain over so that the ventral surface faces upwards. Gently peel back the meningeal tissue from around the hypothalamic region first, then working outwards, peel meningeal tissue as a continuous sheet over the lateral margins of the telencephalic vesicles. Then flip the brain over so that the dorsal surface faces upwards. Continue peeling back the meningeal tissue gently over the telencephalic vesicles so as not to disrupt the meningeal sheet. The cortical neuroepithelium is fragile at this age, and is easily fragmented by careless handling. We find that trans-illumination with a flat panel optic fiber light, set into the cooling plate below the brain tissue, serves to increase the contrast between meningeal and brain tissue. Structures of interest, tissue precursors to the cortex, hippocampus, striatum and diencephalon, for example, can then be microdissected with a microscalpel.

4. To limit the impact of litter-to-litter variations on studies, we typically collate fetal cortical tissue from three to four litters per experiment into a single collection tube

5. The concentration of LIF at 0.15 ng/mL (as with other factors) has been empirically determined in our laboratory. In our hands, a 10-fold increase in LIF

concentration (i.e., 1.5 ng/mL) results in neurosphere cultures becoming adherent to the culture plates, suggesting that greater levels of LIF induce morphological transformation.

6. Cell number can be verified using a standard hematocytometer. Unstained cells are spherical, refract light well, and are easy to visualize and count.

References

1. Jones, K. L., and Smith, D. W. (1975) The fetal alcohol syndrome. *Teratology*, **12,** 1–10.
2. Riley, E. P., and McGee, C. L. (2005) Fetal alcohol spectrum disorders: an overview with emphasis on changes in brain and behavior. *Exp. Biol. Med. (Maywood)*, **230,** 357–365.
3. Clarren, S. K. (1986) Neuropathology in Fetal Alcohol Syndrome, in *Alcohol and Brain Development*. (West, J. R., ed.). Oxford University Press, New York, pp. 158–166.
4. Clarren, S. K., Alvord, E. C., Jr., Sumi, S. M., Streissguth, A. P., and Smith, D. W. (1978) Brain malformations related to prenatal exposure to ethanol. *J. Pediatr.* **92,** 64–67.
5. Sokol, R. J., Delaney-Black, V., and Nordstrom, B. (2003) Fetal alcohol spectrum disorder. *JAMA*, **290,** 2996–2999.
6. Loock C, Conry J, Cook, J. L., Chudley, A. E., and Rosales, T. (2005) Identifying fetal alcohol spectrum disorder in primary care. *CMAJ* **172,** 628–630.
7. Warren, K. R., and Li, T. K. (2005) Genetic polymorphisms: impact on the risk of fetal alcohol spectrum disorders. *Birth Defects Res. A. Clin. Mol. Teratol.* **73,** 195–203.
8. Dobbing, J., and Sands, J. (1979) Comparative aspects of the brain growth spurt. *Early Hum. Dev.*, **3,** 79–83.
9. Bayer, S., and Altman, J. (1995) Neurogenesis and neuronal migration, in *The Rat Nervous System*. (ed 2). Academic Press: San Diego, pp. 1041–1078.
10. Cheema, Z. F., Wade, S. B., Sata, M., Walsh, K., Sohrabji, F., and Miranda, R. C. (1999) Fas/Apo (apoptosis)-1 and associated proteins in the differentiating cerebral cortex: induction of caspase-dependent cell death and activation of, N. F.-kappaB. *J. Neurosci.* **19,** 1754–1770.
11. Cheema, Z. F., Santillano, D. R., Wade, S. B., Newman, J. M., and Miranda, R. C. (2004) The extracellular matrix, p53 and estrogen compete to regulate cell-surface Fas/Apo-1 suicide receptor expression in proliferating embryonic cerebral cortical precursors, and reciprocally, Fas-ligand modifies estrogen control of cell-cycle proteins. *BMC Neurosci.* **5,** 11.
12. West, J. R., and Hamre, K. M. (1985) Effects of alcohol exposure during different periods of development: changes in hippocampal mossy fibers. *Brain Res.* **349,** 280–284.
13. Goodlett, C. R., Marcussen, B. L., West, J. R. (1990) A single day of alcohol exposure during the brain growth spurt induces brain weight restriction and cerebellar Purkinje cell loss. *Alcohol* **7,** 107–114.
14. Bonthius, D. J., and West, J. R. (1990) Alcohol-induced neuronal loss in developing rats: increased brain damage with binge exposure. *Alcohol. Clin. Exp. Res.* **14,** 107–118.
15. Bonthius, D. J., and West, J. R. (1991) Acute and long-term neuronal deficits in the rat olfactory bulb following alcohol exposure during the brain growth spurt. *Neurotoxicol. Teratol.* **13,** 611–619.
16. Hamre, K. M., and West, J. R. (1993) The effects of the timing of ethanol exposure during the brain growth spurt on the number of cerebellar Purkinje and granule cell nuclear profiles. *Alcohol. Clin. Exp. Res.* **17,** 610–622.
17. West, J. R., Hamre, K. M., and Cassell, M. D. (1986) Effects of ethanol exposure during the third trimester equivalent on neuron number in rat hippocampus and dentate gyrus. *Alcohol. Clin. Exp. Res.* **10,** 190–197.

18. Cheema, Z. F., West, J. R., and Miranda, R. C. (2000) Ethanol induces Fas/Apo (apoptosis)-1 mRNA and cell suicide in the developing cerebral cortex. *Alcohol. Clin. Exp. Res.* **24,** 535–543.
19. McAlhany, R. E.J, West, J. R., and Miranda, R. C. (2000) Glial-derived neurotrophic factor (GDNF) prevents ethanol-induced apoptosis and, J. U.N kinase phosphorylation. *Dev. Brain Res.* **119,** 209–216.
20. Mooney, S. M., and Miller, M. W. (2003) Ethanol-induced neuronal death in organotypic cultures of rat cerebral cortex. *Brain Res. Dev. Brain Res.* **147,** 135–141.
21. Mooney, S. M., and Miller, M. W. (2001) Effects of prenatal exposure to ethanol on the expression of bcl-2, bax and caspase 3 in the developing rat cerebral cortex and thalamus. *Brain Res.* **911,** 71–81.
22. McAlhaney, R. E. J, Miranda, R. C., Finnell, R. H., and West, J. R. (1999) Ethanol decreases Glial-Derived Neurotrophic Factor (GDNF) protein release but not mRNA expression and increases, G. D.NF-stimulated Shc phosphorylation in the developing cerebellum. *Alcohol. Clin. Exp. Res.* 23 1691–1697.
23. Luo J, Miller, M. W. (1998) Growth factor-mediated neural proliferation: target of ethanol toxicity. *Brain. Res. Rev.* **27,** 157–167.
24. Miller, M., Jacobs, J., and Yokoyama, R. (2003) Neg, a nerve growth factor-stimulated gene expressed by fetal neocortical neurons that is downregulated by ethanol. *J. Comp. Neurol.* **460,** 212–222.
25. Maier, S. E., Cramer, J. A., West, J. R., and Sohrabji, F. (1999) Alcohol exposure during the first two trimesters equivalent alters granule cell number and neurotrophin expression in the developing rat olfactory bulb. *J. Neurobiol.* **41,** 414–423.
26. Luo J, West, J. R., and Pantazis, N. J. (1996) Ethanol exposure reduces the density of the low-affinity nerve growth factor receptor (p75) on pheochromocytoma (PC12) cells. *Brain Res.* **37,** 34–44.
27. Dohrman, D. P., West, J. R., and Pantazis, N. J. (1997) Ethanol reduces expression of the nerve growth factor receptor, but not nerve growth factor protein levels in the neonatal rat cerebellum. *Alcohol. Clin. Exp. Res.* **21,** 882–893.
28. Pantazis, N. J., Dohrman, D. P., Luo J, Thomas, J. D., Goodlett, C. R., and West, J. R. (1995) NMDA prevents alcohol-induced neuronal cell death of cerebellar granule cells in culture. *Alcohol. Clin. Exp. Res.* **19,** 846–853.
29. Mooney, S. M., Siegenthaler, J. A., and Miller, M. W. (2004) Ethanol induces heterotopias in organotypic cultures of rat cerebral cortex. *Cereb Cortex* **14,** 1071–1080.
30. Hsiao, S. H., DuBois, D. W., Miranda, R. C., and Frye, G. D. (2004) Critically timed ethanol exposure reduces, G. A.BAAR function on septal neurons developing in vivo but not in vitro. *Brain Res.* **1008,** 69–80.
31. Pantazis, N. J., Dohrman, D. P., Goodlett, C. R., Cook, R. T., and West, J. R. (1993) Vulnerability of cerebellar granule cells to alcohol-induced cell death diminishes with time in culture. *Alcohol. Clin. Exp. Res.* **17,** 1014–1021.
32. Pantazis, N. J., Dohrman, D. P., Luo J, Goodlett, C. R., and West, J. R. (1992) Alcohol reduces the number of pheochromocytoma (PC12) cells in culture. *Alcohol,* **9,** 171–180.
33. Marcussen, B. L., Goodlett, C. R., Mahoney, J. C., and West, J. (1993) Developing rat Purkinje cells are more vulnerable to alcohol-induced depletion during differentiation than during neurogenesis. *Alcohol,* **11,** 147–156.
34. Phillips, S. C., and Cragg, B. G. (1982) A change in susceptibility of rat cerebellar Purkinje cells to damage by alcohol during fetal, neonatal and adult life. *Neuropathol. Appl. Neurobiol.* **8,** 441–454.
35. Caviness, V. S., Jr. (1982) Neocortical histogenesis in normal and reeler mice: a developmental study based upon (3H)thymidine autoradiography., *Brain Res.* **256,** 293–302.
36. Takahashi T, Goto T, Miyama S, Nowakowski, R. S., and Caviness, V. S. Jr. (1999) Sequence of neuron origin and neocortical laminar fate: relation to cell cycle of origin in the developing murine cerebral wall. *J. Neurosci.* **19,** 10357–10371.

37. Barnes, D. E., and Walker, D. W. (1981) Prenatal ethanol exposure permanently reduces the number of pyramidal neurons in rat hippocampus. *Brain Res.* **227,** 333–340.
38. Miller, M. (1989) Effects of prenatal exposure to ethanol on neocortical development: II. Cell proliferation in the ventricular and subventricular zones of the rat. *J. Comp. Neurol.* **287,** 326–338.
39. Miller, M. W. (1996) Limited ethanol exposure selectively alters the proliferation of precursor cells in the cerebral cortex. *Alcohol. Clin. Exp. Res.* **20,** 139–143.
40. Miller, M. W., and Nowakowski, R. S. (1991) Effect of prenatal exposure to ethanol on the cell cycle kinetics and growth fraction in the proliferative zones of fetal rat cerebral cortex. *Alcohol. Clin. Exp. Res.* **15,** 229–232.
41. Kotkoskie, L. A., and Norton, S. (1989) Morphometric analysis of developing rat cerebral cortex following acute prenatal ethanol exposure. *Exp. Neurol.* 106 283–288.
42. Kotkoskie, L. A., and Norton S (1988) Prenatal brain malformations following acute ethanol exposure in the rat. *Alcohol. Clin. Exp. Res.* **12,** 831–836.
43. Kotkoskie, L. A., and Norton S (1989) Cerebral Cortical Morphology and Behavior in Rats Following Acute Prenatal Ethanol Exposure. *Alcohol. Clin. Exp. Res.* **13,** 776–781.
44. Bayer, S., Altman, J. (1991) *Neocortical Development.* Raven Press, New York.
45. Dai, M. S., Mantel, C. R., Xia, Z. B., Broxmeyer, H. E., and Lu, L. (2000) An expansion phase precedes terminal erythroid differentiation of hematopoietic progenitor cells from cord blood in vitro and is associated with up-regulation of cyclin E and cyclin-dependent kinase 2. *Blood* **96,** 3985–3987.
46. Uylings H, Van Eden, C., Parnavelas, J., and Kalsbeck, A. (1990) *The Prenatal and Postnatal Development of the Rat Cortex. In The Cerebrtal Cortex of the Rat* (Kolb B., and Tees R. C., eds.). MIT Press, Cambridge. 35–76.
47. Bittman K, Owens, D. F., Kriegstein, A. R., and LoTurco, J. J. (1997) Cell coupling and uncoupling in the ventricular zone of developing neocortex. *J. Neurosci.* **17,** 7037–7044.
48. Tropepe V, Sibilia M, Ciruna, B. G., Rossant J, Wagner, E. F., and van der Kooy, D. (1999) Distinct neural stem cells proliferate in response to EGF and FGF in the developing mouse telencephalon. *Dev. Biol.* **208,** 166–188.
49. Reznikov K, Acklin, S. E., and van der Kooy, D. (1997) Clonal heterogeneity in the early embryonic rodent cortical germinal zone and the separation of subventricular from ventricular zone lineages. *Dev. Dyn.* **210,** 328–343.
50. Soriano, E., Dumesnil, N., Auladell, C., Cohen-Tannoudji, M., and Sotelo, C. (1995) Molecular heterogeneity of progenitors and radial migration in the developing cerebral cortex revealed by transgene expression. *Proc. Natl. Acad. Sci. USA,* **92,** 11676–11680.
51. Anderson, S. A., Kaznowski, C. E., Horn C, Rubenstein, J. L., and McConnell, S. K. (2002) Distinct origins of neocortical projection neurons and interneurons in vivo. *Cereb. Cortex,* **12,** 702–709.
52. Desai, A. R., and McConnell, S. K. (2000) Progressive restriction in fate potential by neural progenitors during cerebral cortical development. *Development* 127, 2863–2872.
53. Nery, S., Fishell, G., and Corbin, J. G. (2002) The caudal ganglionic eminence is a source of distinct cortical and subcortical cell populations. *Nat. Neurosci.* **5,** 1279–1287.
54. Mione, M. C., Danevic, C., Boardman, P., Harris, B., and Parnavelas, J. G. (1994) Lineage analysis reveals neurotransmitter (GABA or glutamate) but not calcium-binding protein homogeneity in clonally related cortical neurons. *J. Neurosci.* **14,** 107–123.
55. Anderson S, Mione M, Yun K, and Rubenstein, J. L. (1999) Differential origins of neocortical projection and local circuit neurons: role of Dlx genes in neocortical interneuronogenesis. *Cereb. Cortex* **9,** 646–654.
56. Donoghue, M. J., and Rakic P (1999) Molecular gradients and compartments in the embryonic primate cerebral cortex. *Cereb. Cortex* **9,** 586–600.
57. Zijlstra M, Bix M, Simister, N. E., Loring, J. M., Raulet, D. H., and Jaenisch R (1990) Beta 2-microglobulin deficient mice lack, C. D.**4**–**8**+ cytolytic T cells. *Nature,* **344,** 742–746.
58. Brazil, J. J., and Gupta, P. (2002) Constitutive expression of the Fas receptor and its ligand in adult human bone marrow: a regulatory feedback loop for the homeostatic control of hematopoiesis. *Blood Cells Mol. Dis.* **29,** 94–103.

59. Aasheim, H. C., Terstappen, L. W., and Logtenberg T (1997) Regulated expression of the Eph-related receptor tyrosine kinase Hek11 in early human B lymphopoiesis. *Blood* **90**, 3613–3622.

60. Hirose, J., Kouro, T., Igarashi, H., Yokota, T., Sakaguchi, N., and Kincade, P. W. (2002) A developing picture of lymphopoiesis in bone marrow. *Immunol Rev*, **189**, 28–40.

61. Bunting, K. D. (2002) ABC Transporters as phenotypic markers and functional regulators of stem cells. *Stem Cells* **20**, 11–20.

62. Miraglia S, Godfrey W, Yin, A. H., Atkins K, Warnke R, Holden, J. T., Bray, R. A., Waller, E. K., and Buck, D. W. (1997) A novel five-transmembrane hematopoietic stem cell antigen: isolation, characterization, and molecular cloning. *Blood* **90**, 5013–5021.

63. Yin, A. H., Miraglia S, Zanjani, E. D., Almeida-Porada G, Ogawa M, Leary, A. G., Olweus, J., Kearney, J., and Buck, D. W. (1997) AC133, a novel marker for human hematopoietic stem and progenitor cells. *Blood*, **90**, 5002–5012.

64. Zhu, G., Chang, Y., Zuo, J., Dong, X., Zhang, M., Hu, G., and Fang, F. (2001) Fudenine, a C-terminal truncated rat homologue of mouse prominin, is blood glucose-regulated and can up-regulate the expression of, G. A.PDH. *Biochem. Biophys. Res. Commun.* **281**, 951–956.

65. Goodell, M. A. (2002) Multipotential stem cells and 'side population' cells. *Cytotherapy* **4**, 507–508.

66. Kim, M., and Morshead, C. M. (2003) Distinct populations of forebrain neural stem and progenitor cells can be isolated using side-population analysis. *J. Neurosci.* **23**, 10703–10709.

67. Goodell, M. A., Brose K, Paradis G, Conner, A. S., and Mulligan, R. C. (1996) Isolation and functional properties of murine hematopoietic stem cells that are replicating in vivo. *J. Exp. Med.* **183**, 1797–1806.

68. Goodell, M. A., Rosenzweig, M., Kim, H., Marks, D. F., DeMaria, M., Paradis, G., Grupp, S. A., Sieff, C. A., Mulligan, R. C., and Johnson, R. P. (1997): Dye efflux studies suggest that hematopoietic stem cells expressing low or undetectable levels of, C. D.34 antigen exist in multiple species. *Nat. Med.* **3**, 1337–1345.

69. Doyle, L. A., Yang W, Abruzzo, L. V., Krogmann T, Gao Y, Rishi, A. K., and Ross, D. D. (1998) A multidrug resistance transporter from human, M. C.F-7 breast cancer cells. *Proc. Natl. Acad. Sci. USA*, **95**, 15665–15670.

70. Alison, M. R. (2003) Tissue-based stem cells: ABC transporter proteins take centre stage. *J. Pathol.* **200**, 547–550.

71. Kim M, Turnquist H, Jackson J, Sgagias M, Yan Y, Gong M, Dean M, Sharp, J. G., and Cowan, K. (2002) The multidrug resistance transporter, A. B.CG2 (breast cancer resistance protein 1) effluxes Hoechst 33342 and is overexpressed in hematopoietic stem cells. *Clin. Cancer Res.* **8**, 22–28.

72. Scharenberg, C. W., Harkey, M. A., and Torok-Storb, B. (2002) The ABCG2 transporter is an efficient Hoechst 33342 efflux pump and is preferentially expressed by immature human hematopoietic progenitors. *Blood* **99**, 507–512.

73. Haubensak W, Attardo A, Denk W, and Huttner, W. B. (2004) Neurons arise in the basal neuroepithelium of the early mammalian telencephalon: a major site of neurogenesis. *Proc. Natl. Acad. Sci. USA*, **101**, 3196–3201.

74. Santillano, D. R., Kumar, L. S., Prock, T. L., Camarillo C, Tingling, J. D., and Miranda, R. C. (2005) Ethanol induces cell-cycle activity and reduces stem cell diversity to alter both regenerative capacity and differentiation potential of cerebral cortical neuroepithelial precursors. *BMC Neurosci.* **6**, 59.

75. Noctor, S. C., Martinez-Cerdeno, V., Ivic, L., and Kriegstein, A. R. (2004) Cortical neurons arise in symmetric and asymmetric division zones and migrate through specific phases. *Nat Neurosci*, **7**, 136–144.

76. Gotz M, Barde, Y. A. (2005) Radial glial cells defined and major intermediates between embryonic stem cells and, CNS neurons. *Neuron* **46**, 369–372.

77. Doetsch F, Caille I, Lim, D. A., Garcia-Verdugo, J. M., and Alvarez-Buylla, A. (1999) Subventricular zone astrocytes are neural stem cells in the adult mammalian brain. *Cell*, **97**, 703–716.

78. Wade, S., Oommen, P., Conner, W., Earnest, D., and Miranda, R. (1999) Overlapping and divergent actions of estrogen and the neurogrophins on cell fate and p-53 dependent signal transduction in conditionally immortalized cerebral cortical neuroblasts. *J. Neurosci.* **15,** 6994–7006.
79. Gary, D. S., and Mattson, M. P. (2001) Integrin signaling via the, P. I.3-kinase-Akt pathway increases neuronal resistance to glutamate-induced apoptosis. *J. Neurochem.* **76,** 1485–1496.
80. Camarillo, C., Kumar, L. S., Bake, S., Sohrabji, F., and Miranda, R. C. (2007) Ethanol regulates angiogenic cytokines during neural development: Evidence from an in vitro model of mitogen-withdrawal induced cerebral cortical neuroepithelial differentiation. *Alcohol. Clin. Exp. Res.* **31,** 324–335.
81. Adachi, J., Mizoi, Y., Fukunaga, T., Ogawa, Y., Ueno, Y., and Imamichi H (1991) Degrees of alcohol intoxication in 117 hospitalized cases. *J. Stud. Alcohol,* **52,** 448–453.
82. Perper, J. A., Twerski, A., and Wienand, J. W. (1986) Tolerance at high blood alcohol concentration: a study of 110 cases and review of the literature. *J. Forensic Sci.* **31,** 212–221.

Part IV
Regulation of Specific Organ Systems in Response to Ethanol

13

Acetaldehyde-induced Barrier Disruption and Paracellular Permeability in Caco-2 Cell Monolayer

R. K. Rao

Summary A significant body of evidence indicates that endotoxemia plays a crucial role in the pathogenesis of alcoholic liver disease. There are several possible factors that may be involved in inducing alcoholic endotoxemia, but increased intestinal permeability to enteric endotoxins appears to be the major contributing factor. In the normal gut, the epithelial barrier function prevents diffusion of toxins across the epithelium. However, the barrier is disrupted in patients with alcoholic liver disease. We showed that acetaldehyde disrupts intestinal epithelial tight junctions and increases paracellular permeability to endotoxins in Caco-2 cell monolayer, the extensively studied model of the differentiated intestinal epithelium. The mechanisms involved in acetaldehyde-induced increase in intestinal permeability to endotoxins can be elucidated in this model of the intestinal epithelium.

Keywords Epithelium; endotoxemia; barrier function; tight junctions; ethanol.

1 Introduction

The evidence for the role of endotoxemia and increased intestinal permeability to endotoxins in alcoholic liver disease (ALD) has been reviewed previously *(1)*. The level of plasma endotoxin is increased in patients with ALD, and a correlation between endotoxemia and the severity of ALD has been established. This correlation is further confirmed by endotoxemia in experimental alcoholic liver damage in rats and mice *(2–4)*. Administration of antibiotics to deplete the colonic luminal source of endotoxins attenuates alcohol-induced endotoxemia and liver damage *(5)*, suggesting that the intestinal luminal microflora as a source of endotoxemia. The three possible mechanisms involved in alcoholic endotoxemia are bacterial overgrowth, alcohol-induced Kupffer cell dysfunction, and an alcohol-induced increase in intestinal permeability to endotoxins; increased intestinal permeability appears as a major contributor to endotoxemia.

In the normal gut, the epithelial barrier function prevents the diffusion of toxins, pathogens and allergens from the gut lumen into intestinal tissue and liver. Alcohol administration increases gastrointestinal permeability to macromolecules both in normal subjects as well as in patients with ALD *(1)*. In some patients with ALD, the increased intestinal permeability persists even after 1–2 wk of sobriety, suggesting that a more sustained barrier dysfunction is present in these patients. Demonstration of increased intestinal permeability in alcoholics showing the symptoms of liver disease and not in those alcoholics without liver disease indicates a close correlation between intestinal permeability and ALD *(6)*. An increased intestinal permeability to macromolecules was also demonstrated in experimental models of alcoholic liver damage *(7)*. Amelioration of alcoholic liver damage in rats by zinc *(8)* or oat bran *(9)* supplementation also associated with attenuation of alcohol-induced intestinal permeability to macromolecules.

Acetaldehyde, the first product of ethanol metabolism, is highly toxic, and it appears to play a key role in cell injury in multiple organs, such as liver, heart, skeletal muscle, lung, and brain. Our recent studies indicated that acetaldehyde disrupts intestinal epithelial barrier function and increases intestinal permeability to endotoxins and other macromolecules *(10–13)*. Acetaldehyde disrupts intestinal epithelial tight junctions (TJ) and adherens junctions (AJ) by a tyrosine kinase-dependent mechanism. Acetaldehyde-induced increase in permeability was mediated by the inhibition of protein tyrosine phosphatases such as PTP1B and increase in tyrosine phosphorylation of TJ and AJ-proteins such as occludin, ZO-1, E-cadherin, and β-catenin. Gastrointestinal mucosal protective factors such as epidermal growth factor (EGF) *(12)* and L-glutamine *(13)* effectively prevents acetaldehyde-induced disruption of TJ and AJ and increase in permeability to endotoxins and macromolecules. Most of these studies were conducted using Caco-2 cell monolayers, an extensively used model of the intestinal epithelium. However, the effects of acetaldehyde on the TJ and AJ and its prevention by EGF and L-glutamine were confirmed in human colonic mucosa *(14)*. Therefore, valuable information can be derived from studies that use Caco-2 cell monolayers as an experimental model of the intestinal epithelium.

2 Materials

2.1 Cell Culture

1. Dulbecco's Modified Eagle's Medium (DMEM; Gibco/BRL, Bethesda, MD) supplemented with 10% fetal bovine serum (FBS, HyClone, Ogden, UT).
2. L-Glutamine, 200 mM, 100× stock solution (Gibco/BRL).
3. Penicillin, 10,000 units/mL–streptomycin, 10,000 μg/mL, 1000× (Gibco/BRL).
4. Gentamycin, 50 mg/mL (1000×, Gibco/BRL).
5. MEM nonessential amino acids solution, 10 mM (100×, Gibco/BRL).
6. Transwell inserts (Costar/Corning, Corning, NY).
7. Trypsin–EDTA: Trypsin solution (0.25%) and ethelenediamine tetraacetic acid (EDTA; 1 mM) from Gibco/BRL.

2.2 Acetaldehyde Exposure

1. Acetaldehyde: 99.5% (A.C.S reagent, Sigma-Aldrich, St. Louis, MO).
2. Sealing tape: 3 M comply indicator tape (Costar).
3. Phosphate-buffered saline (PBS-1): 10 g/L sodium chloride, 0.25 g/L potassium chloride, 1.435 g/L disodium hydrogen phosphate, 0.25 g/L potassium dihydrogen phosphate, pH 7.4 containing 0.134 g/L calcium chloride, 0.1 g/L magnesium chloride, 1.8 g/L glucose, and 6 g/L bovine serum albumin. Diluted from 10× PBS (Sigma-Aldrich, St. Louis, MO; *see* **Note 1**).

2.3 Measuring Transepithelial Electrical Resistance (TER)

1. Millicell-ERS resistance monitoring system and electrodes (Millipore, Billercia, MA).
2. 70% ethanol – prepared by diluting absolute ethanol in order (Mallinckrodt, Phillipsburg, NJ).
3. PBS, pH 7.4 (Gibco/BRL), same as in Subheading 2.2.

2.4 Measuring Paracellular Permeability

1. Inulin-FITC (Sigma, St. Louis, MO), 10 mg/mL in PBS (stock solution).
2. FITC-lipopolysaccharide (LPS), isomer 1 ~98% (HPLC) (Sigma-Aldrich, St. Louis, MO).

3. Fluorescence plate reader: FLx 800 microplate fluorescence reader (Bio-TEK Instruments Inc., Winoonski, VT).
4. 96-well fluorescence plates.

2.5   *Immunofluorescence Staining and Confocal Microscopy*

1. PBS-2: 18.5 g/L sodium chloride, 5.7 g/L disodium hydrogen phosphate, 1.4 g/L sodium dihydrogen phosphate, pH 7.4. Prepared by diluting 10× PBS (Sigma-Aldrich, St. Louis, MO).
2. 3% paraformaldehyde: dissolve 3 g of paraformaldehyde (Sigma-Aldrich, St. Louis, MO) in 100 mL of PBS-2 with a drop of sodium hydroxide heated at 60°C. The pH is re-adjusted to 7.4 (*see* **Note 2**).
3. Acetone:Methanol (1:1): mix equal volumes of acetone and methanol (both from Fisher-Scientific, Fairlawn, NJ) (*see* **Note 3**).
4. Triton X100 (0.2%): prepared by dissolving Triton X100 (Sigma-Aldrich, St. Louis, MO) in PBS.
5. Tris-buffered saline containing Tween 20 (TBST): dilute 10X TBST (20 mM Tris-HCl, 0.15 *M* sodium chloride, pH 8.0, and dissolving 5 ml/L of Tween-20.
6. 4% nonfat milk: dissolve 4 g of blotting grade nonfat dry milk (BIORAD, Hercules, CA) in TBST.
7. Primary antibodies: mouse monoclonal anti-occludin and rabbit polyclonal anti-ZO-1 antibodies (Zymed Laboratories, S. San Francisco, CA).
8. Secondary antibodies: AlexaFlour-488-conjugated goat anti-mouse IgG antibody (2 mg/ml) from INVITROGEN-Molecular Probes (Eugene, OR), and Cy3-conjugated anti-rabbit IgG from Sigma-Aldrich (St. Louis, MO).
9. Mounting fluid: dissolve 25 g DABCO (1,4-Diazabicyclo-[2.2.2]octane; Sigma-Aldrich, cat. no. D-2522) in 25 mL of 10x PBS-2 and mixing 225 mL of glycerol.
10. Nail polish: Nail Enamel/Vernis – clear (REVLON, Mississauga, ON).
11. Glass slides: precleaned Gold Seal Rite-ON microslides (Gold Seal, Portsmouth, NH).
12. Cover slips (Microscope cover glass, Fischer brand, Fisher Scientific, Pittsburgh, PA).
13. Humidifying chamber: you can use a plastic box with wet paper towels on the bottom and two plastic pipets to rest the plate with inserts.
14. Confocal microscope: LSM 5 PASCAL high-performance confocal laser scanning microscope (CARL ZEISS Micro imaging Inc, Thornwood, NY). Others such as BIORAD and Nikon confocal microscopes can be used.
15. Software: LSM 5 PASCAL software is used in both Zeiss LSM 5 PASCAL and BIORAD confocal microscopes. Image J software and LSM reader or Biorad reader plugins can be freely downloaded from NIH web site. Adobe Photoshop can be purchased from Adobe Systems Inc. (San Jose, CA).

2.6   Lactate Dehydrogenase (LDH) Release Assay

1. Cytotoxicity detection kit (Roche Applied Science, cat. no. 11 644 793 001). This kit has 2 reagents, "Catalyst" and "Dye solution." You have to mix them to use. This mixture (assay medium) is stable for several weeks at 4°C.
2. 96-well microplate.

2.7   WST Assay

1. Cell proliferation reagent WST-1 (Roche Applied Science, cat. no. 11644807001). For assay reagent, mix WST-1 reagent and PBS-1 (1:9) just before use. You need 150 μL of assay reagent for cells on 6.5-mm transwells and 400 μL for cells on 12-mm transwells.
2. PBS-1.
3. 96-well microplates.

3   Methods

Caco-2 cell line was derived from colon adenocarcinoma. When grown in culture, Caco-2 cells form a monolayer of epithelial cells, which differentiate into columnar cells and polarize both morphologically and functionally. The phenotypes resemble that of normal ileal epithelial enterocytes. These cells organize TJ and AJ and develop barrier to the diffusion of macromolecules. These cells have been extensively used as a model to study the properties of differentiated enterocytes, the barrier function, and study of oral bioavailability of therapeutic drugs. Studies in our laboratory found this to be a useful model to study acetaldehyde-induced endotoxin absorption *(10–13),* as most of the key observations made using this monolayer were reproduced in human colonic mucosa *(14).*

The barrier function of the epithelium can be evaluated by measuring TER, which evaluates the barrier to ion flux, and by unidirectional flux of extracellular fluid markers such as mannitol and inulin. Extracellular markers are transported through the paracellular space, but not through transcellular route, unless the holes in the epithelium are formed due to dislodging of individual cells. TER can be modulated by changes in both paracellular permeability and transcellular ion permeability through ion channels. Therefore, it is crucial to measure both parameters simultaneously. The integrity of TJ can be evaluated by visualizing the junctional distribution of TJ-specific proteins such as occludin and ZO-1 by immunofluorescence staining and confocal microscopy, followed by densitometric analysis. Disrupted junctional organization of occludin and ZO-1 indicates the disruption of TJ as the cause of increased paracellular permeability. Finally, it is important to confirm that increased permeability is not caused by loss of cell viability. Cell viability can be evaluated by measuring by LDH release into extracellular medium and WST assay to evaluate the mitochondrial activity.

3.1   Caco-2 Cell Monolayers on Transwell Inserts

1. Caco-2 cells are first grown in T75 flasks to achieve 90–95% confluence.
2. Cells are fed with medium every 48 h (*see* **Note 4**).
3. Cells from the flasks are lifted up in trypsin/EDTA solution. Cells are washed in DMEM and resuspended in medium and seeded onto Transwell inserts of different sizes (6.5 mm, 12 mm, or 24 mm diameters). Medium is replaced every 48 hs. TER is monitored in representative wells for the development of barrier function. At 10–14 d after seeding, cells are used for permeability experiments (*see* **Note 5**).
4. Before all experiments cell monolayers are washed two times with PBS-1, and preincubated for 60 min with PBS-1, 0.2 mL and 0.8 mL on upper and lower wells, respectively (*see* **Notes 6** and **7**).

3.2   Acetaldehyde Treatment

1. Prepare acetaldehyde solutions only just few minutes before the use. PBS-1 of volumes required for preparing stock solution of acetaldehyde (10%) and solution for use (0.5%), and chill them at −20°C for 10 min before mixing with acetaldehyde. When you are ready for acetaldehyde treatment, mix 2 mL of acetaldehyde with 18 mL of prechilled PBS-1 and mix quickly to prepare 0.5% acetaldehyde.
2. In a 24-well cluster place transwell inserts with cell monolayers in the middle two rows as shown in Fig. 1. Place 0.5% acetaldehyde solution (1 ml/well) to the wells on the outer rows (*see* **Note 8**).
3. Close the lid and seal by taping on all sides (*see* **Note 9**) and place the plate in an incubator.

3.3   Measuring TER

1. Sterilize chopstick electrodes by dipping in 70% ethanol for 10 min.
2. Remove, air-dry and soak in PBS-1. Keep the electrodes soaked in PBS-1 through the end of experiment.
3. Calibrating electrodes: To test the electrodes, dip it in PBS-1, turn mode switch to "V" and turn power on. Adjust the voltage to 0.0 with a screwdriver by pressing the "Test" button. Turn the mode to "R" and adjust resistance to 0.0.

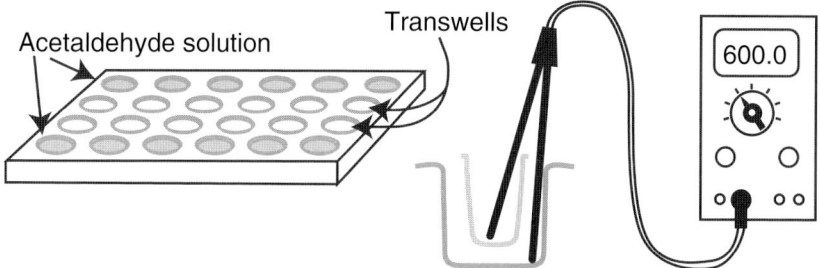

Fig. 1 Diagrammatic representation of 24-well cluster with acetaldehyde solution in outer wells and transwell inserts in inner wells. Diagram also shows the use of Millecel-ERS resistance monitoring meter

4. Insert electrodes through one of the three holes on the wall of the inserts so that the longer electrode sits at the bottom of the lower well and the shorter electrode immersed in buffer in upper well (Fig. 1). Set the mode button to "R" and press "Measure" button to record the TER values (*see* **Note 10**).

5. Normally, TER value can be measured in the same monolayer at different time intervals. However, time course of TER changes during the acetaldehyde treatment cannot be studied in the same monolayer because the plate is sealed during the acetaldehyde treatment. Therefore, TER can be measured only at three time points: before preincubation, after preincubation, and at the end of experiment. A separate monolayer should be used for measuring TER values at different time periods.

6. The polycarbonate membranes at the bottom of transwell inserts pose some level of electrical resistance, which is usually $30\,Ohms.cm^2$. This can be measured in empty wells soaked in PBS-1 for 30 min. This background values should be subtracted from all values.

7. TER values can be presented as $Ohms.cm^2$ by multiplying the recorded values by the surface area of the monolayer, which is $0.33\,cm^2$ for 6.5-mm transwell inserts. TER can also be presented as percent of baseline values to compare the TER of individual monolayers.

8. Acetaldehyde treatment induces gradual decrease in TER, whereas TER in control cell monolayers remains unaffected (Fig. 2A).

3.4 Measuring Paracellular Permeability

1. Stock solution of FITC-inulin is diluted with PBS-1 to obtain a concentration of 0.5 mg/mL and placed in the upper well at the beginning of preincubation (0.2 mL for 6.5-mm transwells), whereas 0.8 mL of PBS-1 without inulin is placed in the lower well.

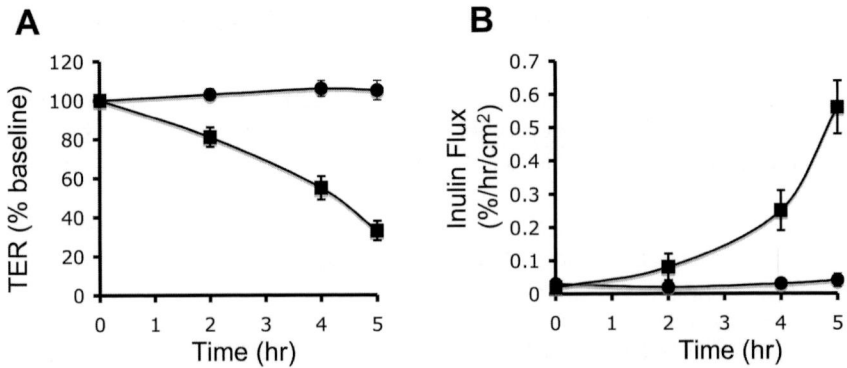

Fig. 2 Acetaldehyde-induced paracellular permeability. Caco-2 cell monolayers were incubated in the absence (●) or presence (■) of acetaldehyde for varying times. TER (**A**) and inulin flux (**B**) were measured. Values are mean ± sem (n = 6)

2. A 50-μL aliquot from the basal well is withdrawn at different time periods and placed in a fluorescence microplate. It is diluted with 50 μL of PBS and fluorescence measured in a 96-well fluorescence plate reader. At the end of experiment, a 10-μL aliquot of buffer from upper well is withdrawn to measure the original amount of inulin administered to upper well.

3. The blank fluorescence reading is measured by using 100 μL of PBS-1. The fluorescence value measured after preincubation is subtracted from values for corresponding monolayer.

4. Inulin flux values can be presented as the absolute flux (μg inulin/ml/hr/cm²) of inulin from upper to lower well using the following formula: Inulin flux (μg inulin/ml/hr/cm²) = $[B \times V_B]/[V \times h \times a]$,

   where B is inulin in aliquot of basal well, V_B is total volume of lower buffer, V is volume of aliquot used, h is time at which flux was measured and a is surface area of the cell monolayer.

   It can also be measured as percent of inulin loaded to the upper well as follows: Percent flux/hr/c m² = $[B \times V_B \times V2 \times h \times a \times 100]/[V1 \times A \times V_A]$,

   where B is inulin in lower buffer aliquot, V_B is volume of lower buffer, V2 is aliquot used from lower buffer, h is hour of incubation, a is surface area of cell monolayer, V1 is aliquot used from upper buffer, A is inulin in upper well, and V_A is volume of upper buffer.

5. Endotoxin permeability can be measured as described previously by replacing FITC-inulin with FITC-conjugated LPS.

6. Acetaldehyde treatment increases inulin flux in a time-dependent manner, whereas the rate of flux remains unaffected in control cell monolayer (Fig. 2B).

3.5   *Immunofluorescence Staining and Confocal Microscopy*

1. At varying times after acetaldehyde treatment cell monolayers are washed in PBS-2 and placed 1 mL of prechilled (−20°C) acetone:methanol (1:1 v/v) onto the cells and allowed to sit at 0°C for 5 min. Aspirate the solvent completely and air dry.

2. Dried samples can be stored at −20°C for several weeks before further process.

3. Rehydrate acetone:methanol-fixed cell monolayers by soaking it in PBS-2 for 15 min. Wash monolayers once with PBS-2.

4. Permeabilization: Place 1 mL of 0.2% Triton X100 solution in PBS-2 (cold), and incubate at 4°C for 5 min. Wash 3X with PBS-2 (1 mL) for 10 min each.

5. Blocking: Cells are blocked to prevent nonspecific binding of antibodies to cell components. Place 1 mL of 4% milk in TBST and incubate at room temperature for 30 min.

6. Incubation with primary antibodies: prepare primary antibodies, anti-occludin, and anti-ZO-1 antibodies, in 4% milk to obtain final concentration of 5 µg/mL and 3 µg/mL, respectively. Place 100 µL of antibody solution on top of the cell monolayer. Incubate in humidifying chamber in dark for 1 h.

7. Aspirate antibody solution and wash cell monolayers in 1% milk three times, 10 min each.

8. Incubation with secondary antibodies: dilute secondary antibodies, Alexafluor-488 conjugated goat polyclonal anti-mouse IgG and Cy3-conjugated goat polyclonal anti-rabbit IgG, in 4% milk (1:100 dilution). Place 100 µl of antibody solution on top of the cell monolayers. Incubate in humidifying chamber in dark for 1 h. Wash tissues with PBS-2 three times, 10 min each.

9. Mounting: cut the membrane at the bottom of transwell inserts with stained cell monolayer and place it on top of a glass slide. Place a drop of mounting fluid on top of membrane. Place the cover slip carefully on top of the mounting fluid, and allow it to spread to all sides. Seal the cover slip with nail polish. Store slides in cold dark area until visualization by microscope.

10. Mounted slides can be stored at 4°C in dark for several months without loosing fluorescence. Use 40x oil immersion objective in a confocal microscope and collect images with a maximum of 500 gain at 10% laser at a 1-µm setting. The laser can be increased if the stain is weak, but keep in mind that higher laser can result in quenching of fluorescence. But, do not increase gain to obtain higher intensity, because it reduces the quality of the image. Store images as pic or lsm files. Stack the sections of images using Image J software and save as RGB tiff files. These files can be opened in Adobe Photoshop software, and processed to optimize the image quality. Using this software you can also superimpose the occludin (green) and ZO-1 (red) stains for co-localization (*see **Note 11***).

11. In untreated control cell monolayers, the occludin and ZO-1 stains co-localize at the intercellular junctions. Acetaldehyde treatment results in redistribution of both occludin and ZO-1 from the intercellular junctions into the intracellular compartments (Fig. 3).

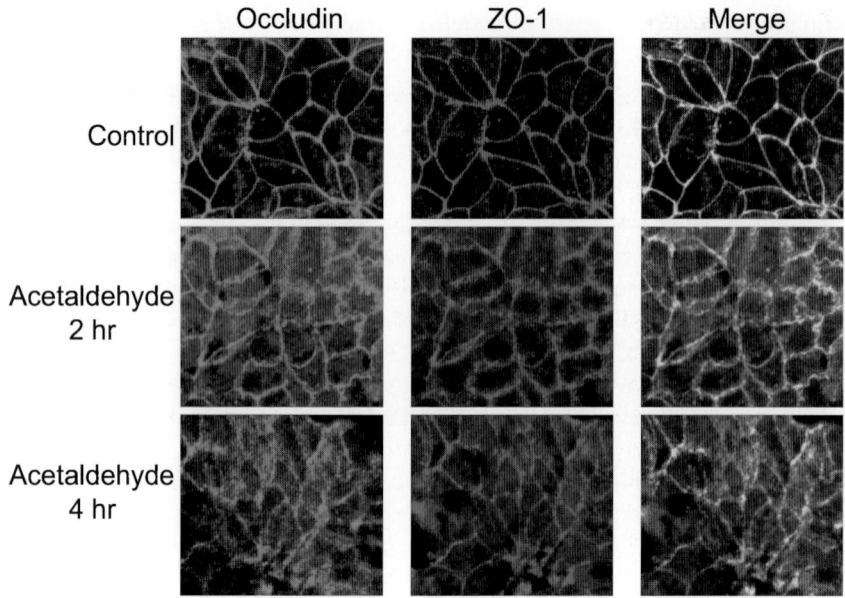

Fig. 3 Acetaldehyde disrupts tight junctions. Caco-2 cell monolayers were incubated with or without (control) acetaldehyde for varying times. Cell monolayers were fixed and double-labeled for occluding and ZO-1 by immunofluorescence method. Images were collected by confocal microscopy

3.6   LDH Release Assay

1. Transfer buffers from both the upper and lower wells to microfuge tubes at the end of experiments or treatments. These samples can be stored at −20°C until the assay.
2. Place 100 µL of assay reagent in a 96-well microplate. Add 100 µL of sample (incubation buffer from the upper well).
3. Prepare blanks using 100 µL of assay reagent and 100 µL of PBS-1. For positive control, use buffer from cells treated with 2% Triton X100 for 10 min.
4. Incubate the microplate at RT for 30 min and measure the absorbance of sample at 490 nm and at 600 nm (as reference to be subtracted from A490).
5. Quantitation and comparison: A490 values obtained can be directly compared between control cells and acetaldehyde-treated cells. Alternatively, percent toxicity can be measured by following formula: A490 for sample divided by A490 for positive control multiplied by 100.

3.7   WST Assay

1. Aspirate buffers from both upper and basal wells of transwell inserts.
2. Place 150 μL or 400 μL of assay medium in transwells of 6.5 mm or 12 mm diameters, respectively.
3. Incubate cells at 37°C for 15 min.
4. Withdraw buffer from upper well into microfuge tubes.
5. Mix solutions and transfer 100-μL solution to 96-well microplate.
6. Measure the absorbance of sample at 450 nm and 600 nm (as reference).
7. Measure blank reading using 100 μL of assay reagent. For positive control, use buffer from cells treated with 2% Triton X100 for 10 min.
8. Quantitation and comparison: A450 values obtained can be directly compared between control cells and acetaldehyde-treated cells. Alternatively, percent toxicity can be measured by following formula: A450 for sample divided by A450 for positive control multiplied by 100.

Acknowledgements   This work was supported by NIH grant R01 AA12307.

4   Notes

1. All solutions should be prepared in Nanopure water (< 18.2 Ohms resistivitity).
2. Paraformaldehyde is a toxic chemical when inhaled. Therefore, preparing 3% paraformaldehyde should be prepared in a fume hood. It dissolves better at alkaline pH and warmer temperature. Therefore, drops of 4 N sodium hydroxide solution should be added and heated at 60°C on a heating magnetic stirrer. Once the solution appears clear, readjust the pH back to 7.4 by using 1 N hydrochloric acid.
3. Acetone:methanol mixture can be stored at −20°C; however, it is important that is is kept in an explosion-proof freezer for safety. Failure to prechill this reagent before use can result in loss of antigenicity of proteins of interest.
4. Caco-2 cells used within 10 passages after thawing. Original Caco-2 cell line and its subclones such as Caco-2$_{bbe}$ can be used for acetaldehyde treatment. However, the sensitivity to certain regulatory factors such as EGF may vary between strains.
5. Caco-2 monolayers develop maximal TER by 7 days after seeding, which is followed during the next week by differentiation into polarized columnar epithelial cells. Cellular responses to many factors depend on the differentiation status of the cells. Therefore, it is important to conduct experiments in a window of time after seeding.
6. EGF, a gastrointestinal mucosal protective factor, prevents disruption of TJ by various injurious factors such as acetaldehyde *(12)* and hydrogen peroxide *(15)*. Serum contains many such mucosal protective factors. Therefore, peincubation for one hour in PBS-1 allows clearance of any serum effects.

7. For 12-mm transwells, 0.8 mL and 1.5 mL, and for 24-mm transwells, 1.5 mL and 2.5 mL of PBS-1 are added to upper and lower wells, respectively. These volumes were chosen to match the level of buffers in upper and lower wells, so that hydrostatic pressure on the cell monolayers is minimized.

8. Small-size transwells of 6.5 mm in diameter are good enough to study the permeability. However, if proteins for biochemical studies are needed one can grow cells in large transwells (12 mm or 24 mm diameters). For 12-mm transwells, 12-well clusters are used. Place transwells in the middle row and acetaldehyde solution in the outer rows. In 24-mm transwells, place transwells in the middle two wells and acetaldehyde in the outer four wells.

9. Acetaldehyde is volatile (boiling point is 18°C) at room temperature and evaporates from solution very rapidly at 37°C. Therefore, it is important to expose cells to vapor-phase acetaldehyde, in a sealed chamber. Acetaldehyde concentration of 0.5% in the reservoir wells results in 400 μM acetaldehyde in the buffer incubating the cell monolayers because of absorption of acetaldehyde vapors.

10. After certain period of incubation, the cells should be equilibrated to room temperature by placing at RT for 5 min. TER tends to increase as the temperature of cell monolayers decrease.

11. The samples double labeled for red and green fluorescence can be scoped simultaneously. However, depending on the filters used, there can be bleeding of fluorescence from one channel to the other channel. In such case, it is safer to take the image with settings for one channel first, and then the other. Always take the image of one with the weaker stain first to minimize the loss of other fluorescence due to quenching.

References

1. Rao R. K., Seth A., and Sheth P. (2004) Role of intestinal permeability and endotoxemia in alcoholic liver disease. *Am. J. Physiol* **286,** G881–G884.
2. Mathurin P., Deng Q. G., Keshavarzian A., Choudhary S., Holmes E. W., and Tsukamoto H. (2000) Exacerbation of alcoholic liver injury by enteral endotoxin in rats. *Hepatology* **32,** 1008–1017.
3. Nanji A. A., Jokelainen K., Fotouhinia M., Rahemtulla A., Thomas P., Tipoe G. L., Su G. L., and Dannenberg A. J. (2001) Increased severity of alcoholic liver injury in female rats: role of oxidative stress, endotoxin, and chemokines. *Am. J. Physiol. Gastrointest. Liver Physiol.* **281,** G1348–G1356.
4. Tamai H., Kato S., Horie Y., Ohki E., Yokoyama H., and Ishii H. (2000) Effect of acute ethanol administration on the intestinal absorption of endotoxin in rats. *Alcohol Clin Exp Res* **24,** 390–394.
5. Anderson, J. M., and van Italie, C. M. (1995) Tight junctions and the molecular basis for regulation of paracellular permeability. *Am. J. Physiol.* **269,** G467–G475.
6. Keshavarzian A., Holmes E. W., Patel M., Iber F., Fields J. Z., and Pethkar S. (1999) Leaky gut in alcoholic cirrhosis: a possible mechanism for alcohol-induced liver damage. *Am J Gastroenterol* **94,** 200–207.

7. Bode C., and Bode J. C. (2003) Effect of alcohol consumption on the gut. *Best Pract Res Clin Gastroenterol* **17,** 575–592.

8. Lambert J. C., Zhou Z., Wang L., Song Z., McClain C. J., and Kang Y. J. (2003) Prevention of alterations in intestinal permeability is involved in zinc inhibition of acute ethanol-induced liver damage in mice. *J Pharmacol Exp Ther* **305,** 880–886.

9. Keshavarzian A., Choudhary S., Holmes E. W., Yong S., Banan A., Jakate S., and Fields J. Z. (2001) Preventing gut leakiness by oats supplementation ameliorates alcohol-induced liver damage in rats. *J Pharmacol Exp Ther* **299,** 442–448.

10. Rao R. K. (1998) Acetaldehyde-induced increase in paracellular permeability in Caco-2 cell monolayer. *Alcohol Clin Exp Res* **22,** 1724–1730.

11. Atkinson K. J., and Rao R. K. (2001) Role of protein tyrosine phosphorylation in acetaldehyde-induced disruption of epithelial tight junctions. *Am J Physiol Gastrointest Liver Physiol* **280,** G1280–G1288.

12. Sheth P., Seth A., Thangavel M., Basuroy S., and Rao R. K. (2004) Epidermal growth factor prevents acetaldehyde-induced disruption of tight junctions in Caco-2 cell monolayer. *Alcohol. Clin Exp. Res.* **28,** 797–804.

13. Seth A., Sheth P., Basuroy S., and Rao R.K. (2004) L-glutamine ameliorates acetaldehyde-induced paracellular permeability in Caco-2 cell monolayer. *Am. J. Physiol.* **287,** G510–G517.

14. Basuroy S., Sheth P., Mansbach C. M., and Rao R. K. (2005) Acetaldehyde induces tyrosine phosphorylation of TJ and AJ proteins and their dissociation from the actin cytoskeleton in human colonic mucosa. *Am. J. Physiol.,* **289,** G367–G37.

15. Sheth P., Atkinson K. J., Gheyi T., Kale G., Giorgianni F., Desiderio D. M., Li C., Naren A. P., and Rao R. K. (2006) Acetaldehyde dissociates PTP1B-E-cadhein-β-catenin complex in Caco-2 cell monolayer by phosphorylation-dependent mechanism. *Biochem. J.* 402, 291–300.

14
Alcohol-induced Oxidative Stress in the Liver

In Vivo Measurements

Gavin E. Arteel

Summary Oxidative stress is increasingly suspected to contribute to the initiation and progression of many disease, including those caused by alcohol exposure. Two major products of reactive oxygen and nitrogen species formation are 4OH-nonenal and 3-nitrotyrosine protein adducts, both of which can be detected by immuno-histochemistry. In the past, immunohistochemical techniques have served largely as qualitative measures of changes. However, coupled with digital capture and analysis of photomicrographs, one can now quantitate treatment-related changes with immunohistochemistry. This chapter summarizes techniques for immunohis-tochemical detection of these products of reactive oxygen and nitrogen species and subsequent image-analysis. Although the methods described herein are based on liver, these techniques have been employed successfully in most tissue types with minor modifications and are therefore broadly applicable.

Keywords Oxidative stress; immunohistochemistry; 4-hydroxynonenal; 3-nitrotyrosine; alcoholic liver disease.

1   Introduction

Early free radical researchers were for the most part chemists (e.g., Henry John Horstman Fenton of the "Fenton" reaction; see *(1)* for review). However, free radical formation was considered for the most part an ex vivo phenomenon. Gershman and Gilbert *(2)* first proposed that radiation-induced cell killing might be mediated by oxygen free radicals, based on the observation that the pattern of toxicity of radiation was quite similar to that hyperoxygen in vivo. In the same year, Harman *(3)* published the "free radical hypothesis of aging." This strength of this hypothesis can be demonstrated by the simple fact that it has endured for almost 50 yr. It was generally thought that free radical formation in the cell was limited to either external stimuli (such as radiation), or random events and did not occur under "normal" circumstances.

A major paradigm shift in free radical research came in 1969, when Fridovich and McCord described the function of SOD as a catalytic reducer of O_2^- to H_2O_2 *(4)*. On the basis of these results, the concept that oxidants are produced by the cell under normal conditions gained hold. The discovery of enzymes that normally produce prooxidants (in contrast to electron leakage from other enzyme systems), such as NAD(P)H oxidases, xanthine oxidases, and myeloperoxidases, further strengthened the case that prooxidants are regularly found in vivo. Discoveries of catalytic functions of the glutathione peroxidases, glutaredoxin peroxidases, thioredoxin peroxidases, and catalases indicated that hydroperoxides are also kept in close check in vivo. It became clear that balance between prooxidants and antioxidants is critical for survival and functioning of aerobic organisms. An imbalance favoring prooxidants and/or disfavoring antioxidants, potentially leading to damage, was coined as oxidative stress by Sies *(5)*.

Reactive oxygen and nitrogen species (ROS and RNS, respectively) can be products of normal cellular metabolism and have potentially beneficial effects (e.g., cytotoxicity against invading bacteria); indeed, there are enzymes in which the key function is to produce ROS/RNS, such as nitric oxide synthases (NOS), NA(D)PH oxidases, and myeloperoxidases. Oxidative stress is proposed to be critically involved in numerous diseases, including alcohol-induced liver disease. In 1966, DiLuzio was the first to characterize lipid peroxidation after chronic exposure to alcohol *(6)*, a finding confirmed by others *(7)*. After the introduction of the spin trapping technique, in vitro *(8–10)* and in vivo *(11)*, evidence for a free radical derived from ethanol, the α-hydroxyethyl free radical was presented.

A general problem of studying prooxidants in vivo is that, because of their inherent reactivity, they generally cannot be measured directly. One is therefore often left with measuring products of the reaction of these molecules with other molecules (e.g., oxidative or nitrative modification of proteins). This indirect detection of "footprints" of ROS/NOS makes it difficult to conclusively identify the parent oxidant. However, coupled with in vivo modulation of free radical production, one can obtain relatively clear data supporting the involvement of specific free radical species involvement in alcohol-induced tissue damage.

This chapter will summarize some techniques that have been used in the study of oxidative stress in alcohol-induced liver injury. Immunohistochemical detection of 4OH-nonenal- and 3-nitrotyrosine-modified proteins will be detailed *(12)*. In the past, immunohistochemical techniques have served largely as qualitative measures of changes but have the advantage of illuminating the location and zonation of the haptene of interest. However, with the advent of digital photography and image analysis, one can now quantitate treatment-related changes in the tissue of interest. Other advantages of these techniques are that they are readily performed in most basic research laboratories and do not require pretreatment of the animals to attain a signal. The latter point implies that historical samples may be analyzed. Although the methods described herein are based liver, these immunohistochemical techniques have been employed successfully in most tissue types with minor modifications.

Although immunohistochemical analysis of ROS/RNS-modified proteins is a useful technique, careful attention must be paid to minimize the effect of variability and artifacts. Obvious possible sources of in vivo variability (e.g., gender, age, dose, etc) should clearly be avoided. Furthermore, variability caused by the processing and staining of the tissue itself should also be minimized. This ex vivo variability is attributable predominantly to the staining itself; it is therefore imperative that all samples to be analyzed are stained simultaneously to avoid interassay inconsistencies. Reasonable measures should also be taken to avoid all possible sources of ex vivo variability, including harvesting and processing of the tissue for immunohistochemistry (*see* **Note 1**).

2   Materials

All solutions should be of analytical grade or purer. Nanopure water is used for preparing all solutions. Glassware should be freed of possible detergent residue by prerinsing with nanopure water.

2.1   *Generation of an In Vivo Positive Control for Staining*

1. Laboratory safety equipment (gloves, mask, protective eyewear, etc.)
2. Lipopolysaccharide (LPS) from *Escherichia coli* (O55:B5; Sigma).
3. Sterile saline: 0.9 % sodium chloride.
4. Needles for injection.
5. Borosilicate glass tubes (12 × 75 mm; Fisher).

2.2 Harvesting of Tissues for Immunohistochemistry

1. Laboratory safety equipment (gloves, mask, protective eyewear, etc.)
2. Rodent anesthetic (e.g., sodium pentobarbital, ketamine/xylazine, etc.).
3. Tools for harvesting tissue.
4. Sterile saline.
5. 10% neutral-buffered formalin (Fisher).
6. 10-mL syringes with plastic infusion needles (16 gage or larger).
7. Razor blades.
8. Plastic tissue cassettes.

2.3 Immunohistochemistry

1. Laboratory safety equipment (gloves, mask, protective eyewear, etc.)
2. Ventilated work area (e.g., chemical fume hood).
3. Magnetic stir plate capable of maintaining a slow stir rate (<100 rpm) and a 1-inch magnetic stir bar (Fisher).
4. $>3\times$ glass staining dishes ($45 \times 75 \times 32$ mm; "Coplin jars") and slide holders (Wheaton).
5. Xylene (Fisher) or less toxic alternative (e.g., CitriSolv; Fisher).
6. 100% and 95% ethanol (Aaper).
7. Phosphate-buffered saline with 0.5% Tween-20 (PBS) ($10\times$ stock) $1.37 M$ NaCl, 27 mM KCl, 100 mM Na_2HPO_4, 18 mM KH_2PO_4 (adjust to pH 7.4), and 5% Tween-20.

   a. Autoclave before storage at room temperature. Sodium azide (0.02%) may be added to the stock to help prevent contamination. Prepare working solution by dilution of one part with nine parts water.

8. 10 mM sodium citrate (pH 6).
9. 30% hydrogen peroxide solution (Fisher).
10. Bovine serum albumin, fraction V (BSA; Sigma).
11. Clear nail polish or Dako Pen (Dako).
12. Slide rack (Fisher). A slide storage box (Fisher) works quite well for this purpose.
13. Rabbit polyclonal anti-4OH-nonenal antibody (Alpha Diagnostic); Rabbit polyclonal anti-nitrotyrosine antibody (Upstate).
14. EnVision, HRP-based staining system (Dako). It is not required, but this system usually gives a good signal to noise ratio and contains both the HRP-conjugated secondary antibody and DAB solutions. Alternatively, HRP-conjugated goat antirabbit antibodies (Vector) and diaminobenzidine (DAB) solution (Dako) may be purchased separately.
15. Aqueous hematoxylin (Dako).
16. Mounting media (Permount; Fisher) and coverslips ($25 \times 50 \times 0.15$ mm; Fisher).

2.4   Digital Image Acquisition and Analysis System (see Note 2)

1. Microscope: Eclipse E600 with a 10× ocular Plan Fluor series objectives (4×, 10×, 20×, 40× and 100×) (Nikon)
2. Camera: Spot RT slider BW/color digital CCD camera (Diagnostic Instruments)
3. Imaging software: MetaMorph image-acquisition and analysis software (Molecular Devices; *see Note 3*).

3   Methods

Both staining protocols are relatively similar and have shared methods. The in vivo methods require prior approval by the local Institutional Animal Care and Use Committee (IACUC).

3.1   Generation of an In Vivo-Positive Control for 4OH-Nonenal and 3-Nitrotyrosine

It is important to have a positive in vivo control for the staining. In our experience, injections of hepatotoxic doses of LPS generate sufficient signals of both 4OH-nonenal and 3-nitrotyrosine to serve as positive controls. Fig. 1 contains representative photomicrographs (10×10 magnification) of a liver from a mouse 24 h after injection with LPS (10 mg/kg intraperitoneally) and immunostained for 4OH-nonenal. The supplier also recommends additional positive and negative controls for 3-nitrotyrosine staining. These can be generated by following manufacturer instructions.

1. Prepare an LPS-injectable solution in sterile saline (injection volume 1-2 mL/kg). Solution must be made in borosilicate glass tubes (LPS binds to plastic and Teflon). **Note:** LPS is highly toxic care should be taken when handling this compound to avoid exposure.
2. Inject the rodent (rat or mouse) with LPS intraperitoneally. This laboratory uses a dose of 10 mg/kg LPS (intraperitoneally) for both species. However, because of batch-to-batch variability in LPS potency, the dose should be titrated in the individual laboratory to achieve serum alanine aminotransferase (ALT) activity of 200–300 IU/L 24 h after LPS injection.
3. Harvest liver 24 h after LPS injection (*see* Subheading 3.2.).

3.2   Harvesting Tissue for Immunohistochemistry

1. Anesthetize the rodent according to your institution's IACUC-approved protocol.
2. Perform a laparotomy to fully-expose the liver.

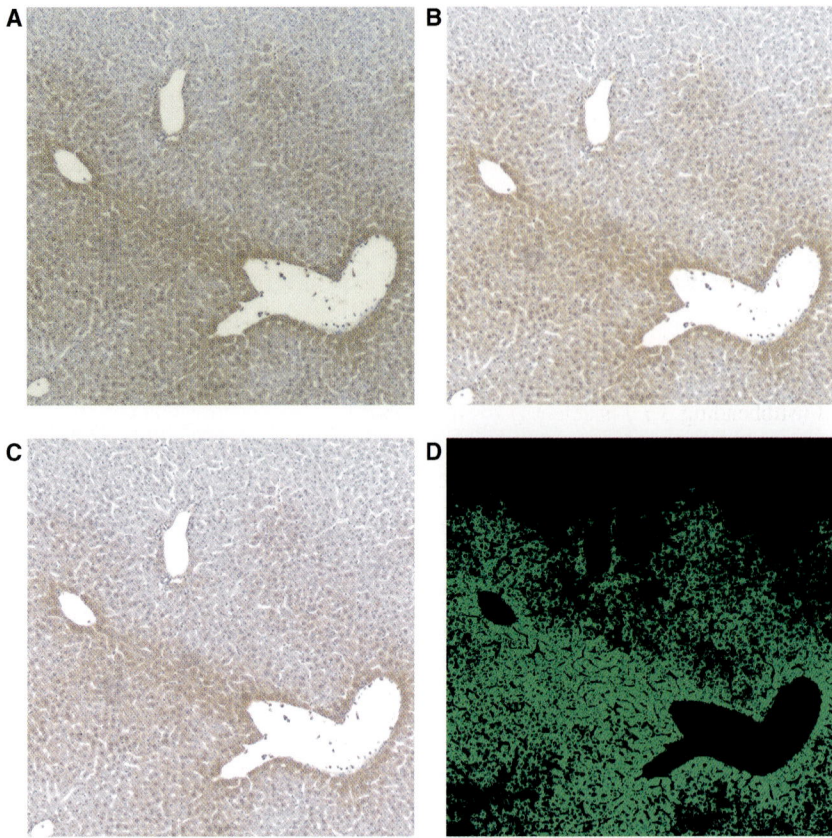

Fig. 1 Effect of LPS on 4OH-nonenal adduct accumulation in mouse liver as determined by quantitative immunohistochemistry. Representative photomicrographs are shown depicting 4OH-nonenal immunostaining of a liver section (100×) of a mouse 24 h after injection with LPS (10 mg/kg intraperitoneally). The same microscope region is shown before (**A**) and after (**B**) white balancing of the image-analysis software. Note that the acellular space in (**A**) is light yellow instead of white. (**C**) is the same as (**B**), but after the blue exposure is increased 10%. Note that the contrast between the DAB chromogen (brown) and the hematoxylin counterstain (blue) is enhanced by digitally increasing the blue exposure. (**D**) shows the microscope region from (**C**), determined by the computer to be within the threshold range for positive staining

3. Exsanguinate the animal by cutting the aorta just before entering the heart.
4. Flush the liver/animal (Optional). Flush the liver with a 10-mL syringe filled with saline by inserting into the portal vein just below the liver. Follow with a 10-mL syringe filled with 10% neutral buffered formalin. This not only helps minimize possible ex vivo haptene formation but also helps retain the architecture of the organ. If other organs are also to be harvested, exsanguination, flushing, and fixing of the whole animal may be achieved by cardiac puncture and perfusion. A larger syringe (e.g., 50 mL) will be required.

5. Cut thin (2- to 4-mm thick) transverse sections of liver (or other organ of interest).
6. Place in tissue cassettes (labeled with pencil) and fix at 4°C for >48 h.
7. Paraffin-embed tissue, section (5- to 6-μm thick), and mount on microscope slides (see **Note 4**).

3.3   *Deparaffinization and Rehydration of Tissue Slides*

At least one sample should be processed in duplicate in this procedure for primary antibody control (*see* Subheading 3.5., step 2.). Normally, the positive in vivo control (Subheading 3.1.) is selected for this control.

1. Slide racks for the Coplin jars have 10 slots (*see **Note 5***). They may hold up to 20 slides if placed "back-to-back" (i.e., tissue side facing out). The jar should be filled with liquid to completely cover the slides (~225 mL). With a stir bar in the bottom of the jar, it is placed on stir plate (set to the lowest RPM). Incubate in each of the following solutions for 5 min:
   a. Xylene (or alternate solution); repeat once
   b. 100% ethanol
   c. 95% ethanol
   d. 70% ethanol
   e. 30% ethanol
   f. Nanopure water
2. Create a buffer dam around tissue (optional). While carefully avoiding drying of the section (do one slide at a time), paint a circle around the tissue section with the Dako pen (or nail polish), carefully avoiding touching the tissue with the solution (>3 mm away from the section). This buffer dam decreases the likelihood of the sample drying-out during antibody incubations. Place samples in a new Coplin jar containing fresh water.
3. Perform antigen demasking (optional). This procedure is strongly recommended for 3-nitrotyrosine immunohistochemistry.
   a. Place slides in Coplin jar containing citric acid solution (Subheading 2.3., step 8.).
   b. Remove metal handle from slide rack. Place jar in a glass dish of water and microwave on high for 2.5 min (700-watt oven). Warning, buffer will get hot; avoid scalding.
   c. Let cool for 3 min. Refill Coplin jar with citrate buffer, if needed.
   d. Repeat steps b-c three times. Let cool in citrate buffer for 20 min.
   e. Rinse sections in Coplin jar containing nanopure water for 5 min. Repeat once.
4. Rinse the samples in a Coplin jar containing PBS-Tween for 5 min. Repeat once.

3.4   Peroxidase and Protein Block of Tissue Sections

1. Make 250 mL of 3% hydrogen peroxide solution diluted in PBS-Tween (*see Note 6*). Pour into Coplin jar.
2. Incubate the slides for 5 min in peroxide solution.
3. Rinse the samples in a Coplin jar containing fresh PBS-Tween for 5 min. Repeat once.
4. Make PBS-Tween containing 1% BSA (PBA). You will need approx 250 μL per slide.
5. Remove the slides from the Coplin jar one at a time. Pour off excess solution and pipette PBA into buffer dam (Subheading 3.3., step 2.) around tissue. Place on slide rack.
6. Incubate slides for 1 h at room temperature. Keep rack covered and hydrated by placing wet paper towels in the bottom of the rack. Add fresh solution as needed to avoid sample drying.
7. Proceed immediately to immunohistochemistry (Subheading 3.5.).

3.5   Immunohistochemical Detection of Haptene Signals

1. While samples are blocking (Subheading 3.4.), prepare the primary antibody solution by diluting the antibody in PBA (~250 μL/slide). Historically, a 1:500 and 1:50 dilution of the 4OH-nonenal and 3-nitrotyrosine antibodies, respectively, were effective. Dilutions may vary on antibody batch.
2. Shake off excess PBA from slides one at a time and replace with primary antibody solution (*see Note 7*). The primary antibody control slide will receive PBA without antibody.
3. Incubate on slide rack (covered and hydrated) for 30 min (4OH-nonenal) or an hour (3-nitrotyrosine). Incubation times may need to be altered to improve signal to noise ratio. Rinse the samples in a Coplin jar containing fresh PBS-Tween for 5 min. Repeat once. Shake off excess PBS-Tween from slides one at a time and replace with secondary antibody solution. Place on slide rack and incubate for 15 min (if using the Dako EnVision kit). Dilutions and incubation times for secondary antibodies from other suppliers should be determined by manufacturer's suggestions.
4. Rinse the samples in a Coplin jar containing fresh PBS-Tween for 5 min. Repeat once.
5. Prepare DAB working solution according to manufacturer's instructions. Place a piece of white paper in the bottom of the slide rack. If the slide rack does not have a cover, make a cover with aluminum foil (DAB is light sensitive).
6. Apply DAB solution to each slide. Cover slides and incubate for 8-10 min. Use the in vivo positive control to determine the length of staining. When the brown DAB precipitate appears to not increase in this slide, stop the incubation. Transfer slides to a Coplin jar containing nanopure water.

7. Rinse Coplin jar by placing under a gentle tap of water for 6 min.
8. Transfer slide rack to a Coplin jar containing hematoxylin. Incubate in solution for 30 seconds while gently moving the slide rack up and down.
9. Return slide rack to Coplin jar containing water. Rinse under tap for 6 min.

3.6   Dehydration and Coverslipping of Slides

1. Incubate the slide rack in each of the following solutions for 3-5 min (see Subheading 3.3., step 1):

   a. Nanopure water
   b. 30% ethanol 100% ethanol
   c. 70% ethanol 95% ethanol
   d. 95% ethanol 70% ethanol
   e. 100% ethanol
   f. Xylene (or alternate solution); repeat once

2. Keep the slides in the second xylene solution (from Subheading 3.6., step 1f) until ready to be coverslipped. Place a thin bead of Permount on one long edge of a coverslip. Gently role the slide onto the coverslip. Capillary action will pull the pieces of glass together. Avoid bubbles over the tissue sections. Should bubbles occur wash off coverslip with xylene solution and recoverslip. Let the slides dry in the fume hood for at least 24 h.

3.7   Image Analysis of Immunohistochemistry

An absolutely critical point in performing quantitative image-analysis of immunohistochemistry is to ensure that all samples to be compared are analyzed simultaneously. It is recommended that before performing image analysis that the slides be critically analyzed under the microscope for any variation in chromogen signal not owing to the in vivo condition. Furthermore, image-analysis should be viewed as a means to quantitate differences between treatment groups, not to create. Specifically, the differences should be readily observable with the eye under the microscope.

For microheterogeneous tissues, such as the liver, a low power magnification (e.g., 100 ×, ocular and objective) is usually sufficient for analysis. For macroheterogeneous tissues (e.g., specific brain regions), a higher magnification may be required to focus on the area of interest. Should staining be located in a specific area within a tissue (for example, the perivenous area of the liver lobule), higher magnification photomicrographs may be also justified. It is recommended, however, that initial attempts be performed at lower magnifications, as this reduces the risk of investigator bias in acquiring the pictures. It is also a good idea to blind analysis of the tissues to reduce the risk of investigator bias.

1. Power-up microscope, camera and image-analysis computer. Turn-on the microscope light source to ~80% of full brightness and allow it to sit for at least 20 min. Most bright-field bulbs are unstable light emitters until they are sufficiently warmed-up. A simple way to validate that the bulb is stable is to take five consecutive photos of the same microscope region and validate that they are consistent in brightness, contrast and color.

2. While the system is warming-up, ensure that the microscope is optimized for taking pictures at the selected magnification (e.g., adjusting the diaphragm height and width). These techniques are routine to experienced microscopists.

3. Make sure that the image analysis software is set to manually determine exposure settings. Otherwise, the software may adjust the exposure between pictures, causing variations in staining color and intensity.

4. Using the image-analysis software, calibrate the exposure settings for the microscope field. First, move the microscope stage to a region of the slide that does not contain the tissue (but does have the coverslip) and defocus the image. Microscopes often have a gradient of light intensity across the field. Most image-analysis programs have a function to account for this gradient and normalize this gradient in captured photomicrographs (in MetaMorph, the command is "get flatfield"). Next, in the same area, determine optimal exposure settings for the digital camera (i.e., "white balance"). Fig. 1A,B compare the same microscope field before (A) and after (B) appropriate white balancing.

5. After performing flatfield and white balance commands, the photomicrographs captured by the software should be similar to those seen with eye. Validate this by capturing a representative image and comparing the digital picture with the image seen with the eye. If necessary, manually adjust the RGB exposure settings.

6. Enhance contrast with digital filtering (optional). Image analysis is much more sensitive to contrast than the human eye. Immunostaining may look very clear by manual analysis but may be fraught with noise during digital analysis. This noise can be alleviated by enhancing the digital contrast. For example, slightly increasing the blue exposure on the RGB setting will enhance the contrast between the DAB chromogen and the hematoxylin counterstain. However, care should be taken if this filtering approach is performed that artifactual signals are not introduced. Fig. 1C is the same microscope region as 1b, but after the blue exposure has been increased 10%.

7. Once steps 1 through 6 are completed, it is imperative that the neither the microscope nor the exposure settings are altered during the image capture phase. Select and capture 10 microscope fields from each section (assuming 10×10 magnification). The regions should be randomly selected with the caveat that staining artifacts (e.g., edges, folds, tears, etc.) should be avoided. Morphologic variations found in the tissue (e.g., very large vessels in the liver lobule) may also be avoided. However, due diligence should be paid to avoid investigator bias in selecting regions. Should a higher magnification be used to focus on a specific region of the tissue, it is recommended that the number of pictures captured be increased.

8. Repeat this step for all tissue sections to be analyzed.
9. Select an image file with strong positive staining for the haptene. Use this image to determine the color, intensity, and saturation ranges for the image-analysis (called "threshold" in MetaMorph). It is recommended that a "conservative" range be initially attempted to reduce noise in the sections. Fig. 1D shows the microscope region from Fig. 1C determined by the computer to be within the threshold range for positive staining.
10. Pool the summary area data from each tissue section to determine the percentage of microscope field positive for the haptene signal. Use this value to determine group means.

3.8   In Vivo Modulation of Free Radical Production Using Genetically Modified Mice

The inherent reactivity of ROS/RNS makes it very difficult to directly measure them under in vivo conditions. Even electron spin resonance spectroscopic analysis of free radicals usually requires a spin-trap to stabilize the radical signal long enough for detection. Therefore, although assessment of changes in ROS/RNS by the techniques described above are useful, these 'footprints' do not give any information on the type(s) and source(s) of reactive species produced that caused these changes.

To date, most work has employed chemical inhibitors of suspected prooxidant enzymes. The effect of these drugs in vivo may not necessarily be specific to inhibition of the enzyme in question. Therefore, caution should be used when basing conclusions on work with pharmacologic inhibitors. One advance towards this end in the alcohol field has been the ability to address key questions at the molecular level by using genetically-altered mice (e.g., knockout/transgenic *(13)*). Such strategies have their own potential pitfalls, such as induction of compensatory mechanisms in knockout mice, which may cause false-positive or -negative results. (*see* **Note 8** for general guidelines on the use of genetically altered mice.)

4   Notes

1. To decrease ex vivo variability, our group always harvests sections for immunohistochemistry from the same liver lobes (left lateral and right medial) and the same location with the lobe (transverse section from the middle of the lobe) to avoid any potential geographic variation in the haptene signals.
2. There are many possible iterations of this system, but the emphasis should be on a system that is able to take high-quality and –resolution brightfield images. Important features of the system include flat-field correction, white balance, and manual control of red/green/blue (RGB) exposures.

3. Windows-compatible personal computer. Most new PCs will have enough speed and RAM to handle MetaMorph. A large hard drive is recommended to accommodate the large microscope files.

4. Microscope sections must be of consistent high quality (e.g., few if any folds or tears). Should the laboratory lack a practiced hand at cutting and mounting tissues, it is highly recommended that the tissues be prepared and processed by a professional histopathology service.

5. A number of newer and more rapid methods to deparaffinize and rehydrate microscope slides are available that use less reagent volumes than Coplin jars. However, it has been the experience of this author that deparaffinization/rehydration can be inconsistent and incomplete with these alternate approaches. These techniques are therefore not recommended for quantitative immunohistochemistry.

6. The Dako EnVision kit supplies a peroxidase block reagent. However, more consistent blocking has been achieved by our group using freshly diluted concentrated hydrogen peroxide in a Coplin jar.

7. A critical component of quantitative immunohistochemistry is consistency, to avoid intra-assay variability. Careful attention should be paid so that the incubation steps for each sample are similar, especially in steps that require you to work with one sample at a time (e.g., Subheading 3.5., steps 2, 5, and 8). A simple solution is to measure the length of time required to add reagent to each sample.

8. The following are general guidelines for usng genetically altered mice in experiments. These guidelines are derived from the Banbury conference on genetic background in mice *(14)*, and have been adopted by many scientific journals:

   a. Genetic background of mouse strain must be clearly reported.

   b. Common strains of mice should be used to aide reproducibility by other laboratories.

   c. The genetically modified strain should be backcrossed into a common background strain.

   d. The strain should be backcrossed into the wild-type strain at least four times to avoid genetic variation at other sites besides the locus of interest.

   e. Genetically modified line should not be maintained by crossing homozygotes, but rather maintained by backcrossing heterzygotes against the wild-type strain. This avoids the potential of genetic drift of the modified strain away from the background strain.

   Results with modified strain should be validated with multiple generations. Further, weight-of-evidence can be increased by comparing pharmacologic targeting of the same protein with the knockout strain.

References

1. Koppenol, W. H. (1993) The centennial of the Fenton reaction. *Free Radic. Biol. Med.* **15,** 645–651.
2. Gerschman, R., Gilbert, D. L., Nye, S. W., Dwyer, P., and Fenn, W. O. (1956) Oxygen poisoning and X-irradiation: a mechanism in common. *Science* **119,** 623–626.

3. Harman, D. (1956) Aging: A theory based on free radical and radiation chemistry. *J. Gerontol.* **11,** 298–300.
4. McCord, J. M., Fridovich, I. (1969) Superoxide dismutase: An enzymatic function of erythrocuprein (hemocuprein). *J. Biol. Chem.* **244,** 6049–6055.
5. Sies, H. (1985) Oxidative stress: introductory remarks, in *Oxidative Stress* (Sies, H., ed) pp. 1–8, Academic Press, London.
6. Di Luzio, N. R. (1966) A mechanism of the acute ethanol-induced fatty liver and the modification of liver injury by antioxidants. *Lab. Invest.* **15,** 50–63.
7. Shaw, S., Jayatilleke, E., Ross, W. A., Gordon, E. R., and Lieber, S. (1981) Ethanol-induced lipid peroxidation: Potentiation by long-term alcohol feeding and attenuation by methionine. *J. Lab. Clin. Med.* **98,** 417–424.
8. Tomasi, A., Albano, E., Biasi, F., Slater, T. F., Vannini, V., and Dianzani, M. U. (1985) Activation of chloroform and related trihalomethanes to free radical intermediates in isolated hepatocytes and in the rat in vivo as detected by the ESR-spin trapping technique. *Chem. Biol. Interact.* **55,** 303–316.
9. Reinke, L. A., Lai, E. K., DuBose, C. M., and McCay, P. B. (1987) Reactive free radical generation in the heart and liver of ethanol-fed rats: Correlation with in vitro radical formation. *Proc. Natl. Acad. Sci. USA* **84,** 9223–9227.
10. Rashba-Step, J., Turro, N. J., and Cederbaum, A. I. (1993) Increased NADPH- and NADH-dependent production of superoxide and hydroxyl radical by microsomes after chronic ethanol treatment. *Arch. Biochem. Biophys.* **300,** 401–408.
11. Knecht, K. T., Bradford, B. U., Mason, R. P., and Thurman, R. G. (1990) *In vivo* formation of a free radical metabolite of ethanol. *Mol. Pharmacol.* **38,** 26–30.
12. McKim, S. E., Gabele, E., Isayama, F., Lambert, J. C., Tucker, L. M., Wheeler, M. D., Connor, H. D., Mason, R. P., Doll, M. A., Hein, D. W., et al. (2003) Inducible nitric oxide synthase is required in alcohol-induced liver injury: studies with knockout mice. *Gastroenterology* **125,** 1834–1844.
13. Arteel, G. E. (2003) Oxidants and antioxidants in alcohol-induced liver disease. *Gastroenterology* **124,** 778–790.
14. Banbury Conference on genetic background in mice. (1997) Mutant mice and neuroscience: recommendations concerning genetic background. *Neuron.* **19,** 755–759.

15
Isolation of Kupffer Cells from Rats Fed Chronic Ethanol

Megan R. McMullen, Michele T. Pritchard, and Laura E. Nagy

Summary   Chronic consumption of ethanol induces hepatic steatosis and inflammation, which can eventually lead to more severe liver injury, characterized by fibrosis and cirrhosis. Recruitment of neutrophils to the liver, as well as activation of Kupffer cells, mediates the inflammatory responses observed after chronic ethanol exposure. Kupffer cells, the resident macrophages of the liver, are critical to the onset of ethanol-induced liver injury. Activation of Kupffer cells leads to an increased production of proinflammatory cytokines, such as tumor necrosis factor-α and also reactive oxygen species, a process mediated in part by changes in lipopolysaccharide-induced TLR4-dependent signal transduction. The isolation and culture of Kupffer cells is an important technique with which one can elucidate the mechanisms that contribute to alcoholic liver injury.

Keywords   Kupffer cell; elutriation; RNA isolation; alcoholic liver disease.

L. E. Nagy (ed.), *Alcohol: Methods and Protocols*
© Humana Press 2008

1   Introduction

The development of alcoholic liver disease (ALD) is characterized by an increase in hepatic steatosis and inflammation, which in turn can lead to fibrosis and then cirrhosis. Kupffer cells, the resident macrophages of the liver, are critical to the onset of ethanol-induced liver injury. Ablation of Kupffer cells prevents the development of steatosis and inflammation in rats chronically exposed to ethanol via intragastric feeding *(1)*. After chronic ethanol exposure, Kupffer cells exhibit increased sensitivity to lipopolysaccharide (LPS) and consequently exhibit enhanced LPS-stimulated inflammatory cytokine production *(2,3)*. Tumor necrosis factor-α (TNF-α) production, resulting from Kupffer cell activation, has been shown to be a critical component in the development of ALD *(4)*. Recent studies have shown that increased sensitivity to LPS after chronic ethanol exposure also leads to an increased production of reactive oxygen species (ROS) *(5,6)*. Isolation of Kupffer cells from rats fed an ethanol-containing diet provides sufficient material for studies using cellular and molecular biology approaches aimed at understanding the effects of ethanol on Kupffer cell function. For example, isolation of mRNA from Kupffer cells from pair- and ethanol-fed rats has shown that LPS-stimulated increases in TNF-α expression can be inhibited by adiponectin *(5)*. In this chapter, we will focus on methods used for isolating Kupffer cells from animals fed the Lieber-DeCarli ethanol diet and their pair-fed controls, as well as how to isolate mRNA from these primary Kupffer cell cultures.

2   Materials

2.1   Surgical Portion of the Procedure

Materials needed for the surgical portion of the Kupffer cell isolation can be found in Table 1. Sources used and catalog numbers associated with them are also included in the table.

2.2   Elutriator

Materials needed for the elutriation portion of the Kupffer cell isolation can be found in Table 2. Also included are the sources used and the catalog numbers associated with them.

2.3   Isolation and Culture of Kupffer Cells

Materials for the isolation and culture of the Kupffer cells can be found in Table 3 along with the sources used and catalog numbers.

Table 1 Materials for Surgical Portion of Procedure

Material	Vendor	Catalog number	Concentration or amount needed
1. Autoclaved water			
2. Hydrogen peroxide			6%
3. 8-channel peristaltic pump	Rainin	7103–058	
4. Four - "butterfly needles" (Blood Collection Infusion Set 19 gauge 3/4 with 12″ tubing)	BD Biosciences	364919	
5. Waterbath at 37°C			
6. Hanks - (Hank's Balanced Salt Solution (w/o Ca^{2+}, Mg^{2+}, $NaHCO_3$))	Mediatech	55–021-PC	250 mL/rat
7. HEPES	Fisher	BP310–500	10 mM
8. EGTA	Sigma	E4378	1 mM
9. Williams E	Sigma	W4125	250 mL/rat
10. Collagenase type IV (from *Clostridium histolyticum*)	Sigma	C5138	50 mg/rat
11. 9″ Pasteur Pipets			
12. 95% O_2 5% CO_2 gas mixture	Praxair	mmoxcd5-k	
13. Heparin	Sigma	H3393	1000 U/mL
14. Ketamine (100 mg/mL)	Henry Schein	995–2949	
15. Xylazine (20 mg/mL)	NLS Animal Health	105650	
16. Acepromazine (10 mg/mL)	Henry Schein	356–7290	
17. Eight 1-mL syringes	Fisher	14–823–2F	
18. Four 26-gauge 5/8-inch needles	Fisher	14–826–6A	
19. Four 21-gauge needles	Fisher	14–826C	
20. Kimwipes			
21. Buttonhole twist thread (2 colors - ~5″)			
22. Two surgical pans (9×13 baking pan)			
23. Two racks (cookie cooling racks - must fit over pan)			
24. Two pairs of sharp/sharp 5 1/2″ scissors	Fisher	13–808–4	
25. One pair of blunt/sharp 5 1/2″ scissors	Fisher	13–808–2	
26. One pair of blunt-point curved 5″ forceps	Fisher	08–887	
27. One pair of curved eye dressing 4 1/2″ forceps	Henry Schein	101–7739	
28. Four 15-mL centrifuge tubes	Sardstedt	52 554 205	
29. Four divided Petri dishes	Fisher	08–757–110	

Item 21 can be atuoclaved in a small box and items 24–27 can be autoclaved together in an autoclaveable box.

Table 2 Materials for Elutriation Procedure

Material	Vendor	Catalog number	Concentration or amount needed
1. Beckman Avanti J25 Centrifuge	Beckman		
2. JE6B Elutriation Rotor and accessories	Beckman		
3. Masterflex L/S Variable Speed Pump	Cole Parmer	EW-77521–40	
4. Masterflex L/S Easey Load II Pump Head	Cole Parmer	EW-77200–60	
5. Autoclaved water			
6. Hydrogen Peroxide			6%
7. Catalase	Sigma	C9322	10–15 mg/100 mL
8. Gey's balanced salt solution			1 L

If working with endotoxin, autoclave and bake all glassware before use.

3   Methods

The following method is designed for the isolation of Kupffer cells from male Wistar rats that were allowed free access to the Lieber-DeCarli liquid diet *(7)* containing 36% of calories from ethanol or pair-fed an isocaloric diet that substituted maltose dextrins for the ethanol. The following procedure has been adapted from methods originally developed by the Friedman and Roll, Knook et al., and Munthe-Kaas et al. *(8–10)*. Kupffer cells are isolated as previously described *(2)*. In brief, livers are perfused with 0.05% collagenase, and the resulting suspension of liver cells is treated with 0.02% pronase for 15 min at 12°C. The resulting cell suspension is centrifuged 3 times at 50 g for 2 min, and the supernatant collected after each centrifugation. The pooled supernatant is then centrifuged at 500 g for 7 min to collect nonparenchymal cells. Kupffer cells are then purified by centrifugal elutriation *(2)*. The following protocol is for isolation of Kupffer cells from 2 pair-fed and 2 ethanol-fed animals; however, the protocol can be scaled up or down according to the quantity of Kupffer cells required.

3.1   Setting up the Peristaltic Pump and Perfusion

All eight channels of the peristaltic pump must be used for isolating Kupffer cells from four rats. Set up the Rainin peristaltic pump per the manufacturer's instructions. It is necessary to check the flow rates of the peristaltic pump before beginning the isolation. For perfusion of two rats, set the flow rate to 25 mL/min and to 50 mL/min for four rats. To sterilize the tubing, first run sterile water through the pump, and then run approx 200 mL of 6% hydrogen peroxide (H_2O_2) through the

Table 3 Materials for Isolation and Culture of Kupffer Cells

Material	Vendor	Catalog number	Concentration or amount needed
1. Waterbath at 37°C			
2. 12°C Shaking Waterbath			
3. Beckman Tabletop Centrifuge			
4. Hanks + (Hank's Balanced Salt Solution (with Ca^{2+}, Mg^{2+}) (w/o $NaHCO_3$))	Sigma	H6136	10 mL/rat
5. CMRL Medium 1066	Invitrogen	11530–037	
6. Antibiotic-Antimycotic	Invitrogen	15240–062	
7. L-glutamine	Invitrogen	25030–081	2 mM
8. Fetal bovine serum certified	Invitrogen	16000–044	10%
9. Pronase	Roche	165 921	30 mg/2 rats
10. DNase I	Roche	104 159	4.5 mg/rat
11. Gey's Balanced Salt Solution			
12. Nycodenz	Accurate Chemical	ACN1002424	11% and 24.6%
13. 3% Acetic acid			
14. Trypan blue			
15. Thirty-50 mL centrifuge tubes	Sardstedt	62 547 205	
16. Six-15 mL centrifuge tubes	Sardstedt	52 554 205	
17. Two-500 mL polypropylene flasks	Fisher	10–041–10D	
18. Two-250 mL polypropylene flasks	Fisher	10–041–10C	
19. Two plastic funnels	Fisher		
20. One pair straight dissecting dressing 4- 1/2″ forceps	Fisher	13–812–39	
21. One pair curved dissecting 4- 1/2″ forceps	VWR	25717–005	
22. Two 3×3″ pieces of Coarse mesh (132 μm)	Sefar Americo (Tetko)	03–132–40	
23. Two 3×3″ pieces of Fine mesh (224 μm)	Sefar Americo (Tetko)	03–225–42	
24. Two - 60-mL syringes	Fisher	13–689–8	
25. Two 18-gauge needles	Fisher	14–826–5G	
26. 5-3/4″ Pasteur pipets			
27. 9″ Pasteur pipets			
28. Rubber bulb	VWR	56311–062	
29. 1.5-mL Eppendorf tubes			
30. Pipet tips (0.5–10 μL, 10–100 μL, 100–1000 μL)			
31. Sterile reagent reservoirs	Fisher	07 200 127	
32. Sterile tissue culture-treated plates (96-well, 24-well, 6-well, chamber slides)			

Items 17–18 should be atuoclaved and items 20–23 can be autoclaved together in an autoclaveable box. If working with endotoxin, autoclave and bake all glassware before use.

pump. Finally, run more sterile water through the pump until all peroxide bubbles are gone from the tubing. While sterilizing the pump, thaw 1 × 10-mL aliquots of 10 mM HEPES and 1 mM ethylene glycol-bis (2-amnioethylether)- N, N, N', N'-tetraacetic acid (EGTA), and bring 1 L of Hanks– (pH 7.4) and Williams E (pH 7.4) to 37°C. When the media are at 37°C, oxygenate each medium with a mixture of 95% O_2 and 5% CO_2 for approx 10 min, and then add 10 mM HEPES and 1 mM EGTA to the Hanks–, mix thoroughly and run through the peristaltic pump. Once the Hanks– is through, stop the pump until you are ready to perfuse the livers. Once the Williams E has been oxygenated, add 200 mg of collagenase, mix by swirling and place in the water bath (*see* **Note 1**).

3.2 Setting up the Elutriator

Before using the elutriator to isolate cells, it is necessary to check the flow rates of the peristaltic pump at the desired rate of centrifugation. Before pumping cells through the elutriator, the tubing should be autoclaved. Set up the elutriation chamber and tubing and cool to 4°C. To prepare the elutriator, turn the pump to the 45 mL/min flow rate and run sterile water through the pump and chamber (~100–200 mL). Next, recirculate 100 mL of 6% H_2O_2 through the tubing and chamber for approx 5 min, with the centrifuge running at 50 g. Make sure the pressure gauge stays under 20 psi at all times. Stop the centrifuge, recirculate 100 mL of catalase (10–15 mg/100 mL) through the tubing, and restart the centrifuge at 24 g for approx 5 more minutes or until all of the peroxide bubbles are gone. Stop the centrifuge again, and then rinse out the tubing and chamber with sterile water. Once the tubing and chamber have been sterilized, run Gey's Balanced Salt Solution (pH 7.4) (GBSS) through the system (let it run all the way through to remove any air bubbles) and then stop the pump until needed.

3.3 Setting Up the Tissue Culture Room

It is helpful to set up needed cell culture supplies before beginning the perfusion or while the perfusion is in progress. Place the following in the cell culture hood:

    2 – 30-mg aliquots of pronase in a 50-mL centrifuge tube
    2 – 4.5-mg aliquots of DNase I in a 15m-L centrifuge tube
    1 – 500-mL bottle of CMRL with antibiotic/antimycotic and 2 mM L-glutamine (CMRL)

Fill each 50-mL centrifuge with 50 mL of CMRL and each 15-mL centrifuge with 3 mL of CMRL.

3.4 Anesthesia

Rats are weighed and the dose of anesthetic is calculated based on diet and body-weight. In our experiments, we use a mixture of xylazine (20 mg/mL), ketamine (100 mg/mL), and acepromazine (10 mg/mL), which expires after 1 mo.

Ketamine:	150 mg (100 mg/mL)	1.5 mL
Xylazine:	30 mg (20 mg/mL)	1.5 mL
Acepromazine:	5 mg (10 mg/mL)	0.5 mL
		3.5 mL

In chow-fed animals, we use a dose of 0.12 mL/100 g bodyweight. Ethanol-fed animals on the Lieber-DeCarli ethanol diet receive a dose of 0.075 mL/100 g bodyweight, whereas their pair-fed controls receive 0.12 mL/100 g bodyweight. Animals receive an intraperitoneal injection in the lower middle abdomen (right or left of midline is fine). Allow 5–10 min for the animals to reach complete sedation. If there is no movement when the foot or base of the tail is pinched, the animal is completely anesthetized.

3.5 Surgery

1. Once the animals are anesthetized, wipe the fur with a Kimwipe wet with water (do not use ethanol) to get rid of any dander, loose fur, or dirt.
2. Using the sharp/blunt scissors and blunt-point curved forceps, peel the abdominal fur by pulling up on the fur at the lower part of the abdomen and cutting up towards the thoracic area, removing enough fur to have a clear surgical field.
3. Remove loose fur from the scissors and animals by wiping with another wet Kimwipe.
4. Using the same sharp/blunt scissors and blunt-point curved forceps, make a horizontal cut into the lower part of the abdomen at midline. Carefully insert the scissors into the opening (do not to puncture the bladder) and make a smooth, V-shaped cut up both sides of the abdomen. Stop the cut at the diaphragm of the animal. Do not puncture the diaphragm.
5. Using a Kimwipe, push the organs to the side and out of the abdominal cavity and place a 15-mL centrifuge tube under the thoracic area to better expose the portal vein.
6. Place two loose ligatures (two different colors of autoclaved buttonhole twist thread) around the hepatic portal vein using the curved eye dressing forceps.

   a. Insert the forceps in the connective tissue near the hepatic portal vein, go underneath the vein and around to the other side. Then, grasp onto one end of the thread and gently pull it through. Pulling too hard on the thread can twist the hepatic portal vein and also cut off circulation to the liver.

   b. After pulling both ligatures through, tie a loose knot in preparation for inserting the perfusion "butterfly" needle and lay the ends of the string out so they are easy to pick up.

7. Inject 0.1 mL of heparin using a 26-gauge, 5/8-inch needle into the posterior vena cava. After the injection, press down on the injection site to stop the bleeding with a Kimwipe, and wait 1min before proceeding.

   a. If you would like to harvest other tissues (skeletal muscle, adipose), this would be the most appropriate/ideal time to do so. The animal will still be alive, and the area for liver perfusion is ready.

   b. To take a blood sample from the animal, use an 18-gauge needle (smaller needles will shear red blood cells resulting in a hemolyzed plasma/serum sample). Insert the needle into the posterior vena cava and pull back slowly. Try to minimize bleeding by applying pressure to the vena cava. The perfusion should be started as soon as possible after collecting the blood.

8. For the perfusion, start the peristaltic pump at the 25 mL/min flow rate, insert the 19-gauge, 3/4-inch perfusion "butterfly" needle behind the rear ligature into the hepatic portal vein and past the front ligature. Tighten the knots of the two ligatures to secure the needle (*see* **Note 2**).

9. As each perfusion needle is secured, cut through the diaphragm and cut the portion of the posterior vena cava in the pericardial cavity. This releases the perfusate and allows it to drain into the pan below.

10. First, perfuse with the Hanks– medium. Stop the pump just before all the media is in the tubing, place the tubing into the William's E + collagenase (*see* **Note 3**) and start the pump again. Do not allow bubbles to get into the liver.

3.6 Isolation of Kupffer Cells

1. After the perfusion, using the second pair of sharp/sharp scissors and the blunt-point curved forceps, remove the liver. Take the forceps and grasp onto a portion of the diaphragm near the liver, at midline. Cut a small "tag" of diaphragm to hold onto while cutting the connective tissue that attaches the liver to other organs. When removing the liver, be careful not to puncture the intestines or stomach, as they are a source of potential bacterial contamination. Once each liver has been removed, place it into a 100mm divided Petri dish. The remainder of the procedure is performed in a laminar flow hood under sterile conditions.

2. Rinse each liver with approx 10 mL of Hanks+.

3. Add 15 mL of CMRL to each liver. Using the straight and curved dissecting forceps, tease each liver slightly.

   a. If the perfusion was successful, the liver will come apart with little effort.

   b. If the liver is hard to tease apart, the perfusion did not work well and Kupffer cell yields will be low.

4. Once the livers have been teased apart, carefully pour the cell suspension from each Petri dish into its appropriately labeled (P or E) 500-mL flask through the plastic funnel. Rinse each Petri dish with 20 mL of CMRL adding the rinse to the flask. Then, add the 50 mL of CMRL + pronase to the flask.

   a. If hepatocytes are required, for experiments, remove them prior to adding the pronase.

5. Pour the 3 mL of CMRL + DNase I into each flask and cover with aluminum foil. At this point, there should be approx 100 mL of media in each flask.
6. Shake the 500-mL flasks for 15 min at 12°C in a shaking water bath (100 rpm).
7. After shaking, filter each suspension through the coarse mesh (224 μm) into each 250-mL flask.
8. Next, filter the suspension from each 250 mL flask through the fine mesh (132 μm) into 4 × 50-mL centrifuge tubes labeled P or E.
9. Centrifuge at 500 g for 7 min (4°C).

   a. This step pellets all cell types: hepatocytes, red blood cells, Kupffer cells, etc.

10. Carefully aspirate off most of the supernatant (leave 1–2 mL over the cells as the pellet is very loose.)
11. Resuspend each of the pellets from two tubes into 10–15 mL of CMRL with antibiotic/antimycotic, 2 mM L-glutamine, and 10% fetal bovime serum (FBS; CMRL 10% FBS) and bring each one to 50 mL. Repeat for the other two tubes, and then the other treatment group.
12. Centrifuge at 50 g for 2 min.
13. Carefully pour approx 25–30 mL of the supernatant from each tube into two sterile centrifuge tubes labeled PA (for pair-fed tube A) or EA (for ethanol-fed tube A) and place them in ice (see **Note 4**).

   a. The volume of media poured off will depend on the size of the pellet. At this step the pellet is very soft and easily disrupted. Pour off as much supernatant as possible without displacing the pellet. The nonparenchymal cells are in the supernatant.

14. Resuspend the pellet in the remaining media in each tube and take the volume up to 50 mL using CMRL 10% FBS.
15. Repeat steps 12–14 two more times.

   a. This step is to separate the Kupffer cells from hepatocytes. After repeating there will be 6 × 50-mL centrifuge tubes for each treatment group.

16. Centrifuge all 12 centrifuge tubes at 500 g for 7 min to pellet nonparenchymal cells.
17. Aspirate off most of the supernatant (the pellet is soft and easily disrupted).
18. Combine the pellets of the six centrifuge tubes from each treatment group into one centrifuge tube with GBSS and then bring the volume up to 50 mL.
19. Once the cells are resuspended in 50 mL of GBSS, load the cells into a 60-mL syringe with an 18-gauge needle attached.

a. First carefully remove the plunger from the syringe, turn it over, and set aside to avoid contamination. Next place the 60-mL syringe (without plunger) into an empty 50-mL centrifuge tube.

b. Carefully, but quickly, pour the cell suspension into the syringe, insert the plunger (some of the suspension will be lost into the centrifuge tube), and invert the syringe so the needle is facing upward.

c. Get rid of as many bubbles as possible and push out any air left in the syringe. Pour the lost cell suspension into a reagent reservoir. Invert the syringe and collect the remaining cells by pulling back on the plunger.

d. Again, remove all the air from the syringe. Replace the cap of the needle onto the syringe and place the syringe in ice.

It is now time to separate the Kupffer cells from other cell types within the cell suspension in the elutriation chamber. Depending on the centrifuge, make sure that the correct rotor is selected on the instrument panel (JE6B) and check that the centrifuge is at 4°C.

20. Turn on the peristaltic pump allowing the GBSS to flow through the tubing into the elutriator and out into the collection bucket. Run the elutriator for approx 1 min and then turn the pump to the 10 mL/min flow rate.

a. Some bubbles may have collected in the tubing near the pressure gauge. Allow these bubbles to pass out of the elutriator into the collection bucket.

21. Once the bubbles are gone, start the centrifuge at 24 g. If there are any bubbles left in the chamber allow them to pass out of the elutriator. If no bubbles are present and the pressure gauge is reading close to 0 psi, increase the speed to 96 g, check again for bubbles and high pressure, and increase the speed finally to 602 g.

22. Load the cells into the elutriation chamber at 10 mL/min. Take one cell-loaded syringe from the ice and invert it to make sure the cells are in suspension. Remove the 18-gauge needle from the end and discard. Invert the syringe and twist it into place on the stopcock making sure it is secure. Turn the valve on the stopcock so the pump starts to pull the cell suspension into the tubing. Once the cells have started to move into elutriator tubing, place the outlet tubing into a 50-mL centrifuge tube.

23. Allow the majority of the cell suspension to be taken into the tubing, return the stopcock to its original position allowing the GBSS to rinse the cells through the elutriation chamber. Remove the syringe, discard it, and allow the first 50-mL centrifuge tube to fill.

24. Change the pump setting to 21 mL/min, and collect another 50 mL of cells; this will elute off any remaining red blood cells, hepatic stellate cells, and endothelial cells.

25. Change the pump setting to 45 mL/min and collect 2 × 50 mL of cells; this will elute off the Kupffer cells.

26. Once you have collected the Kupffer cells, return the outlet tubing to the collection bucket and stop the centrifuge. The cells that were not eluted out of the chamber will wash out into the collection bucket at this time.

27. Allow 100–200 mL of GBSS to wash the elutriation chamber before repeating steps 21–26 for the other cell suspension.

If there is debris left in the chamber, disassemble the chamber and clean out the debris. If left inside the chamber, it may increase the pressure during the next round of elutriation.

28. Centrifuge the Kupffer cells (2P and 2E 50-mL centrifuge tubes) at 500 g for 7 min.
29. Under the tissue culture hood, prepare the nycodenz (ND) gradient for the final purification step.

   a. Add 4 mL of 24.6% ND into 2 × 15-mL centrifuge tubes for each treatment group.
   b. Slowly layer (drop by drop initially) 4 mL of 11% ND onto the 24.6% ND. An interface will be visible between the layers if done correctly.

30. Take the pelleted cells from step 28, aspirate off the GBSS, and resuspend in 10 mL of fresh GBSS.
31. Take 5 mL of cells and slowly layer them onto each ND gradient.
32. Centrifuge the gradients at 1644 g for 15 min WITH NO BRAKE. Using the brake will disrupt the separation of cells within the gradient.
33. After spinning, pair-fed Kupffer cells will be suspended in the 24.6% ND. Anything that pellets at the bottom should be discarded. With cells from rats that were ethanol-fed, two layers of cells can be seen within the gradient. The top layer, within the 11% ND, will be fatty hepatocytes. The bottom layer within the 24.6% ND, is Kupffer cells. Any cells that pelleted through the gradient should be discarded.
34. Under the tissue culture hood, take a 9-inch Pasteur pipet and attach a rubber bulb. Squeeze the rubber bulb and carefully insert the Pasteur pipet through the ND to the Kupffer cells and carefully remove the cells within the 24.6% ND layer.
35. Remove the Pasteur pipet and place the collected cells into a new 50-mL centrifuge tube. Repeat step 34 to remove all of the Kupffer cells from the gradient (*see* **Note 5**).
36. Take the 50-mL centrifuge tube with the Kupffer cells (step 35) and fill to 50 mL with GBSS.
37. Centrifuge the Kupffer cell suspensions at 500 g for 7 min.
38. Aspirate off the GBSS and resuspend the pellet with 10 mL of CMRL 10% FBS.
39. Into a 1.5-mL sterile Eppendorf tube, pipet 70 μL of 3% acetic acid, 10 μL of trypan blue, and 20 μL of Kupffer cells for counting (*see* **Note 6**).
40. Resuspend the cells to your desired density for plating. In our lab, we resuspend the cells to 2×10^6 cells/mL.

   c. We use a plating density of 6.25×10^5 cells/cm^2 or 0.625 mL/cm^2, maintaining a consistent plating density across different types of cell culture plates

41. Wash within 1 h of plating to remove any nonadherant cells

3.7 Isolation of RNA From Primary Kupffer Cells

Analysis of gene expression in isolated Kupffer cells from pair- and ethanol-fed rats is integral to our understanding of the mechanisms that contribute to the pathology associated with chronic ethanol exposure. Using real-time reverse transcriptase polymerase chain reaction (rtPCR), the effects of ethanol on the expression of several genes can be examined rapidly, even with a small number of Kupffer cells. The brief protocol below describes a method to isolate total RNA and synthesize cDNA from rat Kupffer cells for gene expression analysis by real-time rtPCR.

1. Seed six wells of a 96-well plate at a density of 6.25×10^5 cells/cm^2 for each experimental condition as outlined in Subheading 3.6., step 40.
2. The RNeasy Micro kit (Qiagen, Valencia, CA) is used to isolate total RNA from Kupffer cells. The total number of cells per experimental condition is not to exceed 5×10^5 (*see* **Note 7**). Fifty-eight microliters of the lysis buffer, provided with the kit and supplemented with 2-mercaptoethanol by the user, is added directly to each well using a multi-channel pipet. This results in a total of 350 µL of total lysis buffer for a maximum of 5×10^5 cells. Wells of the same experimental condition are combined into a single DNase-RNase-free 1.5-mL microcentrifuge tube before nucleic acid precipitation.
3. The remainder of the RNA isolation, including on-column DNA digestion, proceeds according to the manufacturer's instructions (*see* **Note 8**).
4. RNA yield is small (250–500 ng/sample) when compared with isolating RNA from, for example, whole liver. Therefore, quantify the RNA and test purity using the least amount of sample possible. The NanoDrop instrument (NanoDrop Technologies, Wilmington, DE) is ideal as it only requires 1 µL of undiluted RNA per sample (*see* **Note 9**).
4. We routinely reverse transcribe the RNA immediately to maximize the amount of sample for subsequent analyses. Total RNA is reversed transcribed using Ambion's RETROscript kit (Austin, TX) according to the manufacturer's instructions, using random decamer as primers according to manufacturer's instructions.
5. Finally, cDNA can be diluted 1:2 with sterile, RNase- DNase-free water. This will only increase threshold cycle time by one, but will double the starting material and therefore maximize the number of genes one is able to examine per sample.

Acknowledgments The work in this chapter was supported by National Institute on Alcohol Abuse and Alcoholism Grant no. AA-013868 and AA-011975 (to L.E.N.) and AA-015833 (to M.T.P.).

4 Notes

1. We make use of two surgical pans (similar to a 9 × 13 baking pan), and two wire mesh racks on which to place the four animals. One wire mesh rack is placed on each baking pan, and one pair-fed and one ethanol-fed animal are placed on top

of each wire mesh rack. The purpose of this set up is to allow the 2 L of perfusate (1 L of Hanks– and 1 L of Williams E) to drain from the animal into the surgical pan. This allows for a more sterile field to be maintained around the liver because it is not covered with perfusate and also allows for easy clean up of the area after the perfusion is complete.

2. It may be helpful when first learning the perfusion procedure to insert the butterfly needle into the hepatic portal vein before turning on the peristaltic pump. However, starting the pump before inserting the needle generally allows for a better perfusion of the liver.

   If the needle is inserted before starting the pump, many red blood cells may be left at the edges of the liver, which are hard to separate from the Kupffer cells.

3. It is also important to test different batches of collagenase before starting experiments. We use Collagenase type IV from Sigma (Catalog C5138). If you plan on doing multiple experiments, it is a good idea to call the Reserves Department at Sigma and request a 100-mg tester batch of a lot of collagenase with sufficient reserves to completer all of our experiments. Chow-fed control animals can be used to test the new batch of collagenase. It is important to look at the success of the perfusions, as well as, the yield of Kupffer cells. If there is a problem with either of these, it is likely that the lot of collagenase used will not work for other animals. Call Sigma for another tester batch, if necessary.

4. Centrifuging the cells at 4°C and keeping them in ice, starting at Subheading 3.6., step 13, is important for good quality and yield of Kupffer cells.

5. When taking the ethanol cells off the ND gradient, be careful not to remove the fatty hepatocytes. One trick is to insert the Pasteur pipet just past the fatty hepatocyte layer and then gently squeeze out any remaining air from the rubber bulb. This will push any cells that were in the tip of the pipet out before inserting the pipet down to the Kupffer cell layer.

6. The 3% acetic acid will cause any remaining red blood cells to lyse, and the trypan blue will allow for enumeration of the viable Kupffer cells.

7. The total number of cells from which RNA can be isolated using the RNeasy micro columns is 5×10^5. If you exceed this number of cells, it is likely that RNA yields will suffer in quantity as well as quality. Therefore, determine the number of cells per well and ensure the total number of cells per experimental condition does not exceed 5×10^5. Cell counts can be done immediately before experiment initiation. If the cell culture medium is to be changed before initiation of the experiment, count cells after the media change. If needed, adjust experimental protocol accordingly (i.e., use fewer wells/experimental condition).

8. DNase I used for the on-column DNA digestion step of the RNeasy micro procedure can be dispensed in 10-µL aliquots in 600 µL of RNase- DNase-free tubes and stored at −20°C until use. One of these aliquots is used per sample during RNA isolation.

9. It is also possible to perform the reverse transcription reaction without prior RNA quantification. This is due to the fact that the reagent concentrations for the reverse transcription reaction, using Ambion's RETROscript kit, are suitable for up to 2 µg of RNA, far above what is typically isolated using the small

numbers of Kupffer cells described here. Input sample amounts can be normalized within and between samples by comparing the expression of a house keeping gene to the gene of interest.

References

1. Adachi, Y., Bradford, B. U., Gao, W., Bojes, H. K., and Thurman, R. G. (1994) Inactivation of Kupffer cells prevents early alcohol-induced liver injury. *Hepatology* **20,** 453–460.
2. Aldred, A., and Nagy, L. E. (1999) Ethanol dissociates hormone-stimulated camp production from inhibition of TNFα production in rat Kupffer cells. *Am. J. Physiol. Gastrointest. Liver Physiol.* **276,** G98–G106.
3. Nagy, L. E. (2003) New insights into the role of the innate immune response in the development of alcoholic liver disease. *Exp. Biol. Med.* **228,** 882–890.
4. Thurman RG II (1998). Alcoholic liver injury involves activation of Kupffer cells by endotoxin. *Am. J. Physiol.* **275,** G605–G611.
5. Thakur, V., Pritchard, M. T., McMullen, M. R., and Nagy, L. E. (2006) Adiponectin normalizes LPS-stimulated TNFα production by rat Kupffer cells after chronic ethanol feeding. *Am. J. Physiol. Gastrointest. Liver Physiol.* **290,** G998–G1007.
6. Thakur, V. Pritchard, M. T., McMullen, M. R., Wang, Q., and Nagy, L. E. (2006) Chronic ethanol feeding increases activation of NADPH oxidase by lipopolysaccharide in rat Kupffer cells: role of increased reactive oxygen in LPS-stimulated ERK1/2 activation and TNFα production. *J. Leukoc. Biol.* **79,** 1348–1356.
7. Lieber, C. M., and DeCarli, L. M. (1982) The feeding of alcohol in liquid diets: two decades of application and 1982 update. *Alcohol. Clin. Exp. Res.* **6,** 523–531.
8. Friedman, S. L., and Roll, F. J. Isolation and culture of hepatic lipocytes, Kupffer cells, and sinusoidal endothelial cells by density gradient centrifugation with stractan. *Anal. Biochem.* **161,** 207–218.
9. Knook, D. L., Blansjaar, N., and Sleyster, E. C. (1977) Isolation and characterization of Kupffer and endothelial cells from the rat liver. *Exp. Cell Res.* **109,** 317–329.
10. Munthe-Kaas, A. C., Berg, T., Seglen, P. O., and Seljelid, R. (1975) Mass Isolation and culture of rat Kupffer cells. *J. Exp. Med.* **141,** 1–10.

16
Dendritic Cells in Chronic In Vivo Ethanol Exposure Models

Kevin L. Legge and Annette J. Schlueter

Summary Dendritic cells (DCs) play a key role in the initiation of effective immune responses against infectious agents because they are unique in their ability to provide antigen-specific activation of naïve T cells. To do this, they must acquire antigen and migrate to spleen or lymph node to present the antigen to T cells in association with costimulatory molecules and cytokines. Murine models of chronic EtOH exposure have been developed for dissecting the mechanisms by which EtOH alters immune cell functions. This chapter details methods for assessing DC functions in such models. Methods are presented for 1) the identification and isolation of various DC subsets from spleen, epidermis, and lung, 2) measurement of LC migration out of epidermis and DC migration into peripheral and peribronchial lymph nodes, and 3) measurement of alloantigen presentation in vitro as well as transgenic T-cell activation in vitro and in vivo.

Keywords Dendritic cells; ethanol; immune function; mouse host defense.

1   Introduction

1.1   General Background

As initiators of immune responses, dendritic cells (DCs) play a critical role in the defense against infectious agents. They are uniquely able to activate naive T cells, which are required for the initiation of adaptive immune responses to invading organisms. Optimal DC function requires appropriate development from bone mar-row precursors, migration to peripheral lymphoid and nonlymphoid surveillance locations, maturation (including antigen processing/presentation, costimulatory molecule expression, cytokine production) after encounter with pathogens and migration to lymph nodes and/or spleen to interact with T cells (*1–3*). Langerhans cells (LCs), an epidermal DC resident population, are one well studied example of a peripheral DC subset that participates in antigen transport to and T-cell activation in peripheral LN (*4*). In addition, some DCs are able to suppress or down-regulate immune responses and thus serve an immunoregulatory role (*5*). Ethanol exposure has the potential to interfere with many of these DC functions and thereby increase the organism's susceptibility to infection. The aim of this chapter is to briefly review what is known about the effect of EtOH exposure on DC function, and to outline methodologies that are applicable to studying DC function following chronic EtOH exposure in vivo, in murine models.

1.2   Approaches to Studying DC Function After EtOH Exposure

It is impossible to precisely mimic in vitro the daily variability of EtOH exposure levels, local EtOH metabolite concentrations, or the microenvironment in which DCs normally reside in the intact animal. Thus, it is most relevant to study the effects of in vivo EtOH administration on DC function rather than exposing freshly isolated DC, or those generated from untreated animals, to EtOH in vitro. In some circumstances, no in vivo methods exist to adequately address the DC function under consideration. In these cases, in vitro experiments that measure events initi-ated in vivo or within a few hours of culture preparation (during which time DCs might be expected to maintain the deficits induced by EtOH in vivo) are reasonable surrogates.

1.3   Effect of Chronic EtOH Exposure on DCs

Very few studies have addressed the effect of chronic in vivo EtOH exposure on DC function. Some clues about the effect of chronic EtOH on DC function can be had from studies of unpurified "APC" function from the spleen. Macrophage phagocytosis is reduced by EtOH exposure *(6,7)*. Ag presentation and immunostimulatory cytokine production of APC from EtOH mice is also reduced *(8)* and defects are observed in the afferent phase of DTH in EtOH mice *(9–13)*. EtOH mice also show decreased serum levels of inflammatory cytokines produced by DC and macrophages *(14)*, such as interleukin (IL-6), tumor necrosis factor-α, and IL-1 *(15–17)*. Antigen-presenting cell (APC) functions can be ascribed to a number of different cell types, including monocytes, macrophages, B cells, and DCs. Only one study has specifically investigated the effects of chronic in vivo EtOH exposure on DC function and demonstrated decreased allostimulatory capacity and lower levels of costimulatory molecule (CD86) and major histocompatibility complex (MHC) class II expression on EtOH exposed DC, particularly after inflammatory (CpG) stimulation *(18)*.

1.4 Methods Described in This Chapter

Because of the advantages of an in vivo model of EtOH administration for determining pathophysiologically relevant effects on DC function, this chapter focuses on protocols that rely on this approach. All techniques have been performed after in vivo EtOH administration using the Meadows-Cook model described in the chapter by Coleman, et al. in this volume.

2   Materials

2.1 Identification and Isolation of DCs and LCs

2.1.1   Purification of CD8α+ and CD8α– DCs From the Spleen

1. Dense BSA Solution: 186 mL of phosphate-buffered saline (PBS), 29 mL of 1 *N* NaOH in 65 mL of water. Carefully layer 106 g of bovine serum albumin (BSA; Intergen, Purchase, NY) onto the surface of the solution, cover beaker with foil and place at 4°C overnight. Check density by measuring refractory index; correct BSA density of 1.08 g/mL corresponds to an index of 1.384–1.385 at 25°C. To reduce the refractive index by 0.0005 add 5 mL of PBS or to increase the index by 0.0005, add 5 g of BSA. The pH of the solution should be 7.0–7.4. Filter-sterilize the solution using a self contained filtering unit. Dense BSA may be replaced by Nycoprep 1.077A (cat. no. 10122311).
2. Nycodenz buffer: 9 g NaCl, 0.6055 g Tris, 0.22368 g KCl, 0.11167 g ethylene diamine tetraacetic acid (EDTA); bring to 1.12 L with deionized (DI) water.

3. Nycodenz stock solution: add exactly 15.275 g of Nycodenz powder to 50 mL of DI water. Filter and keep in the dark at 4°C.
4. MACS buffer: PBS, 0.5% BSA, 500 mM EDTA.
5. HT-2 Media: Iscove's DMEM 1X 439 mL, 10% fetal calf serum (FCS) 50 mL, L-glutamine (0.292 mg/mL H$_2$O) 5 mL, Na Pyruvate (0.11 mg/mL H$_2$O) 5 mL, gentamycin sulfate (0.5 mg/mL) 0.5 mL, 2-ME (0.05 mM) 500 μL.
6. Anti-CD8α+ Microbeads (Miltenyi, Auburn, CA, cat. no. 120–000–298).
7. Anti-CD11c+ Microbeads (Miltenyi, cat. no. 130–052–001).
8. LS Columns (Miltenyi, cat. no. 130–042–401).
9. Collagenase.
10. Hanks' Balanced Salt Solution (HBSS, 10X) without sodium bicarbonate, calcium and magnesium.
11. Complete media: Iscove's DMEM 1X, 439 mL, 10% FCS (final concentration) 50 mL, L-glutamine (29.2 mg/mL H$_2$O stock conc.) 5 mL, Na pyruvate (100 mM stock conc.) 5 mL, Pen/Strep (10 μL/mL, 10 μg/mL stock conc.) 5 mL, nonessential amino acids (100X stock conc.) 5 mL, HEPES, 5 mL (1 M stock concentration), 2-ME (89.3 mM stock conc.) 500 μL.

2.1.2 Isolation of LCs for Flow Cytometry From Mouse Ears

1. Trypsin (Amresco, Solon, OH; cat. no. 0458–25G). Prepare 0.5% solution in PBS fresh on the day of the assay.
2. Balanced salt solution (BSS): 10x BSS stock solution 1 (10 g dextrose, 0.6 g KH$_2$PO$_4$, 3.58 g Na$_2$HPO$_4$•7H$_2$O [1.89 g anhydrous], 0.1 g phenol red powder or 20 mL of 0.5% phenol red solution), 10x BSS stock solution 2 (1.86 g CaCl$_2$•2H$_2$O [1.41 g anhydrous], 4 g KCl, 80 g NaCl, 2 g MgCl$_2$•6H$_2$O [1.04 g anhydrous], 2 g MgSO$_4$•7H$_2$O [1.74 g anhydrous]). To make a working solution, combine 50 mL of stock solution #1 with 50 mL of stock solution #2 and bring to 500 mL with deionized water.
3. Ficolite-LM (Atlanta Biologicals, Lawrenceville, GA; cat. no. 140650).
4. Staining buffer: 475 mL BSS, 25 mL bovine calf serum-defined, 0.5 g NaN$_3$
5. 60-mm Petri dish (Falcon; cat. no. 351007).
6. Fine forceps (VWR; cat. no. 25607–856 for straight, 25607–890 for curved).
7. 70-μm filter (Falcon; cat. no. 352350).
8. Glass microscope slides with frosted ends.
9. Dissecting microscope.

2.1.3 Purification of Pulmonary DC Subsets by FACS

1. Iscove's DMEM 1x.
2. Collagenase.
3. DNAse 1.
4. 0.5 M EDTA.

5. Dulbecco's modified Eagle's medium (DMEM).
6. 03–200/48 Nitex Mesh (Sefar, Kansas City, MO; cat. no. F03A0020048I043A02).
7. MACS buffer: PBS, 0.5% BSA, 500 mM EDTA.
8. Polypropylene round-bottom tubes (Falcon, Franklin Lakes, NJ; cat. no. 352063).
9. CD11c MACS Beads (Miltenyi Biotec, Auburn, CA; cat. no. 120–000–322).
10. FCS.
11. Cell Strainer (BD, Bedford, MA; cat. no. 352350).
12. LS Column (Miltenyi Biotec; cat. no. 130–042–401).

2.2 Migration of DCs and LCs

2.2.1 Immunofluorescent Staining of LCs in Epidermal Sheets

1. 0.5 M ammonium thiocyanate: add 19 g of ammonium thiocyanate to 500 mL of 0.1 M phosphate buffer (pH 6.9) or PBS. Store at 4°C.
2. Blocking buffer: 1 mL of goat serum (Invitrogen, Carlsbad, CA; cat. no. PCN5000), 1 mL of 10X PBS, 8 mL of deionized water. Mix, store at 4°C.

   a. Alternate blocking buffer: 8 mL of 100 mM TBS, pH 7.3 (6.06 g of Tris base, 400 mL of deionized H_2O. Adjust pH to 7.3 with concentrated HCl. Adjust volume to 500 mL, then add 0.9 g of NaCl), 1 mL of rat anti-mouse FcγRII/III (clone 2.4G2), 1 mL of goat serum, 0.2 g of BSA (Amresco, Solon, OH; cat. no. 0332–100G), 5 μL of Tween 20.

3. Acetone.
4. Biotinylated or PE conjugated anti-mouse MHC class II.
5. Avidin-Cy5 (Jackson ImmunoResearch Laboratories, Inc., West Grove, PA; cat. no. 016–170–84) or Avidin-Cy3 (Southern Biotechnology Associates Inc. Birmingham, AL; cat. no. 7100–09L) if a biotinylated antibody is chosen in step 4.
6. Petri dishes (60×15 mm, BD Falcon, San Diego, CA; cat. no. 351007).
7. 24-well tissue culture plate (Corning Inc. Life Sciences, Acton, MA, Costar; cat. no. 3524) or 96 well plate (Costar cat. no.3596).
8. Fine forceps (get info from Subheading 2.1.2).
9. Vecta Shield Slide Mounting Media (Vector Labs, Burlingame, CA cat. no. H-1000).
10. Microscope slides (Surgipath Medical Industries, Richmond, IL; cat. no. 00240).
11. Slide cover slips, 22×22 mm.
12. Clear fingernail polish.
13. Drugstore depilatory, e.g., Nair.
14. Dissecting microscope.
15. Fluorescent microscope.

2.2.2   FITC Tracking of LCs and Dermal DCs to Lymph Nodes

1. BSS (See recipe in Subheading 2.1.2.).
2. Liberase blendzyme 3 (Roche Diagnostics, Chicago, IL; cat. no. 1814176) / DNase (Roche Diagnostics, cat. no. 104 159) working solution.

   a. Stock Liberase solution: dissolve Liberase in RPMI 1660 + L-glutamine (Invitrogen, Carlsbad, CA; cat. no. 11875) at 1 mg/mL (*see* **Note 1**).
   b. Stock DNase solution: dissolve DNase in RPMI at 4000 U/mL (*see* **Note 1**).
   c. Working solution: 25 μL of Liberase stock solution, 10 μL of DNase stock solution, 20 μL of bovine calf serum, 945 μL of RPMI. Make fresh on day of use, and keep on ice.

3. Fixative for mouse cells: 12.5 mL of 10x PBS, 2.7 mL of 37% formaldehyde solution, 200 mg of NaN_3. Bring to a final volume of 100 mL.
4. n-butyl phthalate.
5. Acetone.
6. FITC.
7. Cytofix/Cytoperm solution (Becton Dickinson, San Diego, CA; cat. no. 554722).
8. Perm/Wash Buffer 10x (Becton Dickinson, cat. no. 554723). Dilute to 1x before use, according to manufacturer's instructions.
9. Polyclonal unconjugated rabbit anti-mouse langerin (Imgenex, San Diego, CA; cat. no. IMG-5102A).
10. Cy5-conjugated AffiniPure goat anti-rabbit IgG (H+L) (Jackson Immuno-Research, West Grove, PA; cat. no. 111–175–144).
11. Anti-CD11c (and other antibodies to cell surface antigens of interest).
12. 12-well tissue culture tray (Corning Inc. Life Sciences, Acton, MA; cat. no. Costar 3512).
13. 18- and 23-gage needles.
14. Glass vials.
15. 1-mL syringes.
16. Scalpels (1 per lymph node).

2.2.3   Assessment of rDC Migration From the Lungs to the Lymph Nodes by i.n. CFSE Labeling

1. 25 m*M* CFSE stock (Molecular Probes, Eugene, OR, cat. no. C1157): 25 mg dissolved in 1783.6 μL of dimethyl sulfoxide (DMSO; stock solution).
2. Iscove's DMEM 1X.
3. Anti-CD11c APC mAb (Caltag, Carlsbad, CA; cat. no. MCD11c05).
4. FACS buffer: 2% FCS, 0.02% sodium azide in PBS.

2.3 Antigen Presentation by DCs

2.3.1   DC Activation of Naïve CD8 T Cells In Vitro

1. CD8α+ Microbeads (Miltenyi, Auburn, CA, cat. no. 120–000–298).
2. Anti-CD8β mAb biotinylated (Biolegend, San Diego, CA; cat. no. 118004).
3. Strepavidin PerCP (BD, San Jose, CA; cat. no. 554064).
4. CFSE (25 M in DMSO; Molecular Probes, Eugene, OR; cat. no. C1157).
5. Bovine calf serum.
6. Complete medium: Iscove's DMEM 1X, 439 mL; 10% FCS (final concentration), 50 mL; L-glutamine (29.2 mg/mL H_2O stock conc.), 5 mL; Na pyruvate (100 mM stock conc.), 5 mL; Pen/Strep (10 µL/mL, 10 µg/mL stock conc.), 5 mL; nonessential amino acids (100X stock conc.), 5 mL; HEPES, 5 mL (1 M stock conc.); 2-ME (89.3 mM stock conc.), 500 µL.

2.3.1.1   Influenza infection of DCs

1. Infect 1×10^6 purified DC/mL at the desired MOI (5, 10, etc.) by adding virus diluted in Iscove's media without serum to cultured cells.
2. Incubate cells and virus on ice for 30 min to allow virus to attach to the cells and then continue the incubation at 37°C for 30 min to allow infection.
3. Centrifuge cells at 460 g (1600 rpm) for 10 min at 4°C. Remove any remaining virus in the supernatent and resuspend cells in complete media containing serum.
4. Incubate for 18 h at 37°C.

2.3.1.2   CD8 T Cell Activation

1. Purify influenza specific HA$_{529}$ Clone-4 CD8+ T-cells from the spleens of naïve Clone-4 mice using CD8α microbeads.

   a. Prepare single-cell suspension (spleen screen, slides, etc.) in Iscove's DMEM.
   b. Centrifuge at 460 g (1600 rpm) for 10 min at 4°C and resuspend cells in MACS buffer at 90 µL/10^7 cells.
   c. Add 5 µL CD8α+ beads per 10^7 cells and incubate at 4°C for 10 min (refrigerate. not on ice). Add 2 mL of Iscove's DMEM.
   d. Centrifuge at 460 g (1600 rpm) for 10 min at 4°C and resuspend cells in MACS buffer at 10×10^7 cells/mL.
   e. Prewet column with 3 mL of MACS buffer solution.
   f. Place cells over column (max of 5 lungs per column) and collect negative fraction.

   g. Wash with 9 mL of MACS buffer.
   h. Remove column from magnet and elute positive fraction with 6 mL of media
      (2 mL of gravity, 2 mL of gentle plunge, 2 mL of forceful).

2. Centrifuge cells at 460 g (1600 rpm) for 10 min at 4°C and resuspend at 10×10^7
   T-cells/mL in Iscove's DMEM.
3. Label purified T cells with 1.5 µM CFSE. Incubate 10 min at 37°C. At the con-
   clusion of the incubation, quench CFSE by adding an equal volume of calf serum
   (*see* **Note 16**).
4. Purify desired DC populations by MACS or FACS sorting (*see* Subheadings 3.1
   and 3.3) and then infect as described previously.
5. Coincubate purified T cells and DCs at a 5:1 ratio (i.e. 100,000 T celsl:20,000 DCs)
   in 200 µL of complete media at 5% CO_2 at 37°C for 48–72 h (*see* **Note 17**).
6. Harvest cells, stain for CD8β with anti-CD8β biotinylated (clone:CD8b) and
   strepavidin PerCP and measure CFSE dilution by flow cytometry.
7. Method can be combined with staining for T cells activation markers such as
   CD25, CD69, etc.

2.3.2   DC Activation of Antigen-Specific T-Cell Proliferation In Vivo

1. OTI or OTII mice as T-cell source (Jackson Laboratory, Bar Harbor, ME cat.
   no. 003831 or 004194)
2. C57Bl/6 mice as DC source (donor) and as recipient of cell transfers
3. 5 mM stock solution of CFSE (5-(and -6)-carboxyfluorescein diacetate succinimidyl ester, Molecular Probes, Eugene, OR; cat. no. c1157) in DMSO (store at
   −20°C, stable for 6 mo.).
4. 10% FCS in PBS.
5. Super McCoy's medium: 85 mL of McCoy's 5A medium modified 1x with L-
   glutamine (Invitrogen, Carlsbad, CA; cat. no. 16600–082); 2 mL of FCS; 2 mL
   of essential amino acids 50x (Invitrogen, cat. no. 111–30–051); 1 mL of nonessential amino acids 100x (Invitrogen, cat. no. 11140); 1 mL of sodium pyruvate
   100x (Invitrogen, cat. no. 11360); 1 mL of MEM vitamins 100x; 2 mL of peni-
   cillin/streptomycin/glutamine 50x (Invitrogen, cat. no. 5140–122); 50 µL 0.1 M
   2-mercaptoethanol
6. Liberase/DNAse working solution (see recipe in Subheading 2.2.2.).
7. OVA 258–264 peptide (for OTI assays) or OVA 323–339 peptide (for OTII
   assays) stock solution, 2 mg/mL in sterile distilled water (Bio-Synthesis,
   Lewiston, TX; cat. no. 51023–10).
8. 0.5% BSA (Amresco, Solon, OH; cat. no. 0332–100G) in PBS.
9. Anti-CD11c microbeads (Miltenyi Biotec, Auburn, CA; cat. no. 130–052–001).
10. Anti-mouse CD4 and CD8 antibodies, conjugated with a fluorochrome other
    than FITC.
11. 60-mm Petri dish (Falcon; cat. no. 351007).
12. Glass microscope slides with frosted ends.
13. 1 mL of insulin or tuberculin syringes.

14. 1-mL syringe with 23-gage needle.
15. AutoMACS separator (Miltenyi Biotec; *see* **Note 2**).

2.3.3 Allostimulatory Assay of DC Function

 1. BALB/c mouse as source of responder T cells.
 2. C57Bl/6 mouse as source of DCs in first stimulation.
 3. BALB/c, C57Bl/6 and C3H mice as sources of DCs in restimulation.
 4. Super McCoy's medium (recipe can be found in Subheading 2.3.2.).
 5. 0.5% BSA (Amresco, Solon, OH; cat. no.0332–100G) in PBS.
 6. Liberase/DNase working solution (recipe can be found in Subheading 2.3.2.).
 7. Fico-Lite LM (Atlanta Biologicals, Lawrenceville, GA; cat. no.140650).
 8. Anti-CD11c microbeads (Miltenyi Biotec, Auburn, CA; cat. no. 130–052–001).
 9. Anti-biotin microbeads (Miltenyi Biotec, cat. no. 130–090–485).
10. Biotinylated anti-B220, DX5, CD11c, CD11b, and Ter119 antibodies.
11. ^{3}H thymidine (GE Healthcare, Piscataway, NJ; cat. no. 13-TRA120).
12. 96 well plate (Corning Inc. Life Sciences, Acton, MA; Costar, cat. no. 3596).
13. T25 culture flasks (Costar, cat. no. 430639).
14. 60-mm Petri dish (Falcon cat. no. 351007).
15. Glass microscope slides with frosted ends.
16. 1-mL syringe with 23-g needle.
17. AutoMACS separator (Miltenyi Biotec; *see* **Note 2**).
18. Scintillation counter.
19. Cell harvester (Cambridge Technologies, Cambridge, MA) and associated cell harvesting supplies.

3   Methods

3.1   Identification and Isolation of DCs and LCs

3.1.1   Purification of Mature Bulk Splenic DCs

1. Prepare collagenase solutions in HBSS with Ca^{2+} and Mg^{2+}. (Dilute to 100 U/mL. Also make 5 mL of 400 U/mL dilution.).
2. Collect 10 spleens (on Petri dish). Inject 1–2 mL of 100 U/mL collagenase solution into each spleen.
3. Place the injected spleens into Petri dish. Collect the cells that were released in a 50-mL tube containing 10 mL of ice-cold HBSS and keep on ice.
4. Add 6–8 mL of collagenase solution (100 U/mL) into the dish. Tease the spleen into very small pieces. Transfer released cells to collection tube and rinse twice more with 10 mL fresh collagenase (100 U/mL). Pipet up and down vigorously.
5. Add the high concentration collagenase solution (400 U/mL) and incubate at 37°C for 30 min.

6. Pipet up and down vigorously and mash dissociated tissue through the strainer.
7. Pool all the cells, filter through cell strainer and centrifuge at $460\,g$ (1600 rpm).
8. Resuspend cell pellet in dense BSA (no foaming!) Transfer 5 mL of BSA containing cells from spleens into a 15-mL tube.
9. Overlay with 2 mL of Ca^{2+} Mg^{2+}-free HBSS. Spin for 15 min at $2192\,g$ (3300 rpm) without braking.
10. Collect low-density fraction (all of overlaid medium and upper 1 mL of BSA) with Pasteur pipet. Transfer into a 15-mL tube and fill with Ca^{2+} Mg^{2+}-free HBSS, mix well, and centrifuge at $460\,g$.
11. Wash cells again in culture media and centrifuge at $300\,g$.
12. Plate low-density cells into 60-mm tissue culture dish in complete culture medium for 90 min (37°C).
13. Rinse off nonadherent cells using a 10-mL pipet with prewarmed media (HT-2).
14. This wash must be done thoroughly. Pay attention to rims of the dishes. Aspirate the remaining media.
15. Repeat wash.
16. Culture adherent cells overnight (37°C).
17. Next day, collect nonadherent cells (>80% DCs, depends on thoroughness of wash steps 12–13).

3.1.2 Purification of Mature CD8α⁻ and CD8α⁺ DCs from Splenocytes by AutoMACS

1. Follow purification protocol for purification of bulk DCs through differential adherence.
2. Harvest bulk DCs off of the differential adherence plates (wash quadrants 3 times/2 washes pre-warmed media). Spin at $460\,g$ (1600 rpm) 10 min 4°C.
3. Remove supernatant, and resuspend cells in 30 μL of anti-CD8α (Ly-2)-Microbeads. Bring to 3 mL with MACS buffer.
4. Incubate at 4°C, for 15 min (refrigerate, not ice). Spin $460\,g$ (1600 rpm) 10 min 4°C.
5. Resuspend cells in 3 mL of MACS buffer, and run on autoMACS under ";posselds" program.
6. Positive select the CD8α⁻ population with anti-CD11c (N418)-Microbeads by repeating steps 3–5, and keep the CD11c⁺ fraction (*see* **Note 3**).

3.1.3 Manual MACS Purification of Mature CD8α⁻ and CD8α⁺ DCs From Splenocytes

1. Follow purification protocol for purification of bulk DCs through differential adherence.
2. Harvest bulk DCs off of the differential adherence plates (wash quadrants 3 times/2 washes with pre-warmed media). Spin $460\,g$ (1600 rpm) for 10 min at 4°C.

3. Remove supernatant, and resuspend cells in 30 μL of anti-CD8α (Ly-2)-Microbeads. Bring to 3 mL with MACS buffer.
4. Incubate at 4°C (refrigerator, no ice), for 15 min. Spin at 460 g (1600 rpm) for 10 min at 4°C.
5. Prewet LS+ column with 3 mL of MACS buffer solution during the last 3 min of spin.
6. Resuspend cells in 3 mL of MACS buffer, and pass cells through LS column, wash with 9 mL of MACS buffer.
7. Remove column from magnet and elute CD8α+ cells with 6 mL of MACS buffer (2 mL gravity, 2 m L gentle plunge, 2 mL forceful).
8. Repeat steps 3–6 on CD8α⁻ cells (*see* **Note 3**).
9. Pool CD8α⁺ cells, and then positive select the CD8α⁻ population with anti-CD11c Microbeads. Repeat steps 3 through 6, and keep the CD11c⁺ fraction.

3.1.4   Isolation of LCs for Flow Cytometry From Mouse Ears

1. Obtain ears from the mouse. With forceps, split each ear into dorsal and ventral halves.
2. Lay the ears in a 60-mm Petri dish so that the epidermal sides are facing up.
3. Pour 0.5% trypsin in PBS over the ears to submerge the ears in the Petri dish and place them in a 37°C water bath on a platform for 60 min.
4. Rinse the ears in BSS. Under the dissecting microscope, with fine forceps carefully peel off the epidermis taking care to preserve it as a sheet.
5. Prepare a single-cell suspension of the epidermis by gently rubbing it between the frosted ends of two microscope slides. Rinse the slides with BSS.
6. Pass the primary suspension through a 70-μm filter.
7. Transfer the suspension to 15-mL conical centrifuge tube.
8. Fill the tube with BSS. Centrifuge at 400 g at 4°C for 7 min.
9. Resuspend the pellet by adding 3 mL of staining buffer, underlayer with 3 mL of Ficolite.
10. Centrifuge at 1700 g at room temperature for 20 min.
11. Harvest cells from interface, and place them in a new 15-mL centrifuge tube.
12. Fill the tube with BSS. Centrifuge at 400 g, at 4°C for 7 min.
13. Resuspend the pellet. These cells are ready for study by flow cytometry (*see* **Note 4**).

3.1.5   Purification of Pulmonary DC Subsets by FACS

All steps and media to be used in a sterile manner.

1. Remove lungs from naïve or infected mice.
2. Split the lungs into groups of two lungs/tube in Iscove's media.
3. Digest lungs with 1 mg/mL of collagenase and 0.02 mg/mL of DNase 1 in 7.5 mL for 15 min at 25°C in rotator at 18 g (300 rpm).

4. Prepare single cell suspension (spleen screens, slides, etc.). Note: remove debris by passing through Nitex mesh.
5. Continue incubation for another 10 min at 25°C in rotator.
6. Add 500 μL of 0.5 M EDTA, mix by inverting tubes, and incubate at room temperature for 5 min.
7. Centrifuge at 460 g (1600 rpm) for 10 min at 4°C.
8. Remove a small amount of cells to count.
9. Enrich for bulk DC by MACS.

   a. Add 50 μL of anti-CD11c+ microbeads per 10^8 cells and incubate at 4°C for 10 min. (Do not incubate on ice as this will cap Ab-beads on cells leading to contaminating cell populations.)
   b. Centrifuge at 460 g (1600 rpm) for 10 min at 4°C
   c. Resuspend cells in MACS buffer at 10×10^7 cells/mL.
   d. Prewet LS+ column with 3 mL MACS buffer solution.
   e. Place cells over column (max of five lungs per column) and collect negative fraction.
   f. Wash with 9 mL of MACS buffer.
   g. Remove column from magnet and elute positive cells with 6 mL media (2 mL gravity, 2 mL gentle plunge, 2 mL forceful; *see* **Note 5**).

10. Resuspend in Sterile Iscove's +10% FCS at 1×10^7 cell/mL
11. Stain enriched DC with anti-CD11b FITC ($1 \mu L/1 \times 10^6$ cells), anti-CD11c APC $1 \mu L/2 \times 10^6$ cells), anti-I-A/I-E PE ($1 \mu L/2 \times 10^6$ cells) mAbs.
12. Incubate cells for 30 min on ice.
13. Centrifuge at 460 g (1600 rpm) for 10 min at 4°C.
14. Resuspend in Sterile Iscove's +10% BCS at 1×10^7 cell/mL.
15. Run all cells through a cell strainer into new FACS tubes before sorting.
16. Sort into collection tubes of 1 mL of DMEM +10% FCS.
17. Sort on CD11c+ cells, then gate the aDC (CD11c+,I-A/I-E+, CD11b-), iDC (CD11c+,I-A/I-E+, CD11b+), and avelolar Macs (CD11c+, I-A/I-E inter, CD11b moderate (autofluorescent)). (Fig. 1).

3.2   Migration of DCs and LCs

3.2.1   Preparation and Fixation of Epidermal Sheets

1. Obtain ears from the mouse. With forceps, split each ear into dorsal and ventral halves. Lay the ears in a Petri dish so that the epidermal sides are facing up.
2. Apply depilatory (e.g., Nair) to the ears for 20–30 s. Wash with PBS.
3. Submerge the ears in 0.5 M ammonium thiocyanate (in the Petri dish) and place them in the 37°C water bath on a platform for 15–20 min.

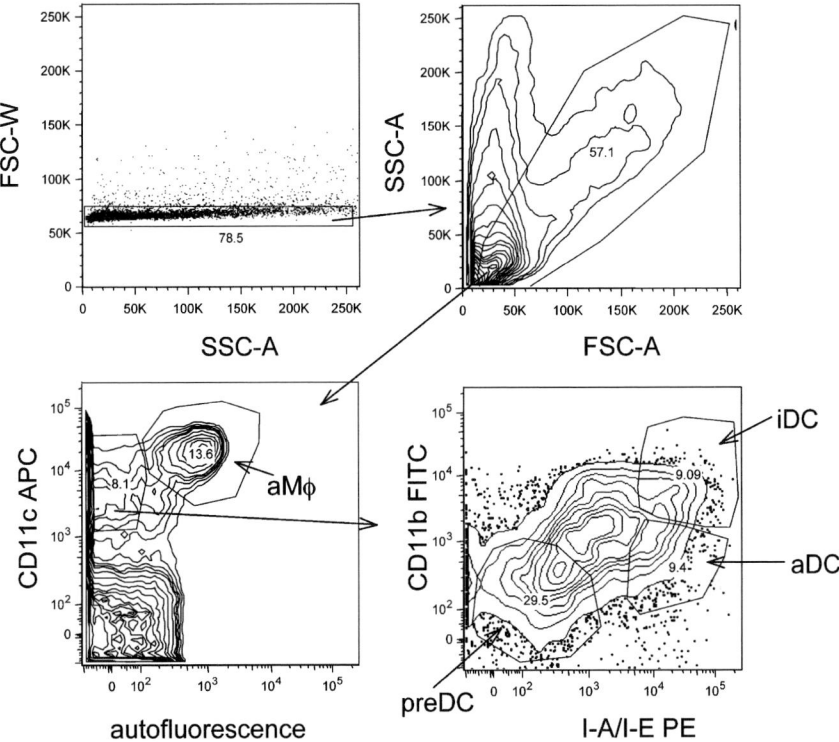

Fig. 1 Gating strategy for FACS purification of pulmonary dendritic cells. Cells are first gated upon a singlet gate (upper left) and then on a large live cell gate (upper right). Next, autofluorescent[+] CD11c[+] cells are sorted (alveolar macrophages) away from the autofluorescent[neg] CD11c[+] cells (lower left). The autofluorescent[neg] CD11c[+] cell population contains interstitial DC (CD11b[+]I-A/I-E[+]), airway and alveolar DC (CD11b[-]I-A/I-E[+]), and precursor DC (CD11b[-]I-A/I-E[-]) cells (lower right)

4. Rinse the ears with PBS. Under the dissecting microscope, with fine forceps, carefully peel the epidermis off as a sheet.
5. Fix the epidermis in ice-cold acetone for 10–20 min. Transfer to PBS for rehydration.

3.2.1.1 Staining the Epidermis for LCs

1. Incubate the tissue for 1 h at room temperature in 500 μL of blocking buffer in a well of a 24-well tissue culture plate or in 200 μL of blocking buffer in a 96-well ELISA plate.
2. Aspirate the blocking buffer.
3. Incubate with 500 μL of biotinylated anti-MHC class II or MHC class II PE direct conjugate at appropriate titer, overnight at 4°C protected from light (*see* **Note 6**).

4. Wash three times in PBS by filling and emptying the well (*see* **Note 7**).
5. Incubate with 500 µL of Avidin-Cy5 or Avidin-Cy3 for 1 h at room temperature (*see* **Note 8**).
6. Wash three times in PBS.
7. Mount on a slide with assistance of fine forceps and the dissecting microscope. Do not use a bright light source during this step.
8. Cover completely with Vecta Shield (about 10–20 µL) and a cover slip. Protect from light.
9. Let dry for 1 h and then seal the four sides of the cover slip to the glass slide with fingernail polish.
10. Store at 4°C until microscopic examination for up to 3 d. For longer storage, freeze slides at −20°C.
11. Examine the slides using a fluorescent microscope. PE fluoresces in the Cy3 channel.

3.2.2   FITC Tracking of LCs and Dermal DCs to Lymph Nodes

All cell washes in this procedure are performed at 4°C at 400 g for 7 min. 0.5% FITC solution preparation:

1. Weigh out 0.001 g of FITC into a glass vial.
2. Add 100 µL of acetone and 100 µL of n-butyl phthalate. Vortex intermittently for 5 min.

3.2.2.1   Animal Application

1. At desired time, (e.g., 12 h before sacrifice) apply 25 µL of the FITC solution to one side of the mouse's abdomen. Apply 25 µL of vehicle alone to the opposite side as a control (*see* Fig. 2).
2. Isolate each inguinal LN. Place each LN individually in 200–400 µL of liberase/DNase working solution. Keep on ice. A 12-well tissue culture tray works well and facilitates identification of individual LNs.
3. Using a scalpel, individually mince each LN in the liberase/DNase working solution.
4. Incubate at 37°C for 30 min.
5. After digestion, prepare a single cell suspension from the LN by passaging the fragments repeatedly in and out of a syringe, initially with an 18-gauge needle followed by a 23-gauge needle.
6. Wash the cell suspension in BSS. Resuspend the cells in Perm/Wash buffer as desired (300–500 µL is recommended for each LN) and perform a cell count.

3.2.2.2   Intracellular Staining

1. Dilute only langerin and goat anti-rabbit secondary antibodies to the appropriate concentration for staining in Perm/Wash buffer.

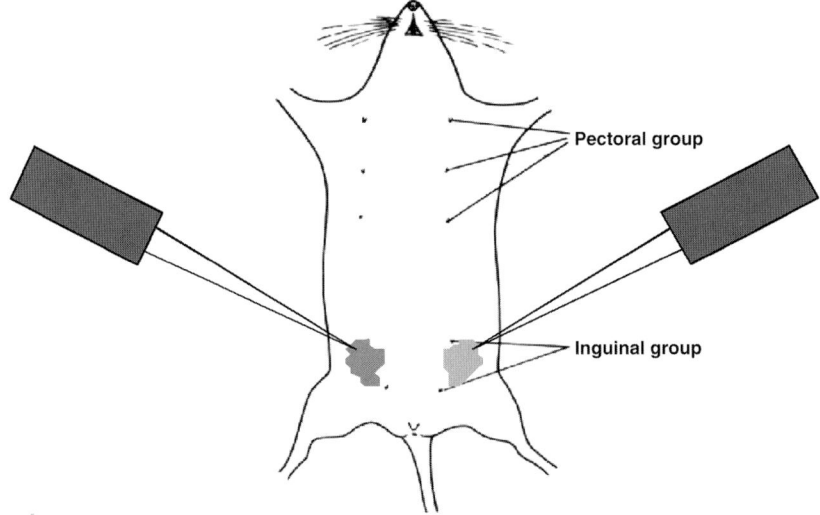

Fig. 2 Location of FITC/vehicle control application.

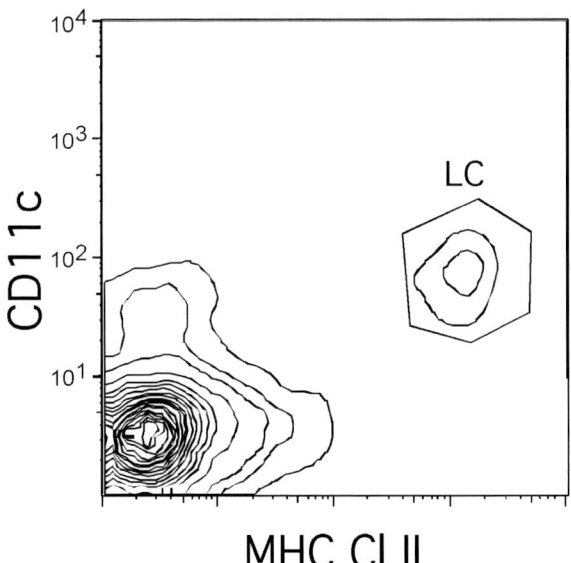

Fig. 3 Results of flow cytometric identification of LC. The gated population of CD11c+ MHC Class II+ cells identifies LC as approximately 3% of the cell suspension

2. Perform surface staining for at least CD11c (to identify DC) and other antigens of interest as usual. Remember to leave the Cy5 channel open for langerin.
3. At final wash, remove as much buffer as possible and resuspend the stained cells in 50–200 μL of Cytoperm/Cytofix solution (it is used undiluted from the manufacturer).

Use 50 µL for 0.6×10^6 cells, and use 200 µL for 3×10^6 cells. Vortex to mix and incubate at room temperature for 5 min.

4. Wash cells twice in Perm/Wash buffer. Add the diluted anti-langerin antibody (from step 1) and incubate at room temperature for 20 min.
5. Wash cells twice in Perm/Wash buffer. Add goat anti-rabbit IgG and incubate cells at room temperature for 20 min.
6. Wash cells twice in Perm/Wash buffer and transfer cells into mouse fix.

3.2.2.3 Alternative Shortened Staining Procedure

This shortened procedure can only be used if the surface antigens are not sensitive to the fixation/permeabilization step (*see* **Note 9**).

1. Dilute all antibodies in Perm/Wash buffer.
2. After counting the cells, pellet them and remove the buffer. Depending upon how many cells are in the pellet, resuspend the cells in Cytoperm/Cytofix buffer (for 3×10^6 cells, use 200 µL). Incubate at room temperature for 5 min (*see* **Note 10**).
3. Wash the cells twice with Perm/Wash buffer. Aliquot cells into staining tubes (if not already done) and add antibodies to surface antigens and langerin. Vortex to mix. Incubate cells for 20 min at room temperature.
4. Wash the cells twice with Perm/Wash buffer and add fluorochrome-conjugated avidin (if biotinylated antibodies have been used) and the goat anti-rabbit IgG. Incubate at room temperature for 20 min.
5. Wash cells twice with Perm/Wash buffer and transfer cells into mouse fix (*see* **Note 11**).

3.2.3 Assessment of rDC Migration From the Lungs to the Lymph Nodes by i.n. CFSE Labeling

1. Prepare working CFSE solution from stock by diluting to an 8 m*M* solution with Iscove's DMEM Media (16 µL of CFSE + 34 µL media /mouse; *see* **Note 12**).
2. Anesthetize 1–4 mice (*see* **Note 13**).
3. Hold mouse vertical to 45°. Administer 50 µL of CFSE i.n. alternating back and forth between nostrils (*see* **Note 14**).
4. Allow mice to recover.
5. Six hours later, i.n. administer DC maturation and/or migration stimulus (virus, inflammatory agent, etc.).
6. At various times after stimulus introduction (i.e., 6, 12, 18, 24 h), remove the lung draining peribronchial and mediastinal lymph nodes. Prepare a single-cell suspension and stain cells for CD11c+ expression with 1 µL/well of anti-CD11c APC mAb (Caltag/Invitrogen, cat. no. MCD11c05) per 1×10^6 cells/100 µL FACS buffer.

7. Analyze cells on flow cytometer (*see* **Note 15**). Monitor and calculate the number and percentage of recruited DC (i.e., CFSE+CD11c+ cells) in the lymph nodes.

3.3 Antigen Presentation by DCs

3.3.1 DC Activation of Antigen-Specific T-Cell Proliferation In Vivo

All steps must be performed with sterile technique. All wash steps are performed at 4°C for 7 min at 400 g.
Day 1

1. Prepare a single-cell suspension of splenocytes from OTI or OTII mice, wash in PBS, count, and resuspend at 5×10^7/mL in PBS.
2. Mix 5 mM CFSE stock solution:cell suspension 1:10 (to achieve a final concentration of 5 μM CFSE), and incubate at 37°C for 10 min.
3. Wash three times in cold (4°C) 10% FCS in PBS.
4. Inject 5×10^7 CFSE-labeled splenocytes into each recipient C57Bl/6 mouse intravenously using an insulin syringe (*see* **Note 18**).

Day 2

1. Remove spleen from donor C57Bl/6 mice (EtOH or H_2O treated), being careful not to disrupt the capsule. Place into a 60 mm Petri dish. Inject 1 mL of Liberase/DNase working solution into the spleen, and incubate 20 min in a 37°C water bath, supported by a platform.
2. Prepare a single-cell suspension of splenocytes in the Petri dish by rubbing the spleen between the frosted ends of two microscope slides, and aspirate the suspension through a 1-mL syringe with a 23-g needle to break up remaining cell clusters.
3. Collect the cells in a 15-mL centrifuge tube. Wash with 0.5% BSA in PBS, and count the cells.
4. Label with CD11c+ microbeads as follows:

   a. Resuspend the cells in 400 μL of 0.5% BSA in PBS/10^8 cells.
   b. Add 100 μL of anti-CD11c microbeads per 10^8 cells, mix well, and incubate at 4°C (in the refrigerator, not on ice) for 15 min.
   c. Wash with 0.5% BSA in PBS.
   d. Resuspend cells in 500 μL of 0.5% BSA in PBS/10^8 cells.

5. Purify the DC on the autoMACS following the manufacturer's instructions (using "POSSELD" selection mode).
6. Wash the DCs once in 0.5% BSA in PBS and resuspend in Super McCoy's medium at 5×10^6/mL.
7. Pulse the purified DC with OVA 258–264 (for OTI) or OVA 323–339 (for OTII) stock solution to achieve a final concentration of 100 nM.

8. Incubate at 37°C for 2 h, wash in PBS, and resuspend at 5×10^7/mL in PBS.
9. Transfer 5×10^6 peptide-pulsed DC intravenously into each recipient using insulin syringe.

Day 3

1. Prepare a single cell suspension of recipient splenocytes as in step 1 of day 1.
2. Stain with anti-CD8 (OTI) or anti-CD4 (OTII) using a fluorochrome conjugate other than FITC (this channel is needed to collect the CFSE fluorescence).
3. Run samples on flow cytometer.

3.3.2   Allostimulatory Assay of DC Function

All steps in this procedure must be performed under sterile conditions. All washes are performed in 0.5% BSA in PBS, $400 g$ at 4°C for 7 min unless otherwise noted. Preparation of T cells:

1. Prepare a single-cell suspension of splenocytes from BALB/c mouse. Wash and resuspend in 3 mL of 0.5% BSA in PBS.
2. Underlayer the cell suspension with 3 mL of Fico-Lite LM. Centrifuge the cells at $1700 g$ for 20 min at room temperature. Remove the cells from the interface. Wash and resuspend the cell pellet in 1 mL of 0.5% BSA in PBS.
3. Incubate splenocytes with appropriate concentrations of biotinylated anti-B220, DX5, CD11c, CD11b, and Ter119 antibodies for 20 min on ice.
4. Wash three times. Resuspend cell pellet in 80 µL of 0.5% BSA in PBS per 10^7 cells. Add 20 µL of anti-biotin microbeads per 10^7 cells. Mix well and incubate for 15 min at 4°C in the refrigerator (not on ice).
5. Deplete the antibody-coated cells on the autoMACS according to the manufacturer's instructions using the "DEPLETE" mode.
6. Wash the cells, resuspend in Super McCoy's medium, and count.

Preparation of DC for initial stimulation:

1. Purify DCs from C57Bl/6 mouse spleen by using anti-CD11c microbeads on the autoMACS according to steps in Subheading 3.3.1, day 2, steps 1–5.
2. Wash the cells. Resuspend the purified DCs in Super McCoy's medium, and count the cells. Put DCs and T cells into a T25 culture flask at a 1:10 ratio of DC to T cells (the density is generally around 1.1×10^6 cells/mL, 10 mL per flask). Incubate in 5% CO_2 at 37° C.

Restimulation of T cells

1. On day 3 of culture, harvest the cells. Wash 3 times and resuspend in 400 µL of 0.5% BSA in PBS per 10^8 total cells.
2. Deplete the DCs from the cell suspension. Use anti-CD11c microbeads to label the DCs as in step 1 of "Preparation of DCs for Initial Stimulation" in this protocol,

but run the autoMACS in "DEPLETE" mode to remove rather than purify the DCs from the cell suspension.

3. Resuspend the re-purified T cells in Super McCoy's medium and place into a new T25 culture flask. Rest these cells for 2 d by incubating in 5% CO_2 at 37°C.

4. On day 3, purify DCs from the spleens of C57BL/6, C3H or BALB/c mice as in step 1 of "Preparation of DCs for Initial Stimulation" in this protocol.

5. Resuspend the cells in Super McCoy's medium, count, and place into culture in a 96-well plate at graded doses (e.g., 1000, 5000, and 10,000 cells) with 1×10^5 T cells/well (see **Note 19**).

6. Incubate in 5% CO_2 at 37°C for 3 d.

7. Measure T-cell proliferation by adding 1 µCi of ^{3}H-thymidine/well and incubating for an additional 18 h.

8. Harvest the cells using a cell harvester and measure the radioactivity incorporated into the divided cells with a scintillation counter using standard methods.

4   Notes

1. Liberase and DNase stock solutions are stable at 4°C for 1 wk or at −20°C for at least 2 mo.

2. DC purification can also be done using the LS Column (Miltenyi Biotech, Auburn, CA; cat. no. 130–042–401) as described in Subheading 3.1.5

3. CD11c+, CD8α-fraction can be further enriched for CD11c+ CD4+ DC by using the Miltenyi Biotec Multisort Release Kit to remove anti-CD11c Microbeads, stain with anti-CD4 Microbeads and positively select CD4+ cells.

4. Generally approx 1–3% of the cell suspension is LC, as identified by CD11c+ MHC Class II+ cells by flow cyotmetry (Fig. 3).

5. Negative fraction can be used for unstained and comps (be sure to lyse RBC using HBSS).

6. anti-CD11c (clone N418) PE or biotin conjugates have also been used in our hands to label LCs in the ear. In general the biotinylated antibodies give brighter staining with less background than the direct conjugates. PE has a tendency to photobleach readily.

7. Avoid excessive handling of the epidermal sheets as they are quite friable.

8. Avidin-Cy3 is brighter than avidin-Cy5 in our hands. The latter is useful primarily if multicolor staining is performed.

9. Antigens that have been tested for sensitivity and found to be unaffected are epitopes recognized by the following antibodies: 4B12 (CCR7), C34–3448 (CCR5), GI-1 (CD86), 1C10 (CD40), and N418 (CD11c).

10. Performing the permeabilization in bulk will aid in the cell recovery if that is a concern.

11. LCs are identified in LN as FITC+langerin+, and dermal DC as FITC+CD11c+ langerin- (Fig. 4).

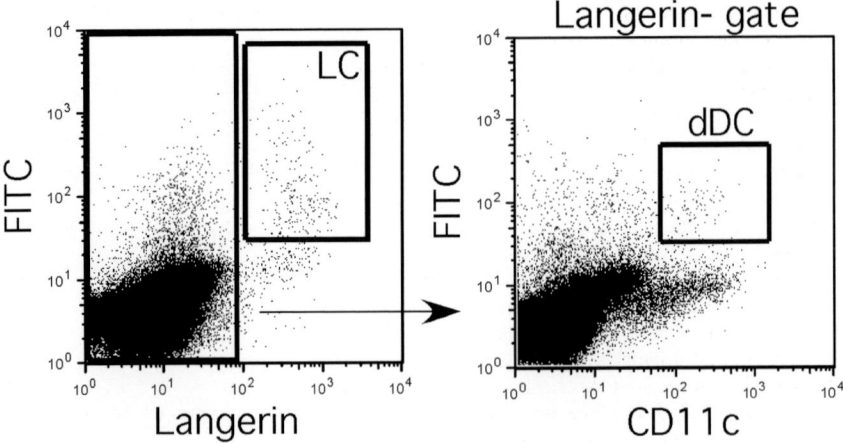

Fig. 4 Gating strategy for LC and dDC in draining LN following FITC skin painting. Recently immigrated LC are identified as langerin+FITC+, and dDC are langerin-CD11c+FITC+.

12. CFSE solution will be cloudy with some precipitate. However, it will still label cells.
13. Mouse must be anaesthetized enough to allow inhalation of CFSE without coughing or swallowing it.
14. Be careful not to get onto hands, skin, or other areas of mouse or those DCs will label and you will observe CFSE+ DC in the appropriate non-lung draining lymph nodes.
15. There is substantial bleed-over of CFSE into the FL2 channel. Careful compensation is required.
16. To measure T cell cytokine production, you can choose to skip the CFSE labeling steps.
17. Supernatants can be used for ELISA.
18. Intravenous injections are easily accomplished by accessing the retro-orbital plexus immediately behind the medial aspect of the eye. Insulin (or tuberculin) syringes work best because there is no void volume left in the syringe (all the cells are transferred).
19. Restimulation with C57Bl/6 DC measures the allogeneic response of the T cells, restimulation with BALB/c DC measures the syngeneic response, and restimulation with C3H DC measures the third party (primary) response.

References

1. Shortman, K. (2002) Mouse and human dendritic cell subtypes. *Nat. Rev. Immunol.* **2**, 151–161.
2. Randolph, G. J., Angeli, V., and Swartz, M. A. (2005) Dendritic-cell trafficking to lymph nodes through lymphatic vessels. *Nat. Rev. Immunol.* **5**, 617–628.

3. de Jong, E. C., Smits, H. H., and Kapsenberg, M. L. (2005) Dendritic cell-mediated T cell polarization. *Spring. Sem. Immunol.* **26**, 289–307.

4. Valladeau, J., Saeland, S. (2005) Cutaneous dendritic cells. *Seminars in Immunology* **17**, 273–283.

5. Steinman, R. M., Hawiger, D., Nussenzweig, M. C. (2003) Tolerogenic dendritic cells. *Ann. Rev. Immunol.* **21**, 685–711.

6. Bagasra, O., Howeedy, A., Kajdacsy-Balla, A. (1988) Macrophage function in chronic experimental alcoholism. *Immunol.* **65**, 405–409.

7. Castro, A., Lefkowitz, D. L., Lefkowitz, S. S. (1993) The effects of alcohol on murine macrophage function. *Life Sci.* **52**, 1585–1593.

8. Mikszta, J., Waltenbaugh, C., Kim, B. S. (1995) Impaired antigen presentation by splenocytes of ethanol-consuming C57BL/6 mice. *Alcohol* **12**, 265–271.

9. Waltenbaugh, C., Peterson, J. D. (1997) Ethanol impairs the induction of delayed hypersensitivity in C57BL/6 mice. *Alcohol* **14**, 149–153.

10. Wang, K., Busker-Mannie, A. E., Hoeft, J., Vasquez, K., Miller, S. D., Melvold, R. W., Waltenbaugh, C. (2001) Prolonged Hya-disparate skin graft survival in ethanol-consuming mice: correlation with impaired delayed hypersensitivity. *Alcohol. Clin. Exp. Res.* **25**, 1542–1548.

11. Zisman, D. A., Strieter, R. M., Kunkel, S. L., Tsai, W. C., Wilkowski, J. M., Bucknell, K. A., Standiford, T. J. (1998) Ethanol feeding impairs innate immunity and alters the expression of Th1- and Th2-phenotype cytokines in murine *Klebsiella* pneumonia. *Alcohol. Clin. Exp. Res.* **22**, 621–627.

12. Peterson, J. D., Vasquez, K., Waltenbaugh, C. (1998) Interleukin-12 therapy restores cell-mediated immunity in ethanol-consuming mice. *Alcohol. Clin. Exp. Res.* **22**, 245–251.

13. Waltenbaugh, C., Vasquez, K., Peterson, J. D. (1998) Alcohol consumption alters antigen-specific Th1 responses: mechanisms of deficit and repair. *Alcohol. Clin. Exp. Res.* **22**, 220S–223S.

14. Cumberbatch, M., Dearman, R. J., Kimber, I. (1996) Constitutive and inducible expression of interleukin-6 by Langerhans cells and lymph node dendritic cells. *Immunol.* **87**, 513–518.

15. Jerrells, T. R., Sibley, D. (1995) Effects of ethanol on cellular immunity to facultative intracellular bacteria. *Alcohol. Clin. Exp. Res.* **19**, 11–16.

16. Standiford, T. J., Danforth, J. M. (1997) Ethanol feeding inhibits proinflammatory cytokine expression from murine alveolar macrophages *ex vivo. Alcohol. Clin. Exp. Res.* **21**, 1212–1217.

17. D'Souza, N. B., Nelson, S., Summer, W. R., Deaciuc, I. V. (1996) Alcohol modulates alveolar macrophage tumor necrosis factor-alpha, superoxide anion, and nitric oxide secretion in the rat. *Alcohol. Clin. Exp. Res.* **21**, 156–163.

18. Lau, A. H., Abe, M., Thomson, A. W. (2006) Ethanol affects the generation, cosignaling molecule expression, and function of plasmacytoid and myeloid dendritic cell subsets in vitro and in vivo. *J. Leukoc. Biol.* **79**, 941–953.

17

Formation and Immunological Properties of Aldehyde-derived Protein Adducts following Alcohol Consumption

Geoffrey M. Thiele, Lynell W. Klassen, and Dean J. Tuma

Summary Most ingested ethanol is eliminated from the body through oxidative metabolism in the liver. Alcohol dehydrogenase is the enzyme that is most important in the oxidation of ethanol to acetaldehyde. However, it has also been demonstrated that cytochrome P4502E1 also can contribute to this process. However, this is not the only aldehyde that is produced after chronic ethanol consumption because oxidative stress and lipid peroxidation can be induced in the liver, which results in the production of malondialdehyde and 4-hydroxy-2-nonenal. These aldehydes are highly reactive and have the ability to react with (adduct) many macromolecules to alter their structure and play a major role in the derangements of hepatic function. Therefore, the formation of these types of adducts in the liver has been proposed as key events leading to the development and/or progression of alcoholic liver disease. In this chapter, methods for the production and detection of these modified proteins will be discussed.

Keywords Aldehydes; acetaldehydye (AA); malondialdehyde (MDA), 4-hydroxy-2-nonenal (HNE); malondialdehyde-acetaldehyde (MAA); adduct; lipid peroxidation; oxidative stress.

1   Introduction

Most ingested ethanol is eliminated from the body through oxidative metabolism in the liver. Alcohol dehydrogenase is the enzyme that is most important in the oxidation of ethanol to acetaldehyde. However, it has also been demonstrated that cytochrome P4502E1 can also contribute to this process *(1)*. However, this is not the only aldehyde that is produced after chronic ethanol consumption because oxidative stress and lipid peroxidation can be induced in the liver, which results in the production of malondialdehyde (MDA) and 4-hydroxy-2-nonenal (HNE) *(2,3)*. These aldehydes are highly reactive and have the ability to react with (adduct) many macromolecules to alter their structure and play a major role in the derangements of hepatic function *(4-6)*. Therefore, the formation of these types of adducts in the liver has been proposed as key events leading to the development and/or progression of alcoholic liver disease (ALD). In this chapter, methods for the production and detection of these modified proteins will be discussed.

Adducts of proteins with acetaldehyde, the first metabolite of ethanol, have been described in a number of studies. Acetaldehyde forms adducts primarily via binding to reactive lysine residues of preferred target proteins *(5,7–10)*. Although the data on the reactivity of acetaldehyde at physiologically relevant concentrations have remained controversial, it appears that, among acetaldehyde-exposed proteins, those with abundant amounts of reactive lysine residues are readily modified even at low concentrations of acetaldehyde under appropriate reducing conditions *(11,12)*. On the other hand, even in the absence of reducing agents, stable cyclic imidazolidinone structures are generated as a result of a reaction between acetaldehyde and the free alpha-amino group of the aminoterminal valine of hemoglobin *(13,14)*.

Aldehydic products of lipid peroxidation, such as MDA and HNE, also form Schiff's base adducts with proteins. MDA is a highly reactive dialdehyde originating from nonenzymatic lipid peroxidation of a variety of unsaturated fatty acids as a result of lipid peroxidation that occurs during phagocytosis by monocytes or from arachidonic acid catabolism in thrombocytes *(15,16)*. HNE is produced by the free radical-mediated oxidation of long-chain polyunsaturated fatty acids, and can react with the sulfhydryl, lysine and/or histidine residues of macromolecules through a Michael addition reaction *(15–17)*. Oxidative modification of proteins with MDA and HNE have been demonstrated to occur in vivo on arterial vessel walls of atherosclerotic lesions *(18,19)*. Similar epitopes have also been found from the liver specimens of patients with alcoholic liver disease (ALD) *(20)* and from animals subjected to experimental iron overload *(21,22)*.

Tuma and coworkers have demonstrated the formation of hybrid adducts with acetaldehyde and malondialdehyde, designated as MAA adducts, in livers of ethanol fed rats *(23)*. Such hybrid adducts may act in a synergistic manner and may also be involved in the mechanisms for stabilization of protein adducts in vivo *(23)*. In addition, the appearance of hydroxyethyl radicals, a reactive species resulting from ethanol during its oxidation in the presence of iron, has also been described from liver microsomes of ethanol-fed animals *(24,25)*. Therefore, the formation of these types of adducts in the liver has been proposed as key events leading to the development and/or progression of ALD. In this chapter, methods for the production and detection of these modified proteins will be discussed.

2 Materials

2.1 Preparation of 5X Chelator Solution

1. A 5X stock chelator solution containing; sodium phosphate (Na_2HPO_4), phytic acid (PA), diethylenetriaminepentaacetic acid (DTPA), is prepared at pH 7.2 as follows *(26)*:

   10 mM PA = 9.24 g phytic acid (FW = 923.83)/liter of deionized H_2O
   10 mM DTPA = 3.94 g DTPA (FW = 393.35)/liter of deionized H_2O
   0.1 M Na_2HPO_4 = 14.2 g of sodium phosphate (Na_2HPO_4; FW = 141.96)/L of deionized H_2O

2. The buffer is placed on ice and nitrogen bubbled through the solution for 30 min.
3. The solution is sterile filtered through a 0.2-micron filter and stored at 4°C until used.

2.2   Synthesis of Malondialdehyde

1. Malonildialdehyde-bis-dimethyl acetal is purchased from Aldrich Chemical (10838-3).
2. In a 125-mL Ehrlenmeyer flask, combine 5 g of malonildialdehyde-bis-dimethyl acetal, 5 mL of deionized H_2O, and 2.5 mL 1 N HCl.
3. Cover the flask with parafilm, place on a rotary shaker at room temperature for 1 hour at low speed.
4. Cool the flask on ice for 15 min, add 6 N NaOH dropwise (swirl the flask) until the pH reaches 10.0. NOTE: The pH will start at about 1.0 and will rise very slowly at first. As it begins to become more basic the pH will change more rapidly. Also, the color of the solution will turn brownish.
5. Add ice-cold acetone to the 100-mL mark (precipitates the sodium salt) and let set on ice for approx 1 h.
6. Attach a 4.25-cm Buchner funnel with a Whatman #1 Filter paper to a 500-mL side arm flask. Connect the side-arm flask to a vacuum source, pre-wet the filter paper with ice-cold acetone, filter the precipitate, rinse with cold acetone.
7. Scrape the precipitate on the filter paper into a 25-mL Erhlenmeyer flask and dissolve the precipitate in as little deionized H_2O as possible (approx 1–1.5 mL), by adding a few drops at a time.
8. Add ice-cold acetone slowly until the precipitate barely remains in solution. This should be about the same amount as the deionized H_2O added previously, and the solution should be slightly cloudy. This is a critical step and it is imperative that no more acetone than necessary is added.
9. Cover with parafilm and place at −20°C for 1.5 h. Do not it allow to go overnight.
10. Filter as in step 6, scrape the precipitate into a glass container, and dry under vacuum for 2–6 h or overnight.
11. Store at −20°C in a parafilm sealed glass vial away from light.
12. This procedure should result in a theoretical yield of 2.85 g, but our experience is that the actual yield is much less than this (5–10 times less). The product is the sodium salt of MDA, with a MW of 94 as determined by mass spectrometry. Other procedures can be done to give higher yields, but are more expensive, use much more acetone, and the purity is not significantly greater.

2.3   Synthesis of Hexyl-MAA

1. Hexyl-MAA (1-hexyl-4-methyl-1,4-dihydropyridine-3,5-dicarbaldehyde), a synthetic analog of MAA, is synthesized via the Hantzsch reaction as previously reported (27).
2. Briefly, a 1.0 M solution of MDA (10 mL, 0.01 mol) is transferred to a 100-mL round bottom flask, hexylamine added (1.1 mL, 0.81 g, 0.008 mol), the pH

adjusted to 4.0 with concentrated HCl (5–10 drops), and then the AA (0.28 mL, 0.22 g, 0.005 mol) added.

3. The flask was capped and allowed to stir at 50°C for 2–4 h and the solvent removed *in vacuo*.

4. Column chromatography afforded the product as a yellow solid and recrystallization using 95% ethanol afforded the product as yellow needles.

2.4   Synthesis of N-ethyl lysine (NEL)

1. Poly-L-lysine is purchased from the Sigma Chemical Co., adjusted to 1.2 mg/mL, and AA (5 mM) and NaCNBH3 (30 mM final concentration) added.

2. The reaction is allowed to continue at 37°C for 72 h as described previously *(12)*.

3. The purity of the NEL prepared by this method is greater than 99%, as determined with high-performance liquid chromatography *(12)*.

4. Unmodified lysine, should be used as a control and is prepared in the same manner as above except that no AA is added to the initial reaction mixture.

5. The samples are dialyzed, lyophilized, and subjected to hydrolysis at 115°C for 22 h.

6. The samples are stored at −20°C in sealed amber glass ampules with very little head space.

2.5   Preparation of Lysine Sepharose 4B-CL Beads

1. Sepharose 4B-CL are purchased from the Sigma Chemical Co., washed, and activated using 1 mM HCl.

2. To 1 mL of swollen beads, 25 mg of either Nε-t-BOC-L-lysine or Nα-t-BOC-L-lysine (Sigma Chemical Co.) in bicarbonate buffer, pH 8.3, is added and the suspensions are rotated for 2 h at 37°C, and then overnight at 4°C with 0.2 M glycine to block residual active groups.

3. The beads are washed with 10 volumes of bicarbonate buffer, and the blocking groups are removed by treatment with 50% (v/v) trifluoroacetic acid/dichloromethane, and neutralized with 5% (v/v) diisopropylethylamine in dichloromethane for 45 min.

4. The beads are washed with another 10 vol. of buffer, leaving one set of beads with a lysine attached at the ε-amino group and the α-amino group free (S4Bα) and another set of beads where the α-amino group was attached leaving the ε-amino group available for binding (S4Bε).

5. The amount of lysine bound to the Sepharose 4B-CL was determined by assessing the number of free amino groups bound to the beads using the ninhydrin method *(28)*. Using this procedure, it was determined that 1–2 μmol of lysine

was bound to each milliliter of beads, regardless of whether the attachment was through the α-amino or the ε-amino group.

6. These beads (20 mL) are modified under reducing (100 mM AA, with 600 mM NaCNBH$_3$) or nonreducing (500 mM AA) conditions for 16 h at 37°C. For MAA-modified beads, 10 mM MDA and 10 mM AA in chelators are added for 72 h at 37°C (*29*).

7. Beads are washed twice with 100 mL of 0.1 M sodium phosphate buffer, pH 7.2, and dialyzed against this same buffer overnight at 4°C with constant stirring.

2.6   Method of Habeeb for Determining Lysine Modification (30)

1. The macromolecule concentration should be adjusted to 0.6–1.0 mg/mL
2. Place 1.0 mL of macromolecule solution in a 5-mL tube, add 1 mL of 4% NaHCO$_3$, 1 mL of 0.1% picrylsulfonic acid (TNBS), and incubate at 40°C for 2 h.
3. Add 1.0 mL of sodium dodecyl sulfate (SDS), 0.5 mL of 1 N HCl, and read at visible wavelength of 355 nm.
4. Use 1.0 mL of H$_2$O as the blank and the unmodified macromolecule as the negative control.
5. Calculate:

$$\%\text{lysine modification} = \frac{\text{Modified macromolecule at 355 nm}}{\text{Unmodified macromolecule at 355 nm}} \times 100$$

2.7   Determination of the Fluorescence/Macromolecule (F/M) Ratio for MAA Adducted Macromolecules (27)

1. Using a Fluorometer, adjust using the following settings:

   Excitation wavelength = 398
   Emission wavelenth = 460

2. Blank on 0.1 M phosphate buffer.
3. Dilute the MAA adducted sample to 1 mg/mL and determine the intensity.
4. Calculations for F/M:

$$\text{Fluorescent Binding} = \frac{F}{k} \times \frac{\text{Dilution Factor}}{\text{Concentration of Protein}} = \frac{\text{nmoles}}{\text{mg}}$$

Where F = 398/460 and k = Constant = arbitrary fluorescence units/nmole hexyl-MAA. Calculate the weight of the macromolecule adducted in nmoles/mg. Using the following formula, calculate the amount of fluorescence on a macromolecule:

$$F/M = \frac{\text{nmoles/mg of fluorescent binding}}{\text{nmoles/mg of macromolecule}}$$

5. F/M should be at least 1.0 if it is properly adducted. If not, change the MDA:AA ratio during the modification of the macromolecule to achieve this level.

2.8   Casein (2%) Blocking Buffer

1. To 2.5 liters of ddH$_2$O, add 5 mL of NaOH, 80 g of casein, and 0.4 g of thimerosol.
2. Stir for approx 6 h or overnight at room temperature.
3. The solution will remain translucent, should q.s. to 4.0 liters, and the pH adjusted to 7.0–7.5. If it is necessary to use HCl to bring the pH down, the casein may begin to clump a little but will dissolve if stirred for a while.
4. Can be stored at 4°C for 1 wk.

2.9   5X Tris (0.125 M) Buffer

To make 2 liters of this buffer, the following are needed:

1. 30.25 g of Tris (Fisher BP 154-1)
2. pH to 7.4 and q.s. to 2 liters
3. Store at 4°C.

2.10   Casein Wash Buffer

To make 1 liter of casein wash buffer, the following are needed:

1. 250 mL of casein (2%) blocking buffer.
2. 200 mL of 5X Tris buffer.
3. 550 mL of ddH$_2$O.

2.11   Peroxidase Substrate for ELISAs

0.4 mg/mL of 1-phenylendiamine, 0.4 µl/mL hydrogen peroxide (30%), 5.1 mg/mL citric acid, 6.1 mg/mL anhydrous Na$_2$HPO$_4$, pH 5.0.

2.12   Immunization Schedules

To produce antibodies to the different adducts animals are immunized with
adducted macromolecules:

1. Emulsified in Complete Freund's Adjuvant at week 1. The process is repeated at
   weeks 3 and 5, but using Freund's Incomplete Adjuvant. Different animals are
   injected as follows:

Rabbits:	2 mg/mL	intraperitoneally (ip)
Rats:	500 μg/mL	ip
Mice:	25 μg/mL	ip

2. In the case of MAA-adducted proteins, it is possible to immunize mice and rats
   weekly for 5 wk in the *absence* of adjuvants:

Rats	500 μg/mL	ip
Mice	25 μg/mL	ip

3. The immunization schedule for T-cell proliferation assays consists of immunization
   of five animals per group once with 50 μg in the foot pad with either the nonadducted
   or adducted macromolecule in an emulsion of complete Freund's adjuvant (50 μL)
   and assaying 7 d later. Immunization in the foot pad is a standard protocol that
   enhances the absolute number of T cells for use in proliferation assays.

2.13   Antibodies to Aldehydes (*see Note 1*)

Antibodies to aldehyde modified proteins have been described by many individuals.
Importantly, with the exception of three of these antibodies, the epitope specificity
has not been adequately described. The reduced adduct has been shown to form on
lysine residues and antibody responses to this adduct can be specifically inhibited with
N-ethyl-lysine (29). Antibody responses to the MAA adduct can be inhibited
with the hexyl-MAA construct, which contains the 1-hexyl-4-methyl-1,4-dihydro-
pyridine-3,5-dicarbaldehyde moiety (23). Finally, the antibody to HNE has been
shown to specifically react with the 4-hydroxy-2-nonenal Michael adduct.
    The antibodies to MDA are less well-characterized, and no specific epitope has
been identified. Indeed, in our hands, these antibodies primarily react with the
MAA adduct (Thiele, et al., unpublished observation) as demonstrated by direct
ELISA and Competitive Inhibition ELISA using hexyl-MAA as the inhibitor.
    Similarly, antibodies to the nonreduced adduct have not been characterized with
respect to their epitope specificity. Most investigators have used high concentrations of AA to "adduct" a macromolecule and then screened them against the
adducted and non-adducted form of this macromolecule. Thus, they have produced
an antibody that recognizes the "adducted" form and not the non-adducted form of
the macromolecule. Interestingly, many of these antibodies recognize other "adducted"
macromolecules. Thus, these antibodies have been designated as recognizing
the non-reduced AA adduct. However, no epitope has ever been determined, and

the best that can be said about these antibodies is that they recognize something on nonreduced AA macromolecules that is not found on the nonadducted form of the macromolecule. This "something" could be a Schiff base, aggregates of protein, or some other epitope not yet identified.

1. AA under reducing conditions: Also, known as *N*-ethyl-lysine (NEL) *(29)*. Available (from Geoffrey M. Thiele, Ph.D.) as a rabbit anti-NEL (affinity purified) or a monoclonal mouse anti-NEL (IgG2b isotype). Must be detected using a isotype specific assay.
2. MDA: Multiple sources of this antibody can be purchased as: 1) Anti-malondial-dehyde-modified proteins (cat. no. 442730; Calbiochem, Darmstadt, Germany).
3. AA under nonreducing conditions. This antibody is not currently available *(29)* but can be produced as described by many investigators. However, great caution as to its specificity should be noted, as outlined previously.
4. MAA: Available (from Geoffrey M. Thiele, Ph.D.) as a rabbit anti-MAA (affinity purified) or a monoclonal mouse anti-MAA (IgG1 isotype). Must be detected using a isotype specific assay *(31)*.
5. HNE: Multiple sources of this antibody can be purchased: 1) anti-4-hydroxy-2-nonenal Michael adducts, reduced (cat. no. 393207; Calbiochem, Darmstadt, Germany); 2) monoclonal anti-4-hydroxynonenal antibody (cat. no. MAB3249; R&D Systems, Inc., Minneapolis, MN).

2.14 SDS-Polyacrylamide Gel Electrophoresis (SDS-PAGE) (32)

1. Separating buffer (4X): $1.5 M$ Tris-HCl, pH 8.7, in 0.4% SDS. Store at room temperature.
2. Stacking buffer (4X): $0.5 M$ Tris-HCl, pH 6.8, in 0.4% SDS. Store at room temperature.
3. Thirty percent acrylamide/bis solution (37.5:1 with 2.6% C) and N,N,N,N'-tetramethyl-ethylenediamine (TEMED, Bio-Rad, Hercules, CA; *see* **Note 2**).
4. Ammonium persulfate: Prepare 10% solution in water and use immediately.
5. Running buffer (5X): $125 mM$ Tris-HCl, $960 mM$ glycine, 0.5% (w/v) SDS. Store at room temperature.
6. Prestained molecular weight markers: Kaleidoscope markers (Bio-Rad, Hercules, CA).

2.15 Western Blot for the Detection of Adducts

1. Setup buffer: $25 mM$ Tris-HCl (do not adjust pH), $190 mM$ glycine, 20% (v/v) methanol.
2. Transfer buffer: Setup buffer plus 0.05% (w/v) SDS.
3. Supported nitrocellulose membrane from Millipore, Bedford, MA and 3 MM Chromatography paper from Whatman, Maidstone, UK.

4. Tris-buffered saline with Tween (TBS-T): Prepare 10X stock with 1.37 M NaCl, 27 mM KCL, 250 mM Tris-HCl, pH 7.4, 1% Tween-20. Dilute 100 mL with 900 mL of water before use.
5. Blocking buffer: 5% (w/v) nonfat dry milk in TBS-T
6. Antibody dilution buffer: TBS-T.
7. Secondary antibody: anti-isotype specific antibody produced in either a goat or rabbit is preferential (BD Biosciences, San Jose, CA).
9. Tertiary antibody: Anti-goat or rabbit conjugated to horse-radish peroxidase (Sigma Chemical Company, St. Louis, MO).
10. Enhanced chemiluminescent (ECL) reagents from Kirkegaard and Perry (Gaithersburg, MD) and Bio-Max ML film (Kodak, Rochester, NY).

3 Methods

3.1 Generation of Malondialdehyde-Acetaldehyde Adducts

1. Prepare a 0.1 M sodium phosphate buffer, pH 7.2, and place the buffer in ice. Slowly bubble nitrogen through the buffer for 30 min (23). All macromolecules and stock solutions are made with this buffer as follows:

   10 mM MDA = 0.94 mg in 1 mL of buffer
   10 mM AA = Make a 1 M stock solution first
       56 μL of AA (17.9 M) in 944 μL of buffer.
       Dilute the 1 M stock AA solution 1:100.
   Macromolecule = 2 mg/mL

2. Combine these materials in the smallest tube possible in order to keep the head space to a minimum. Make a final solution in 1.6 mL to make MAA adduct under the conditions of a 1:1 ratio of MDA (1 mM; see Note 3):AA (1 mM):

10 mM MDA	160 μL
10 mM AA	160 μL
2 mg/mL of macromolecule	800 μL
5X Chelators	320 μL
0.1 M Phosphate, pH 7.2	160 μL

3. The tubes are sealed with parafilm, covered with aluminum foil, and placed on a rotator at 37°C for 72 h. The solution should turn light yellow, indicating the formation of the dihydropyridine product on the macromolecule.
4. Unbound aldehydes are removed by overnight dialysis at 4°C against 0.1 M Na_2HPO_4 buffer, pH 7.2.
5. To assure the proteins are not aggregated, all proteins are centrifuged at 100,000 g for 30 min or passed through a 0.2 μM syringe filter, and protein concentrations reassessed using the method of Groves et al. (33).

6. The presence of MAA adducts are evaluated by measuring the fluorescence intensity at 398/460 nm (excitation/emission) wavelength pairs according to Tuma et al. *(27)*.
7. Some macromolecules do not label very well at 1 m*M* MDA and 1 m*M* AA (1:1). It may be necessary to change this ratio and this can easily be done by altering the amount of MDA or AA added to the solution. It is up to the investigator to determine which concentrations are optimum for their studies.

3.2 Generation of Malondialdehyde Adducts

1. Prepare a 0.1 *M* sodium phosphate buffer pH 7.2 and place the buffer in ice. Slowly bubble nitrogen through the buffer for 30 min. All macromolecules and stock solutions are made with this buffer as follows:

   10 m*M* MDA = 0.94 mg in 1 mL of buffer
   Macromolecule = 2 mg/mL

2. Combine these materials in the smallest tube possible to keep the head space to a minimum. To make a MDA adduct at 1 m*M* (*see* **Note 3**) in 1.6 mL, add the following:

10 m*M* MDA	160 μL
2 mg/mL of macromolecule	800 μL
5X Chelators	320 μL
0.1 *M* phosphate, pH 7.2	320 μL

3. The tubes are sealed with parafilm, covered with aluminum foil, and placed on a rotator at 37°C for 72 h. The solution should turn light yellow, indicating the formation of the dihydropyridine product on the macromolecule.
4. Unbound aldehydes are removed by overnight dialysis at 4°C against 0.1 *M* Na_2HPO_4 buffer, pH 7.2.
5. To assure the proteins are not aggregated, all proteins are centrifuged at 100,000 X g for 30 min, or passed through a 0.2-μ*M* syringe filter, and protein concentrations reassessed using the method of Groves et al. *(33)*.
6. The presence of MDA adducts was evaluated by measuring the fluorescence intensity at 399/471 nm (excitation/emission) wavelength pairs according to Cominacini et al. *(34)*.

3.3 Generation of Reduced Acetaldehyde Adducts

1. Prepare a 0.1 *M* sodium phosphate buffer, pH 7.2, and place the buffer in ice. Slowly bubble nitrogen through the buffer for 30 min. Prepare all stock solutions with this buffer.

2. Acetaldehyde-modification of macromolecules is performed under reducing conditions by incubating the following materials in a small parafilm sealed tube at 37°C overnight on a rotator (*12*):

2 mg/mL of macromolecule	800 µL
50 m*M* AA	160 µL
300 m*M* sodium cyanoborohydride	160 µL
0.1 *M* sodium phosphate buffer, pH 7.2	480 µL

Finally, the samples are dialyzed overnight against a 0.1 *M* sodium phosphate buffer, pH 7.2.

3. A variation on the production of the reduced adduct is to incubate 1 mL of a 1.2 mg/mL solution of the macromolecule with 240 m*M* AA (14 µL of AA stock = 17.9 *M*) for 1 h at 37°C on a rotator. This is followed by adding 186 µL of a 0.645 *M* sodium cyanoborohydride solution to the tube for 30 mins at 37°C on a rotator. Finally, the samples are dialyzed overnight against a 0.1 *M* sodium phosphate buffer, pH 7.2.

4. There have been many more variations on this procedure, but using antibodies to the Reduced adduct (*N*-ethyl lysine) has demonstrated that all of these methods will generate the reduced adduct. Whether there are other products associated with these adducted macromolecules is unknown.

5. To assure the proteins are not aggregated, all proteins are centrifuged at 100,000 X g for 30 min, or passed through a 0.2 µM syringe filter, and protein concentrations reassessed using the method of Groves et al. (*33*).

3.4   Generation of Nonreduced Acetaldehyde Adducts

1. Prepare a 0.1 *M* sodium phosphate buffer, pH 7.2, and place the buffer in ice. Slowly bubble nitrogen through the buffer for 30 min. Prepare all stock solutions with this buffer.

2. Acetaldehyde-modification of macromolecules is performed under non-reducing conditions by incubating the following materials in a small parafilm sealed tube at 37°C overnight on a rotator (*5*):
   Finally, the samples are dialyzed overnight against a 0.1 *M* sodium phosphate buffer, pH 7.2.

2 mg/mL of macromolecule	800 µL
10 m*M* AA	160 µL
5X Chelators	320 µL
0.1 *M* sodium phosphate buffer, pH 7.2	320 µL

3. Under nonreducing conditions, we have shown that 5 m*M* AA incubated for 5 d at 37°C produces an adduct similar to that produced by 100 m*M* AA overnight at 37°C. There have been many more variations on this procedure. All have

resulted in various products, but it is unknown whether there are other products associated with these adducted macromolecules. Additionally, this adduct has no identified structure which makes it impossible to identify without any doubt that the adduct exists.

4. To assure the proteins are not aggregated, all proteins are centrifuged at 100,000 g for 30 min, or passed through a 0.2-μM syringe filter, and protein concentrations reassessed using the method of Groves et al. *(33)*.

3.5 Generation of Hydroxyethyl Radical Adducts

1. Hydroxyethyl radical adducts are prepared by incubating 1 mg/mL of macromolecule for 30 min at 37°C with; 100 mmol/L ethanol, 50 μmol/L Fe(NH$_4$)$_2$(SO$_4$)2, 0.1 mmol/L EDTA, and 0.1 mmol/L H$_2$O$_2$ in PBS, pH 7.4, as previously described *(24)*.

2. Because oxidation of many macromolecules can occur during the formation of hydroxyethyl radicals, oxidized-macromolecules were prepared by incubating them under the same conditions as for the formation of hydroxyethyl radicals, but without ethanol.

3.6 Determination of Macromolecule Modification by ^{14}C Acetaldehyde

To determine the optimum concentrations of AA to use in the modification of acromolecules, ^{14}C acetaldehyde can be used to detect maximum binding *(35)*.

1. To determine AA binding under nonreduced conditions, the macromolecule is rehydrated in 0.1 M sodium phosphate buffer (pH 7.4) to 1 mg/mL and incubated with 1 mM ^{14}C-AA at 37°C in polypropylene tubes that are sealed to minimize the loss of volatile radioactivity as previously described 4. Because these adducts are prepared under many different conditions, this concentration of ^{14}C-AA should be used as a tracer.

2. To determine AA binding under reducing conditions, the ^{14}C-AA is again used as a tracer in the presence of reducing agents as outlined above.

3. To determine the optimum conditions to produce MAA adducts, 1 mM ^{14}C acetaldehyde is incubated with varying concentrations of MDA ranging from 0.5 to 8.0 mM. All incubations are performed in phosphate buffer (0.1 M, pH 7.4) in the presence of the chelators diethylenetriamine-pentaacetic acid (DTPA) and phytic acid (Aldrich Co., Milwaukee, WI) at 10 mM each to prevent oxidative reactions from occurring.

4. At the end of the incubation period, the reaction mixture is exhaustively dialyzed against 0.1 M sodium phosphate buffer for 24 h at 4°C. Radioactivity was measured in the retentate and data expressed as nmoles AA bound per milligram of

macromolecule. The concentration of protein was determined using the method of Groves et al. *(33)*.

3.7   Methods for Stabilizing Nonreduced AA Adducts for Detection

As noted, nonreduced AA adducts are not stable and no known structure has been identified for which there is a chemical analog. This means that any procedure to detect nonreduced adducts will cause the adduct to deteriorate. However, it is possible to stabilize the nonreduced adduct by adding sodium cyanoborohydride or MDA to the sample to produce the reduced or MAA adduct, respectively. Both of these adducts can be detected by monoclonal antibodies that are currently available by; ELISA, Competitive ELISA, Western Blotting, etc.

1. Stabilization by sodium cyanoborohydride. If AA is present in the sample, then it is possible to detect this modification by adding 30 mM sodium cyanoborohydride overnight at 37°C (*see* **Note 3**). This will convert the nonreduced AA adduct to the reduced adduct (NEL) and the antibody RT1.1 can be used for detection.
2. Stabilization by MDA. If AA is present in the sample, then it is possible to detect this modification by adding 1 mM malondialdehyde overnight at 37°C (*see* **Note 4**). This will convert the non-reduced AA adduct to the MAA adduct and the antibody to MAA can be used for detection.

3.8   Absorption and Elution of Polyclonal Antibodies From Sepharose 4B-CL Lysine Beads Modified with Nonreduced AA Adducts, Reduced AA Adducts, or MAA Adducts

1. Antiserum from animals immunized with the different adducts is added to modified Sepharose 4B-CL lysine beads and incubated with constant shaking for 1 hour at 37°C *(29,31)*.
2. The beads are washed with 10 volumes of 0.1 M sodium phosphate buffer, pH 7.4.
3. The bound antibody is eluted with 0.1 M acetic acid and dialyzed immediately against 0.02 M Tris containing 0.01 M glycine, pH 8.2.
4. Eluted materials are tested for antibody titers to an unmodified macromolecule (negative control) or the same macromolecule modified with; AA under nonreducing and reducing (5/30 and 240/100) conditions or MAA.

3.9   Measurement of Circulating Antibody Responses in Mice by ELISA

1. An indirect ELISA can be used to determine circulating antibody levels as previously described *(23,29,36,37)*. Briefly, carrier macromolecules and adducted macromolecules are diluted to 20 μg/mL in sodium bicarbonate buffer, pH 9.6.

2. Additionally, purified myeloma isotype standards (IgG1, IgG2a, IgG2b, IgG3, and IgM) are serially diluted down the appropriately labeled ELISA plate(s) in sodium bicarbonate buffer (pH 9.6) starting at 1280 ng/mL (6.4 μL in 5 mL of coating buffer, 1B and 2B) and ending with 20 ng/mL (1H and 2H) to construct a standard curve.

3. ELISA plates (Immulon IV, Dynatech, Chantilly, VA) are coated in duplicate with 100 μL of the test materials in columns: 3 and 4, 5 and 6, 7 and 8, 9 and 10, 11 and 12.

4. The plate(s) are incubated at 37°C overnight, washed three times with casein wash buffer, and add 200 μL of casein blocking buffer to block the nonspecific binding sites on the plate. Incubate at 37°C for 30 min.

5. Each plate is washed with casein wash buffer and 100 μL of test sera is added where appropriate and incubated at 37°C. At the end of 45 minutes the plates are washed three times with casein wash buffer and 100 μL of rabbit anti-mouse isotype (IgG1, IgG2a, IgG2b, IgG3, 1:1000; IgM, 1:1,000; Zymed, San Francisco, CA) is added.

6. After a 45-min incubation at 37°C, the plates are washed three times with casein wash buffer, and 100 μL of:

   A. Alkaline phophatase-labeled goat anti-rabbit IgG (H & L; 1:10,000) is added (Jackson Laboratories, West Grove, PA); or,

   B. Peroxidase-linked goat anti-rabbit IgG (H & L; 1:10,000) is added (Jackson Laboratories, West Grove, PA), and incubated for 45 min at 37°C.

7. The plates are washed three times with casein wash buffer and:

   A. 100 μL of p-nitrophenyl phosphate (substrate) added, and the plates read on an ELISA Reader at 405 nm.

   B. 150 μL of peroxidase substrate for ELISAs were added. After 15 minutes the reaction was stopped by adding 50 μL of 2 N H_2SO_4. Absorbances were read on an ELISA Reader at 490 nm.

8. Where appropriate a standard curve is extrapolated from each plate and relative antibody levels are determined. Isotype standards on each plate are utilized in order to standardize the comparison between plates and between assay dates. Although this method of calibration allows consistency between groups, direct coating of purified Igs onto plastic leads to the potential overestimation of antigen specific antibody concentrations *(38,39)*. Therefore, calculations of antigen concentration are not intended to indicate actual concentrations of

specific antibodies found in serum, but are a relative measure of that isotype of antibody present.

9. Alternatively, the quantification of antibody can been determined in a number of other ways including; titers *(40)* and international units *(41)*. The use of titers for simple comparisons of antibody concentrations have been reported to be misleading because of differences in sensitivity of ELISAs *(38)*.

3.10   Measurement of Circulating Antibody Responses in Humans by ELISA

1. Plates are prepared as above, except the standards are human; IgG, IgA, or IgM.
2. After the incubation of the plates with serum, the plates are washed and incubated with alkaline phosphatase- or peroxidase-labeled goat anti-human; IgG (dilution 1:5,000), IgA (dilution 1:5,000) or IgM (dilution 1:5,000)(Dako S.P.A., Milano, Italy), for 45 min at 37°C.
3. The plates are washed and developed as outlined previously.

3.11   Measurement of Circulating Antibody Responses in Rabbits by ELISA

1. Plates are prepared as noted previously, except the standards are Rabbit IgG.
2. After the incubation of the plates with serum, the plates are washed and incubated with alkaline phosphatase- or peroxidase-labeled goat anti-rabbit IgG (dilution 1:5,000; Jackson Laboratories, West Grove, PA), for 45 min at 37°C.
3. The plates are washed and developed as outlined previously.

3.12   Analysis of the Antibody Specificity by Competitive ELISA

A competitive ELISA is used as previously described *(23,36)* to determine the specificity of the antibodies raised against MAA-adducted or NEL-adducted macromolecules.

1. ELISA plates (Immulon IV) are coated (bicarbonate buffer, pH 9.6) with 100 μL of solid phase macromolecules (approximately 20 μg/mL) overnight at 37°C.
2. 125 μL of diluted serum (see note below) from the species being tested is added to RIA vials in a 96 well platform containing an equal volume of various inhibitors (hexyl-MAA, NEL, etc.) adjusted to different concentrations (125 μL), sealed and placed at 4°C overnight. **Note**: This dilution of serum is used as it

was determined to be the amount necessary to result in 1/2 maximum antibody activity in the direct ELISA (1.0 O.D. after 30 min) as performed previously.

3. The next morning, the coated wells are washed three times with casein wash buffer to remove unbound protein, and blocked with casein blocking buffer for 30 min at 37°C.

4. The plates are washed with casein wash buffer and 100 μL of samples from step 2 are added to the corresponding wells of the coated plates and incubated for 45 min at 37°C.

5. After washing 3 times with casein washing buffer, 100 μL of an enzyme labeled antispecies antibody is added. The ELISA is developed as described above for the direct ELISA in order to obtain optical density measurements.

6. Results for each inhibitor concentration are expressed as percent inhibition, which was calculated using the following formula:

$$\% \, inhibition = \frac{(ODmax - BKG) = (ODsample - BKG)}{(ODmax - BKG)} \times 100$$

Where Odmax = OD in the absence of inhibitor;
BKG = OD from nonspecific absorption of assay reagents; and
ODsample = OD resulting from a given concentration of test samples

Note: This procedure can be adapted to any adduct for which the structure is known and an inhibitor can be synthesized.

3.13   T-Cell Proliferation Assays

1. Single-cell suspensions of popliteal lymph nodes or spleen cells are isolated 7 d after the animals are immunized in the foot pad (fp) or intraperitoneally (ip) with 25 μg of adducted macromolecules in Freund's complete adjuvant (FCA) *(26,42,43)*.

2. Cells are washed 1 time in DMEM-10, centrifuged at 200 g and decanted.

3. To isolate T cells, the Miltenyi Magnetic Activated Cell Sorting (MACs™) system is used.

  A. Briefly, the lymph node or spleen cells are brought up in a MACs™ separation buffer (2 mM Na₂EDTA, 0.5% BSA, Ca²⁺ and Mg²⁺ free HBSS without phenol at pH 7.4), passed through a 30-m preseparation mesh filter (Miltenyi Biotec, Auburn, CA) to get rid of debris, counted on a hemocytometer, stained with anti-Thy1.2 magnetic beads, and isolated using a MIDI-MACs™ isolation column as per the manufacturer's protocol.

  B. Purity of isolated cells is determined to be greater than 95% pure as determined using anti-CD3 FITC on flow cytometry.

4. To determine proliferative responses of the T cells, the unadducted and adducted macromolecule is sterile filtered and added to the appropriate wells of a 96-well

plate at twice the final desired concentration and serially diluted down the plate starting at 1000 μg/mL and going to 250 μg/mL (final concentration).
5. Isolated T cells (1×10^5 cells/well) and irradiated (2000 Rads) spleen cells (5×10^5 cell/well) are added to each well of the 96 well plate that contained the appropriately diluted antigen.
6. Plates are incubated for 48 to 72 h at 37°C, in 5% CO_2, pulsed with 1 μCi/well of [^{3}H] thymidine (New England Nuclear) for 16 h, harvested on a 96-well harvester (Tomtec, Orange, CT), and read on a scintillation counter (Betaplate, Wallac, Turku Finland).
7. In order to compare proliferative responses, a measure of the relative increase in [^{3}H]thymidine incorporation over background [^{3}H]thymidine counts is made and referred to as stimulation index (SI):

$$SI = \frac{CPM_{(Exp)} - CPM_{(Bkg)}}{CPM_{(Con)} - CPM_{(Bkg)}}$$

3.14 Production of Monoclonal Antibodies

The monoclonal antibodies to NEL (RT1.1) and MAA have been reported previously (29,31).

1. Spleens from animals whose serum showed antibody to the adducted macromolecule are removed and homogenized by mechanical disruption.
2. A single cell suspension is prepared and mixed with an equal concentration of mouse myeloma P3.NS-1-AG4-1 (ATCC).
3. Using the method of Kohler and Milstein (44), these cells were fused and placed in 96-well plates at 1×10^5 cells/well.
4. Selections were performed using hypoxanthine-aminopterine-thymidine (HAT) followed by hypoxanthine-thymidine (HT) in the medium.
5. Wells are monitored for growth and tested for reactivity to the nonadducted or adducted macromolecule of interest using an ELISA system.
6. Positive wells are subcloned twice, using limiting dilution, to ensure monoclonality.

3.15 Immunoprecipitation of Adducted Macromolecules

1. Briefly, antibody to the various adducts are added to the different samples and incubated at 4°C overnight.
2. Antigen-antibody complexes will be removed by adding Staph Protein G (SpG) for 2 h at 37°C. The SpG will be pelleted, washed three times with PBS, and resuspended in SDS-run buffer.
3. These supernatants will be separated by SDS-PAGE and transferred to nitrocellulose as outlined previously.

4. Western blot analysis can be performed to detect the macromolecules that are adducted by utilizing antiadduct antibody or anti-macromolecule antibodies.
5. Alternatively, silver staining is performed to detect all macromolecules precipitated by the antibody and unique to modification by the various adducts.
6. In some cases, it may be useful to stain the nitrocellulose for content (Coommassie Brilliant Blue), and then cutting out the appropriate bands and send them out for protein sequencing to identify the macromolecules that were recovered.

3.16   SDS-PAGE

1. These instructions assume the use of a Bio-Rad Mini-Gel apparatus (Bio-Rad, Hercules, CA), but can be modified for other formats. It is critical that the glass plates are scrubbed clean with detergent and rinsed extensively with distilled water.
2. Prepare a 10% separating gel (lower gel) by mixing 3.75 mL of 4X separating buffer, with 5 mL of acrylamide/bis solution, 6.25 mL of water, 50 μL of ammonium persulfate solution, and 10 μL of TEMED. This should be enough to pour two separating gels leaving enough space for the stacking gel. The gel should be overlayed with isobutanol and allowed to polymerize for 30 min.
3. The top of the gel should be washed with water.
4. Prepare a stacking gel by mixing 1.25 mL of 4X stacking buffer with 0.65 mL of acrylamide/bis solution, 3.05 mL of water, 25 μL of ammonium persulfate solution, and 5 μL of TEMED. Pour the stacking gel, insert the comb, and let polymerize for about 30 minutes.
5. Prepare the running buffer by diluting 250 mL of the 4X running buffer with 750 mL of water.
6. Once the stacking gel has set, remove the comb, wash the wells with running buffer.
7. Add the running buffer to the upper and lower chambers of the gel unit and load 25 μL of each sample to a well. Include one well for prestained molecular weight markers.
8. Complete the assembly of the unit and connect to a power supply. The gel can be run at 100 Volts or 20 mA for about 1–2 h or until the blue dye front reaches the bottom of the gel.

3.17   Western Blot for the Detection of Adducts

1. The samples separated by SDS-PAGE are transferred to nitrocellulose electrophoretically. These directions assume the use of a Bio-Rad Mini-Gel transfer system. A tray of setup buffer is prepared that is large enough to lay out a transfer

cassette with its pieces of foam and with two sheets of 3 MM paper submerged on one side. A sheet of the nitrocellulose is cut just larger than the size of the separating gel is laid on the surface of a separate tray of distilled water to allow the membraned to wet by capillary action. The membrane is then submerged in the setup buffer on top of the 3 MM paper.

2. The SDS-PAGE unit is disconnected from the power supply and disassembled. The stacking gel is removed and discarded and one corner cut from the separating gel to allow its orientation to be tracked. The separating gel is then laid on top of the nitrocellulose membrane.

3. Two further sheets of 3 MM paper are wetted in the setup buffer and carefully laid on top of the gel, ensuring that no bubbles are trapped in the resulting sandwich. The second wet foam sheet is laid on top and the transfer cassette closed.

4. The cassette is placed into the transfer tank such that the nitrocellulose membrane is between the gel and the anode. It is vitally important to ensure this orientation or the proteins will be lost from the gel into the buffer rather than transferred to the nitrocellulose.

5. The lid is put on the tank and the power supply activated. Transfers can be accomplished at either 30 V overnight or 100 V for 1 h.

6. Once the transfer is complete the cassette is taken out of the tank and carefully disassembled, with the top sponge and sheets of 3 MM paper removed. The gel is carefully removed to leave the membrane with the colored molecular weight markers clearly visible.

7. The nitrocellulose is incubated in 50 mL of blocking buffer for 1 h at room temperature on a rocking platform, at which time the blocking buffer is removed and the gel is washed two times with TBS-T.

8. The antibody to the different types of adducts are added at a dilution of 1:1000 in TBS-T/2% BSA for 1 h at room temperature on a rocking platform.

9. The primary antibody is removed and the membrane is washed three times with 50 mL of TBS-T.

10. The secondary antibody is freshly prepared at a 1:10,000 if it is not labeled, 1:20,000 if it is labeled, and added to the membrane for 1 h at room temperature.

11. If a tertiary antibody (labeled) is necessary, it is prepared at a 1:20,000 and added to the membrane for 1 hour at room temperature.

12. This last antibody is discarded and the membrane washed five times for 10 min each with TBS-T. During the final wash, 2-mL aliquots of each portion of the ECL reagent is warmed separately to room temperature and the remaining steps are done in a dark room under safe light conditions. Once the final wash is removed from the blot, the ECL reagents are mixed together and then immediately added to the blot, which is then rotated by hand for 1 minute to ensure even coverage.

13. The blot is removed from the ECL reagents, blotted with Kim-Wipes, and then placed between the leaves of an acetate sheet protector that has been cut to fit into an X-ray film cassette.

14. The acetate containing the membrane is placed in an X-ray film cassette with film for a suitable exposure time, typically a few minutes.

Acknowledgments The authors would like to thank Monte S. Willis, Michael J. Duryee, Karen C. Easterling, Carlos D. Hunter, and Bartlett C. Hamilton III for their valuable technical assistance. Also, we would like to thank all of those who have worked on various aspects of these procedures throughout the years, and are too numerous to mention individually. This work has been supported by The Alcohol Center at the Omaha VA Medical Center, Department of Veterans Affairs; VA Merit Reviews, Department of Veterans Affairs; and NIH/NIAAA grants, R01 AA10435 (to Thiele), R01 AA07818 (to Klassen), R01 AA04691 (to Tuma).

4 Notes

1. The antibodies to NEL or MAA are available upon request from Dr. Thiele. The other antibodies outlined have been used by our group successfully. There are others that are available, and as long as their specificity has been defined, could be substituted. Finally, all antibody concentrations suggested are based on those utilized in our laboratories. Each laboratory will have to determine which concentrations work best for them.
2. TEMED is best stored at room temperature in a desiccator. It rapidly declines in quality (gels take longer to polymerize) once opened and this should be taken in to account when purchasing this product.
3. It is extremely important to note that the addition of MDA at levels higher than 2-3 mM begin to produce MAA on their own. This is most likely due to the break down of MDA to AA. Thus, higher levels of MDA will generate MAA and antibodies to MDA will not detect the MDA adduct. It is also been shown that a number of anti-MDA antibodies react with MAA adducts (inhabitable by hexyl-MAA) as they were prepared using extremely high levels of MDA (> 10 mM in most cases).
4. We try to use a 5-fold molar excess in performing these types of stabilization. The problem is, this is only an estimate as to how much AA is present. Therefore, each laboratory will have to determine these levels for their particular assay.
5. We try to use a 5-fold molar excess in performing these types of stabilization. The problem is that this is only a guess as to how much AA is present in the unknown sample. Therefore, each laboratory will have to determine the appropriate levels for their particular assay. However, it is important to note that the addition of MDA at levels higher than 2-3 mM begin to produce MAA on their own. Thus, higher levels of MDA will generate false positive levels of MAA.

References

1. Tsutsumi, M., Lasker, J. M., Shimizu, M., Rosman, A. S., and Lieber, C. S. (1989) The intralobular distribution of ethanol-inducible P450IIE1 in rat and human liver. *Hepatology*. **10**, 437–446.

2. Niemela, O., Parkkila, S., Pasanen, M., Viitala, K., Villanueva, J. A., and Halsted, C. H. (1999) Induction of cytochrome P450 enzymes and generation of protein-aldehyde adducts are associated with sex-dependent sensitivity to alcohol-induced liver disease in micropigs. *Hepatology*. **30**, 1011–1017.

3. Cederbaum, A. I. (2001) Introduction-serial review: alcohol, oxidative stress and cell injury. *Free Radic. Biol. Med*. **31**, 1524–1516.

4. Tuma, D. J. (2002) Role of malondialdehyde-acetaldehyde adducts in liver injury. *Free Radic. Biol. Med*. **32**, 303–308.

5. Tuma, D. J., and Sorrel, M. F. (1995) The role of acetaldehyde adducts in liver injury. in *Alcoholic Liver Disease: Pathology and Pathogenesis* (Hall, P., ed) pp. 89–99, Edward Arnold, London.

6. Niemela, O. (2001) Distribution of ethanol-induced protein adducts in vivo: Relationship to tissue injury. *Free Radic. Biol. Med*. **31**, 1533–8.

7. Lieber, C. S. (1988) Metabolic effects of ethanol and its interaction with other drugs, hepato-toxic agents, vitamins, and carcinogens: a 1988 update. *Semin. Liver Dis*. **8**, 47–68.

8. Donohue, T. M., Jr., Tuma, D. J., and Sorrell, M. F. (1983) Acetaldehyde adducts with proteins: binding of [14C]acetaldehyde to serum albumin. *Arch. Biochem. Biophys*. **220**, 239–246.

9. Israel, Y., Hurwitz, E., Niemela, O., and Arnon, R. (1986) Monoclonal and polyclonal antibodies against acetaldehyde- containing epitopes in acetaldehyde-protein adducts. *Proc. Natl. Acad. Sci. U.S.A*. **83**, 7923–7927.

10. Stevens, V. J., Fantl, W. J., Newman, C. B., Sims, R. V., Cerami, A., and Peterson, C. M. (1981) Acetaldehyde adducts with hemoglobin. *J. Clin. Invest*. **67**, 361–369.

11. Jennett, R. B., Sorrell, M. F., Saffari-Fard, A., Ockner, J. L., and Tuma, D. J. (1989) Preferential covalent binding of acetaldehyde to the alpha-chain of purified rat liver tubulin. *Hepatology*. **9**, 57–62.

12. Tuma, D. J., Newman, M. R., Donohue, T. M., Jr., and Sorrell, M. F. (1987) Covalent binding of acetaldehyde to proteins: participation of lysine residues. *Alcohol. Clin. Exp. Res*. **11**, 579–584.

13. San George, R. C., and Hoberman, H. D. (1986) Reaction of acetaldehyde with hemoglobin. *J. Biol. Chem*. **261**, 6811–21.

14. Fowles, L. F., Beck, E., Worrall, S., Shanley, B. C., and de Jersey, J. (1996) The formation and stability of imidazolidinone adducts from acetaldehyde and model peptides. A kinetic study with implications for protein modification in alcohol abuse. *Biochem. Pharmacol*. **51**, 1259–67.

15. Esterbauer, H., Schaur, R. J., and Zollner, H. (1991) Chemistry and biochemistry of 4-hydrox-ynonenal, malonaldehyde and related aldehydes. *Free Radic. Biol. Med*. **11**, 81–128.

16. Palinski, W., Yla-Herttuala, S., Rosenfeld, M. E., Butler, S. W., Socher, S. A., Parthasarathy, S., Curtiss, L. K., and Witztum, J. L. (1990) Antisera and monoclonal antibodies specific for epitopes generated during oxidative modification of low density lipoprotein. *Arteriosclerosis* **10**, 325–335.

17. Stadtman, E. R. (1992) Protein oxidation and aging. *Science* **257**, 1220–1224.

18. Steinberg, D., Parthasarathy, S., Carew, T. E., Khoo, J. C., and Witztum, J. L. (1989) Beyond cholesterol. Modifications of low-density lipoprotein that increase its atherogenicity. *N. Engl. J. Med*. **320**, 915–24.

19. Haberland, M. E., Fong, D., and Cheng, L. (1988) Malondialdehyde-altered protein occurs in atheroma of Watanabe heritable hyperlipidemic rabbits. *Science* **241**, 215–218.

20. Niemela, O., Parkkila, S., Yla-Herttuala, S., Halsted, C., Witztum, J. L., Lanca, A., and Israel, Y. (1994) Covalent protein adducts in the liver as a result of ethanol metabolism and lipid peroxidation. *Lab. Invest*. **70**, 537–546.

21. Parkkila, S., Niemela, O., Britton, R. S., Brown, K. E., Yla-Herttuala, S., O'Neill, R., and Bacon, B. R. (1996) Vitamin E decreases hepatic levels of aldehyde-derived peroxidation products in rats with iron overload. *Am. J. Physiol*. **270**, G376–384.

22. Houglum, K., Filip, M., Witztum, J. L., and Chojkier, M. (1990) Malondialdehyde and 4-hydroxynonenal protein adducts in plasma and liver of rats with iron overload. *J. Clin. Invest*. **86**, 1991–1998.

23. Tuma, D. J., Thiele, G. M., Xu, D., Klassen, L. W., and Sorrell, M. F. (1996) Acetaldehyde and malondialdehyde react together to generate distinct protein adducts in the liver during long-term ethanol administration. *Hepatology* **23**, 872–880.

24. Clot, P., Bellomo, G., Tabone, M., Arico, S., and Albano, E. (1995) Detection of antibodies against proteins modified by hydroxyethyl free radicals in patients with alcoholic cirrhosis. *Gastroenterology* **108**, 201–207.

25. Moncada, C., Torres, V., Varghese, G., Albano, E., and Israel, Y. (1994) Ethanol-derived immunoreactive species formed by free radical mechanisms. *Mol. Pharmacol.* **46**, 786–791.

26. Willis, M. S., Klassen, L. W., Tuma, D. J., Sorrell, M. F., and Thiele, G. M. (2002) Adduction of soluble proteins with malondialdehyde-acetaldehyde (MAA) induces antibody production and enhances T-cell proliferation. *Alcohol. Clin. Exp. Res.* **26**, 94–106.

27. Tuma, D. J., Kearley, M. L., Thiele, G. M., Worrall, S., Haver, A., Klassen, L. W., and Sorrell, M. F. (2001) Elucidation of reaction scheme describing malondialdehyde-acetaldehyde-protein adduct formation. *Chem. Res. Toxicol.* **14**, 822–32.

28. Friedman, M. (2004) Applications of the ninhydrin reaction for analysis of amino acids, peptides, and proteins to agricultural and biomedical sciences. *J. Agric. Food Chem.* **52**, 385–406.

29. Thiele, G. M., Tuma, D. J., J.A., M., Wegter, K. M., McDonald, T. L., and Klassen, L. W. (1998) Monoclonal and polyclonal antibodies recognizing acetaldehyde-protein adducts. *Biochem. Pharmacol.* **56**, 1515–1523.

30. Habeeb, A. F. (1966) Determination of free amino groups in proteins by trinitrobenzenesulfonic acid. *Anal. Biochem.* **14**, 328–336.

31. Thiele, G. M., Miller, C., Miller, J. A., McDonald, T. L., Tuma, D. J., and Klassen, L. W. (1996) Polyclonal and Monoclonal antibodies to an epitope made by the combination of malondialdehyde and acetaldehyde. *Clin. Exp. Res. Suppl.* **20**, 119A.

32. Laemmli, U. K. (1970) Cleavage of structural proteins during the assembly of the head of bacteriophage T4. *Nature* **227**, 680–685.

33. Groves, W. E., Davis, F. C., Jr., and Sells, B. H. (1968) Spectrophotometric determination of microgram quantities of protein without nucleic acid interference. *Anal. Biochem.* **22**, 195–210.

34. Cominacini, L., Garbin, U., De Santis, A., Campagnola, M., Davoli, A., Pasini, A. F., Faccini, G., Pasqualini, E., Bertozzo, L., Micciolo, R., et al. (1996) Mechanisms involved in the in vitro modification of low density lipoprotein by human umbilical vein endothelial cells and copper ions. *J. Lipid Media.t Cell Signal.* **13**, 19–33.

35. Tuma, D. J., and Sorrell, M. F. (1985) Covalent binding of acetaldehyde to hepatic proteins: role in alcoholic liver injury. *Prog. Clin. Biol. Res.* **183**, 3–17.

36. Xu, D., Thiele, G. M., Kearley, M. L., Haugen, M. D., Klassen, L. W., Sorrell, M. F., and Tuma, D. J. (1997) Epitope characterization of malondialdehyde-acetaldehyde adducts using an enzyme-linked immunosorbent assay. *Chem. Res. Toxicol.* **10**, 978–986.

37. Xu, D., Thiele, G. M., Beckenhauer, J. L., Klassen, L. W., Sorrell, M. F., and Tuma, D. J. (1998) Detection of circulating antibodies to malondialdehyde-acetaldehyde adducts in ethanol-fed rats. *Gastroenterology* **115**, 686–692.

38. Gupta, R. K., and Siber, G. R. (1995) Method for quantitation of IgG subclass antibodies in mouse serum by enzyme-linked immunosorbent assay. *J. Immunol. Methods* **181**, 75–81.

39. Mattila, P. S. (1985) Quantitation of antibody isotypes in solid-phase assays. Comparison of myeloma protein and monoisotypic antibody standards. *J. Immunol. Methods* **83**, 43–53.

40. Esparza, I., and Kissel, T. (1992) Parameters affecting the immunogenicity of microencapsulated tetanus toxoid. *Vaccine* **10**, 714–720.

41. German-Fattal, M., Bizzini, B., and German, A. (1987) Immunity to tetanus: tetanus antitoxin and anti-BIIb in human sera. *J. Biol. Stand.* **15**, 223–230.

42. Willis, M. S., Klassen, L. W., Tuma, D. J., and Thiele, G. M. (1997) Different levels of protein adduction by alcohol metabolites induce antibody and T cell responses specific to the carrier protein in a dose response manner. *Hepatology* **26**(4, Pt. 2), 254A.

43. Willis, M. S., Klassen, L. W., Tuma, D. J., and Thiele, G. M. (1998) T cell immunogenicity induced by malondialdehyde-acetaldehyde protein adduction without the use of adjuvant. *Hepatology* **28**, 639A-639A.

44. Kohler, G., and Milstein, C. (1975) Continuous cultures of fused cells secreting antibody of predefined specificity. *Nature* **256**, 495–497.

18
Assessment of Natural Killer (NK) and NKT Cells in Murine Spleens and Livers

Michael R. Shey and Zuhair K. Ballas

Summary Natural killer (NK) cells are part of innate immunity. NK cells have been assigned numerous functions, including the ability to serve as a bridge between innate and adaptive immunity. In evaluating NK cell function, two pathways need to be examined: their ability to kill certain tumors spontaneously and their ability to secrete cytokines, interferon-gamma (IFN-γ), in particular. Although NK cells are distinct from T lymphocytes, a new lymphocyte subset, termed NKT cell, has been described. NKT cells express surface markers that are unique to NK cells (e.g., NK1.1) as well as markers that are unique to T cells (e.g., CD3). Most NKT cells recognize glycolipids and are thought to play an important immunoregulatory role. This chapter will detail the methodology needed for examination of NK and NKT cells in mice.

Keywords Natural killer; NK; NKT; Chromium-release assay; Cytotoxicity; Flow cytometry; cytoplasmic staining; cytokine secretion; ADCC; reverse-ADCC (R-ADCC).

L. E. Nagy (ed.), *Alcohol: Methods and Protocols*
© Humana Press 2008

259

1   Introduction

Chronic alcohol consumption is associated with several abnormalities in the immune system, including increased frequency and severity of certain infections *(1)*. In addition, chronic alcohol consumption is associated with an increased risk for cancers of various organs *(2,3)*. Natural killer (NK) cells are immune surveillance cells that inhibit the development and growth of certain tumors *(4–7)*. A closely related lymphocyte subset, the so-called NKT cell, is also believed to serve a regulatory function in various diseases *(8)*. There is a large body of literature examining the effect of alcohol on NK cell function. There is no consensus on the effect of alcohol on NK cells, and the relationship seems to differ depending on whether one examines acute or chronic alcohol consumption and whether alcoholic liver disease is present *(9–12)*. There is very little literature examining the effect of chronic alcohol consumption on NKT cells. This chapter is intended to detail the methodology used to examine splenic and hepatic NK and NKT cells in mice and is not intended as a comprehensive review of these two cell subsets.

NK cells are part of the innate immune system and thus do not require any specific sensitization for the expression of their function *(4–7)*. NK cells are lymphocytes that do not belong to the T- or B-cell lineages; they are often referred to as large granular lymphocytes, although not all NK cells are large. NK cells have several functions, most of which can be assigned to their ability to secrete cytokines or their ability to exert cytotoxicity *(4–7)*. NK cells are a major source of interferon gamma (IFN-γ). NK cells are also able to kill certain tumor target cells directly with no need for prior sensitization. Indeed, it is this killing ability that earned NK cells their name *(4–7)*.

NKT cells are a newly defined lymphocyte subset that shares characteristics of NK cells and T cells. NKT cells express certain markers unique to NK cells but also express some T-cell markers, including CD3 and a restricted pattern of TCR expression. NKT cells also are a major source of cytokines, including IFN-γ, and IL-4. NKT cells, however, are quite heterogeneous. Some NKT cells subsets are thymically derived whereas others arise from an extrathymic pathway. Most, although not all, NKT cell subsets recognize glycolipids and are thus unique among the various lymphocyte lineages *(8,13)*.

A recently described cell, termed interferon-producing killer dendritic cells (IKDCs) *(14)*, shares phenotypic and functional qualities with both NK cells and DCs and can be identified by surface expression of both CD11c and CD49b. IKDCs resemble plasmacytoid DCs in that they express both B220 and moderate levels of MHC II but may be differentiated from this DC subset by their lack of Ly-6g. These cells are beyond the scope of this chapter and will not be covered.

This chapter will detail the methodology for the isolation of NK and NKT cells from murine spleen and liver as well as the methodology employed to examine their number and function. Examination of NK and NKT cells can be readily and easily accomplished if one pays attention to some of the pitfalls detailed in this chapter. We hope that the methods described below will simplify the task of investigators seeking to examine NK and NKT cells both of which are proving to be of vital importance in health and disease.

2 Materials

2.1 Purification of NK and NKT From Murine Spleens and Livers

1. Minimal Essential Media (Gibco cat. no. 11095-080), Hank's Balanced Salt Solution (Gibco cat. no. 14175-095), and RPMI 1640 (Gibco cat. no. 11875-093). RPMI 1640 is supplemented (cRPMI-5) with penicillin-streptomycin (final concentration 100 U/mL and 100 μg/mL, respectively), L-glutamine (final concentration 2 mM), gentamicin (final concentration 50 μg/mL), 2-mercaptoethanol (2-ME) (final concentration 5×10^{-5} M), and 5% heat-inactivated fetal calf serum. The above ingredients are filter-sterilized (0.2 μm) before use, stored at 4°C, and used within 30 d.
2. 60 × 15-mm sterile tissue culture dishes (Falcon #35-3002).
3. Tissue grinder (Thomas Scientific cat. no. 3431-D70).
4. 100-μm cell strainer (Falcon #352360).
5. Barrel from a 10-mL syringe (BD cat. no. 309604).
6. 0.83% NH$_4$Cl in 1 mM Trizma Base (RBC lysis buffer).
7. 40% Percoll (Sigma cat. no. 4937) made isotonic via addition of 10x PBS containing 100 U/mL heparin. For optimal separation of murine lymphocytes, osmolality should be adjusted to 330 mOsm/Kg H$_2$O *(15)*.

2.2 Flow Cytometry Examination of Surface and Cytoplasmic Markers

1. Single-cell suspension of splenocytes or hepatic mononuclear cells (as described below).
2. Staining buffer: Hank's Balanced Salt Solution (Gibco cat. no. 14175-095) with 1% bovine serum albumin (Sigma cat. no. 7030) and 0.1% sodium azide. pH 7.2–7.4.
3. 96-well round-bottom tissue culture plate (Corning Costar #3799).
4. 24-well flat-bottom tissue culture plate (Corning Costar #3524).
5. 5 mL polystyrene 12 × 75-mm tubes (Falcon #352054).
6. Monoclonal antibodies 2.4G2 (α-CD16/32) unlabeled, prepared from hybridoma culture supernatant (ATCC cat. no. HB-197), PK136 (α-NK1.1), PE and APC (BD Biosciences cat. no. 553165 and 550627), DX5 (α-CD49b) APC (eBioscience cat. no. 17-5971), XMG1.2 (α-IFN-γ) PE (BD Biosciences cat. no. 554412), and C363.29b (α-CD3ε) FITC (Beckman Coulter cat. no. 73191).
7. Murine CD1d-PBS57 (α-GalCer analog) APC (NIH Tetramer Facility, Atlanta, GA).
8. Phorbol 12-myristate 13-acetate (Sigma; cat. no. P8139) and Ionomycin (Sigma; cat. no. I3909).
9. BD Cytofix/Cytoperm Plus Golgi Stop (BD Biosciences cat. no. 554715).
10. LincoPlex Mouse Cytokine Panel (Linco Research cat. no. MCYTO-70K-PMX16).

2.3   Cytotoxicity Assays

1. RPMI 1640 (Gibco cat. no. 11875-093) supplemented with penicillin-streptomycin (final concentration 100 U/mL and 100 µg/mL, respectively) and 5% heat-inactivated calf serum (Assay Medium)
2. 96 well Styrene V-bottom plates (Fisher Scientific cat. no. 14-245-72).
3. 2% acetic acid Chromium-51 Sodium Chromate radionuclide (^{51}Cr; Perkin Elmer cat. no. NEZ03001).

3   Methods

3.1   Purification of NK and NKT From Spleen and Liver

The following should be performed in a Tissue Culture laminar flow hood using aseptic technique.

3.1.1   Isolation of Splenocytes

1. Remove spleens and place in 10 mL of MEM.
2. Gently grind spleens with homogenizer and centrifuge cell suspension for 5 min at 200 g to pellet the cells.
3. Remove supernatant and resuspend cells in 2 mL of RBC lysis buffer for 45 s to lyse red blood cells. Immediately bring volume to 10 mL with cRPMI-5 and flash spin to remove aggregates.
4. Centrifuge cell suspension for 5 min at 200 g and resuspend cells in 10 mL of cRPMI-5 to wash the cell pellet.
5. Centrifuge cell suspension for 5 min at 200 g and resuspend cells in 10 mL of cRPMI-5 for counting.

3.1.2   Isolation of Hepatic Mononuclear Cells

1. Perfuse liver via the portal vein with 10 mL of cold HBSS. Alternatively, if serum or PBMC are desired, thorough bleeding of the mouse prior to sacrifice will adequately remove peripheral blood from the organ (*16*) (*see* **Note 1**).
2. Mince liver and press through a 100-µm cell strainer using the barrel of a 10-mL syringe.
3. Centrifuge cell suspension for 5 min at 200 g and resuspend cells in 20 mL of HBSS to wash the cell pellet.
4. Repeat wash then centrifuge cell suspension at 30 g for 5 min to pellet hepatocytes.

5. Transfer supernatant to a new 50-mL conical tube and centrifuge for 5 min at 200 g.
6. Resuspend cell pellet in 20 mL of 40% isotonic Percoll containing 100 U/mL heparin. Centrifuge for 20 min at 800 g.
7. Wash the cell pellet with HBSS and proceed with RBC lysis as detailed in the splenocyte isolation protocol.

Expected recoveries vary as a function of age, strain and immune status of the mice used, but typical yields range from $7–10 \times 10^7$ splenocytes and $2–5 \times 10^6$ hepatic mononuclear cells per mouse. With proper technique cell viabilities should exceed 90%.

3.1.3 Enrichment of NK and NKT Cells From Splenocytes and Hepatic Mononuclear Cells

Splenic NK and NKT are relatively rare populations comprising roughly 3–10% and 1–2% of lymphocytes in normal animals, respectively. Although their frequency is increased in preparations of hepatic mononuclear cells (10–15% NK and 15–30% NKT), enrichment of these cells is often required for functional analyses. The enrichment of NK cells has been facilitated by the availability of antibodies coupled to magnetic particles available from a variety of sources including Miltenyi Biotec (Auburn, CA) and Stem Cell Technologies (Vancouver, BC). Two general strategies are employed, each offering distinct advantages. Positive selection using a magnetic particle-coupled monoclonal antibody directed against a cell-surface marker specifically expressed by the cell type of interest (in this case CD49b or NK1.1) usually offers greater purity and yield. The concern with positive selection is that engagement of a surface marker might alter the activation status of the cells. Negative selection, in which a cocktail of monoclonal antibodies are used to deplete unwanted cell types (B, T, DC and macrophages; *see* **Note 2**), will leave the investigator with un-manipulated NK cells. Depending upon the antibodies used, NKT cells may be the major contaminant in NK-cell preparations. One can use the differential expression of CD5 to further separate these two populations. Using CD5 rather than CD3 will avoid potential activation of NKT cells via crosslinking of the CD3 complex. In situations where highly purified (>99%) populations are desired (e.g., RT-PCR or western blot), sorting by flow cytometry is the method of choice. One can maximize cell recovery, and minimize the time needed on the sorter, by pre-enriching the cells of import by paramagnetic bead selection (*see* **Note 3**).

3.2 Surface and Cytoplasmic Markers for NK and NKT Cells

3.2.1 Flow Cytometry Staining for Cell-Surface Molecules

1. Dilute cells to 1×10^7/mL with staining buffer and add 100 µL to desired wells of 96-well tissue culture plate.

2. Add 100 μL of staining buffer to each well, mix gently, and centrifuge at 200 g for 30 s to pellet the cells.
3. Remove supernatant by quickly inverting and flicking plate.
4. Add 10 μL of 2.4G2 [1 mg/mL] to block Fc receptor-mediated binding and 20 μL of each fluorochrome-conjugated antibody. Optimal concentrations for staining should be empirically determined. Incubate 10 mins at room temperature. If CD1d tetramer staining is being performed, increase incubation to 30 min.
5. Wash cells twice with staining buffer and resuspend in 200 μL of staining buffer. Transfer to 12 × 75 tubes and bring final volume to 500 μL with staining buffer. Analyze immediately on flow cytometer. Detector voltages should be set with unstained cells and compensation values set with cells stained with a single fluorochrome.

3.2.2   Intracellular Staining for IFN-γ

1. In a 24-well flat-bottom plate, add 5×10^6 cells to 2 mL of cRPMI-5. Add Golgi Stop (monensin) to a final concentration of 2 μM. Incubate alone (for unstimulated control) or with PMA [10 ng/mL] and ionomycin [200 ng/mL] for 5 h at 37°C in a 5% CO_2-humidified incubator. Another alternative for stimulation is to use YAC-1, the standard mouse NK target cell, for stimulation. Lymphocytes should be cultured with YAC-1 at a ratio of 10:1.
2. Harvest cells, wash and count then proceed with cell surface staining as detailed previously, followed by intracellular staining as described in the BD Cytofix/Cytoperm Kit manual.
3. Flow cytometry has proved to be an invaluable tool for dissection of the immune system and NK and NKT cell research is no exception. The choice of fluorochromes for particular applications will vary depending upon the instrumentation available to the investigator, but the methods are routine and conjugated monoclonal antibodies are widely available from a variety of vendors.

NK and NKT cells can be identified by their expression of several surface and cytoplasmic markers (see **Note 4**). The principal marker for NK and NKT cells is NK1.1 (CD161), which is recognized by the monoclonal antibody PK136. The expression of this antigen is restricted to certain mouse strains such as C57BL/6, SJL, and NZB. In all strains, CD49b (mAb DX5) may be used to identify NK cells with the caveat that it is not expressed by all NK cells (17). As shown in Figure 1, when one examines NK1.1 vs. DX5 staining in an NK1.1+ strain (C57BL/6 in this case), there are some NK1.1+ cells that do not express DX5.

To differentiate between NK and NKT cells during flow cytometric analysis, one should stain with both NK1.1 and CD3 mAbs so that these populations can be gated upon for subsequent evaluation of the expression of other markers of interest. A host of other molecules are associated with NK cell function, including the Ly-49 and NKG2D receptors. Like other lymphocytes, NK and NKT cells constitutively

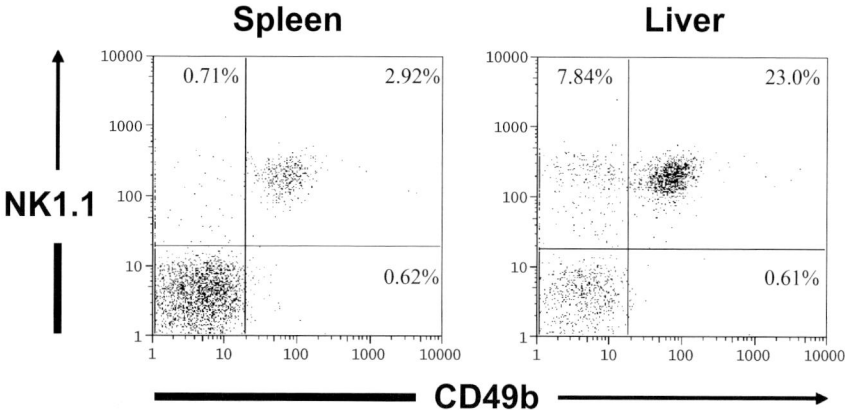

Fig. 1 NK1.1 Vs DX5 expression on NK cells. Splenocytes and hepatic mononuclear cells were isolated from C57BL/6 mice and stained with mAbs specific for CD3, NK1.1, and CD49b (DX5). Lymphocytes were gated on CD3(–) events to exclude NKT cells. The majority of NK cells coexpress NK1.1 and CD49b, but there is a significant proportion which is negative for expression of CD49b, particularly in the liver

express a variety of adhesion molecules (CD11a, CD11b and, in a subset, CD11c) and cytokine receptors (such as the beta and gamma chains of the IL-2 receptor, CD122, and CD132, respectively). NK cells also contain intracellular molecules which are essential for their cytotoxic function (perforin and granzymes). Both perforin and granzymes can be identified by cytoplasmic staining coupled with surface marker staining and flow cytometry analysis *(4–7)*.

By definition, NK cells are CD3- whereas NKT cells are CD3+ (Fig. 2). Also shown in Fig. 2 is that the percentage of NK and NKT cells is much greater in the liver than the spleen, although the absolute numbers may not necessarily be so. NK cells do not express T-cell receptors whereas NKT cells have a restricted expression of certain TCR haplotypes. NKT cells are not homogeneous and may be broadly classified into four groups based upon phenotypic and functional criteria *(13)*. Classical (or Type I) NKT cells are the predominant subset found in murine liver and express an invariant TCR alpha chain (Vα14-Jα18). They are restricted by the CD1d glycolipid antigen presenting molecule and are highly reactive with the lipid antigen α-galactosylceramide (α-GalCer), which is derived from the marine sponge *Agelas mauritanius*. CD1d tetramer reagents complexed with this molecule (obtained from the NIH Tetramer Facility-Emory University Vaccine Center, Atlanta, GA) are particularly useful for the identification of this subset. Staining with this tetramer is shown in Fig. 3. Of note in Fig. 3 is the subset that binds the tetramer but does not express NK1.1 on its surface. The exact nature of this subset is not known but may reflect an NKT cell that either down-modulated or never expressed NK1.1 *(8,18,19)*.

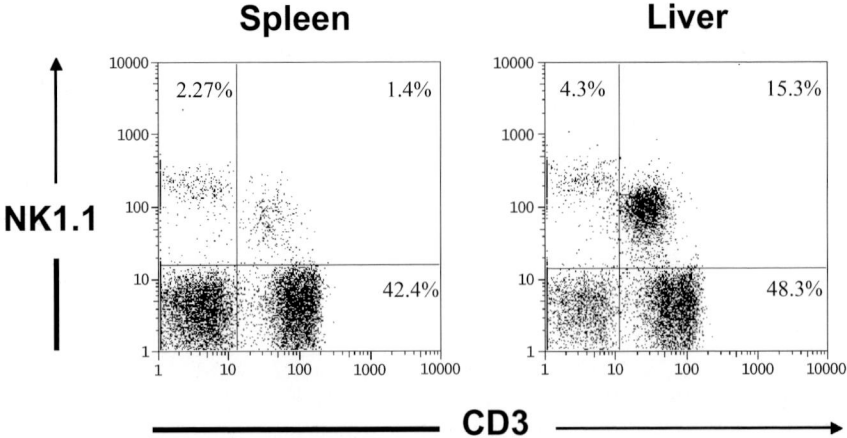

Fig. 2 NK1.1 Vs CD3 Expression on spleen and liver mononuclear cells. Splenocytes and hepatic mononuclear cells were isolated from C57BL/6 mice and stained with mAbs specific for CD3 and NK1.1. Double positive NKT cells are detectable in both organs, but are greatly enriched in the liver

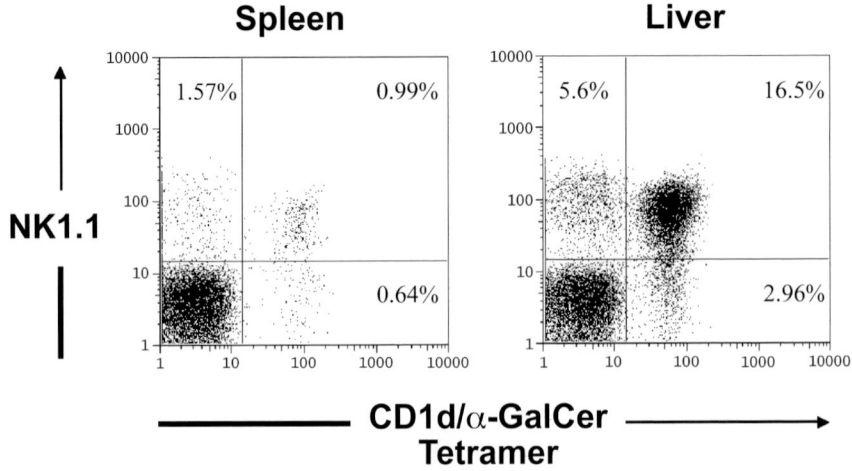

Fig. 3 NKT cells bind CD1d/α-GalCer tetramers. Splenocytes and hepatic mononuclear cells were isolated from C57BL/6 mice and stained with mAbs specific for CD3, NK1.1 and the CD1d/α-GalCer tetramer. Lymphocytes were gated on CD3(+) cells. Note the significant number of tetramer positive NK1.1 negative cells suggesting that some NKT cells may down-modulate or not express NK1.1 on their surface

3.3 Functional Assessment of NK and NKT Cells

NK cells are so named for their ability to lyse susceptible target cells with no need for stimulation. The prototypical target for murine NK cells is the YAC-1 lymphoma. While NK cells can kill directly ex vivo, they may also be stimulated to higher lytic activity through short term in vitro culture with a variety of stimuli including cytokines and certain CpG oligodeoxynucleotides (ODN). Long-term culture of murine splenocytes(>3-5 days) with Interleukin-2 results in the induction of lymphokine-activated killer (LAK) cells, which demonstrate high lytic activity against NK-sensitive targets and also acquire the ability to lyse NK-resistant targets such as EL4 lymphoma and P815 mastocytoma. LAK are a heterogeneous population derived from both NK and T lymphocyte precursors (15,20–22).

NK cells express Fc-γRIIIA and mediate antibody-dependant cellular cytotoxicity (ADCC) by binding the Fc portion of an IgG molecule whose F(ab)$'_2$ region is bound to its specific epitope on a target cell. A variation upon this theme is redirected-ADCC (R-ADCC) in which the antibody's F(ab)$'_2$ region binds a cell-surface molecule on NK or NKT cells whereas its Fc portion binds FC receptors on the target cell. Since NKT cells express CD3 on their surface, R-ADCC using the mAb 145-2C11 specific for the CD3ϵ chain has been successfully employed to analyze NKT cell-mediated lytic activity (23–25).

Environmental factors impact endogenous levels of fresh NK activity and mice housed in barrier facilities typically have lower levels of splenic NK activity (26). In situations in which endogenous NK activity is low, it may be advantageous to augment responses through the enrichment of NK cells by depletion of B cells or the addition of stimulatory cytokines such as IL-2 to the 4 hr [51]Chromium (^{51}Cr)-release assay (see Note 2).

3.3.1   [51]Cr Labeling of Target Cells

1. Wash target cells twice in assay media centrifuging for 5 min at 200 g after each wash.
2. Aspirate supernatant after the second wash and resuspend cells in 50 µL of assay media.
3. Add 200 µL of [51]Cr (1 millicurie/mL).
4. Incubate at 37°C and 5% CO_2 -humidifier-air for 30 min.
5. Wash three times in assay media and resuspend at 1×10^5/mL after the final wash. Target cells will be plated in 100 µL of volume, thus giving 1×10^4 target cells/well.

3.3.2   Preparation of Effector Cells

1. Wash effector cells in assay media and count cells.
2. Resuspend cells in the appropriate volume for the desired effector: target (E:T) ratio (e.g. 1×10^7/mL for a 100:1 ratio; the targets being at a constant 1×10^4 cells/well).

3.3.3   Plating (Typically Performed in Duplicates)

1. Add 100 µL of assay media to wells in which a 2-fold serial dilution of effectors will be performed.

2. Add 200 μL of effectors to the wells which will contain the highest E:T ratio and perform serial dilution leaving 100 μL per well.
3. Add 100 μL of target cells per well for a final volume of 200 μL per well. Determination of maximal and spontaneous release is performed by adding 100 μL of target cells to 2% acetic acid or assay media, respectively.
4. Centrifuge at 300 g for 30 s to facilitate cell-cell contact.
5. Incubate 4–6 h at 37°C in 5% CO_2-humidified-air atmosphere.
6. After incubation, centrifuge plate at 300 g. Carefully harvest 100 μL of supernatant and transfer to borosilicate centrifuge tubes and measure radioactivity (expressed as counts-per-minute, CPM).

3.3.4 Data Analysis

Percent specific lysis is calculated using the following formula:

$$Percent\ Specific\ Lysis = [(experimental\ CPM-spontaneous\ CPM)]/[(maximum\ release\ CPM-spontaneous\ CPM)] \times 100$$

The calculation of lytic units (LUs) is a convenient method to quantify lytic activity and is useful for comparison of the lytic activity among various populations. The method for calculation of LU involves plotting the percentage of specific lysis against the log_{10} of the various E:T ratios. The linear portion of the log_{10} [E:T] versus specific lysis curve is selected and the data entered into a computer for calculation by a standard linear regression program. The E:T ratio required to effect a given percent specific lysis (commonly 30%) of the targets is obtained from the regression line. Because the number of targets is a constant, one can calculate from this ratio the number of effectors needed to give 30% specific lysis; this is defined as one LU. It is conventional to present the data as the number of LU per million effector cells.

An example of a ^{51}Cr-release assay is shown in Fig. 4. Unfractionated splenocytes have a detectable but low cytotoxic activity; this activity is increased upon depletion of B cells. The activity can be further amplified by the addition of IL-2 directly to the wells of the microtiter plate. IL-2 in this short-term stimulation enhances the killing per NK cell but does not expand the spectrum of susceptible target cells. Another approach useful in the assessment of NK cell function and responsiveness is to culture B cell-depleted splenocytes overnight either with IL-2 or stimulatory CpG ODN (Fig. 5). NK cells lose their cytotoxic activity in overnight culture unless

Fig. 5 Short-term stimulation of NK cell activity by IL-2 and CpG. Splenocytes were isolated from naïve C57BL/6 mice and cultured overnight with rIL-2 (200 u/mL), control CpG oligodeoxynucleotide 2243 (10 μg/mL) or activating CpG oligodeoxynucleotide 2336 (10 μg/mL). The cultured cells were harvested, counted, and used as effectors in a standard 4-h ^{51}Cr-release assay to determine lytic activity. Splenocytes cultured in the absence of stimulation rapidly lose endogenous NK activity, while the addition of either IL-2 or activating CpG enhance lytic activity against YAC-1 tumor targets

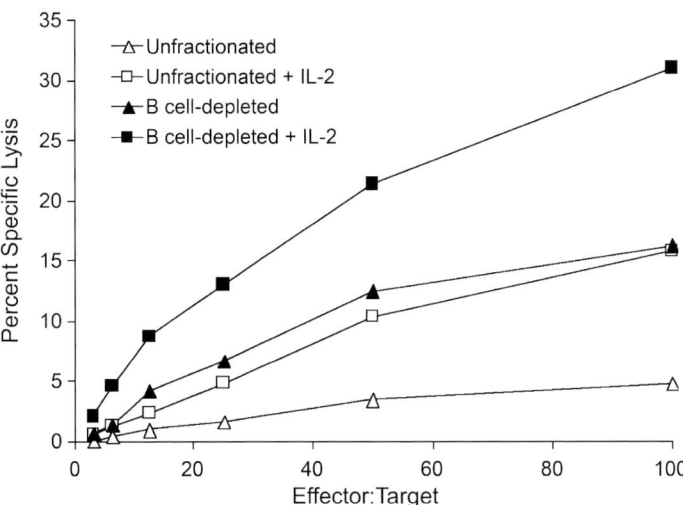

Fig. 4 Cytotoxic activity of freshly obtained unfractionated and B cell-depleted splenocytes. Splenocytes were isolated from naïve C57BL/6 mice and either used unfractionated or depleted of B cells using magnetic beads coated with goat-anti-mouse IgM and goat-anti-mouse IgG. A standard ^{51}Cr-release assay was performed with rIL-2 [1000 U/mL] being added, where indicated, immediately prior to the addition of radiolabeled YAC-1 target cells

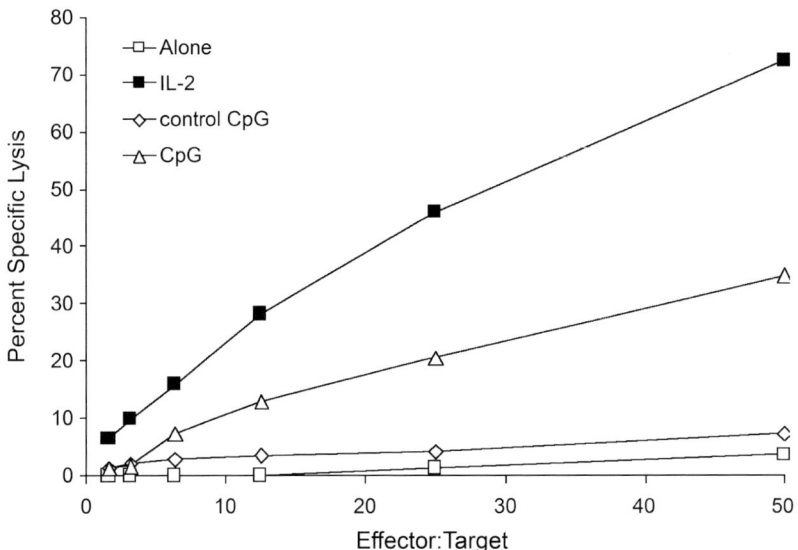

stimulated. IL-2 stimulates NK cells directly while CpG ODN stimulated DC which then secrete cytokines that stimulate NK cell activity *(21,22)*.

3.4   Cytokine Production

The production of various cytokines, particularly IFN-γ and IL-4, is another important function of NK and NKT cells *(4–8,27)*. Various methods can be employed to analyze cytokine secretion including intracellular cytokine staining (Section 3.2.2) and the enzyme-linked immunosorbent assay (ELISA). Intracellular cytokine staining is a flow cytometry-based assay and allows the investigator to assess the production of cytokines by particular cells as defined by simultaneous staining of the cell surface for particular markers (such as NK1.1 and CD3). The analysis of intracellular cytokines by flow cytometry relies upon the blockade of protein transport by inhibitors of the Golgi apparatus such as monensin and brefeldin A. These inhibitors are included in short-term cultures with various NK and NKT agonists such as PMA/Ionomycin, anti-CD3 mAbs and α-GalCer. After 4–6 h of culture, cells are stained for cell surface markers then fixed and permeabilized for staining of intracellular markers. Flow cytometry is then performed to analyze expression of the cytokine of interest by particular cell types. Careful consideration should be given to the selection of fluorochromes for intracellular cytokine staining. Investigators will often select the brightest fluorochromes such as phycoerythrin (PE) and allophycocyanin (APC) for the anticytokine mAb because the intensity may be low and the cytokine-producing cells infrequent. Additionally, the fixation/permeabilization process may degrade the intensity of the surface staining hence the use of bright mAb conjugates is essential. An example of surface and cytoplasmic staining of NK cells is shown in Fig. 6, where one can clearly see that PMA/Ionomycin stimulation induces the cytoplasmic expression of IFN-γ, which is restricted to NK1.1+ cells.

Another approach to examine cytokine secretion by NK and NKT cells is to measure cytokine content in the culture supernatants of stimulated cells. ELISA yields quantitative information about the amount of cytokine present in culture supernatants as well as in sera and other biological fluids. ELISA is a standard assay in many immunological studies and commercial kits with detailed instructions are readily available. It relies upon the availability of a pair of mAbs directed against distinct epitopes on the analyte of interest. The capture mAb is bound to the surface of a microtiter well, the sample to be analyzed is added to the wells and the presence of the analyte is detected with the second (detection) mAb which is usually labeled with a dye that can be assessed objectively. A recent modification in this methodology is the multiplex suspension array technology *(28)* in which capture mAbs are linked to microparticles that are complexed with varying amounts of fluorescent dyes. After incubation with the sample and a cocktail of detection antibodies, quantitation of numerous analytes can be performed in a manner analogous to flow cytometry *(28)*. This method is particularly useful when limited quantities of the sample are available.

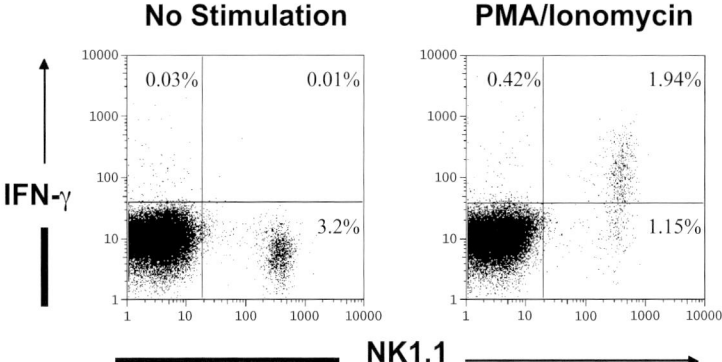

Fig. 6 Cytoplasmic IFN-γ expression in NK cells. Splenocytes from C57BL/6 mice were cultured in the presence of monensin (2 μ*M*) either without stimulation or with PMA (10 ng/mL) and ionomycin (200 ng/mL) for 5 h. They were then harvested and surface-stained for CD3 and NK1.1. Following fixation and permeabilization the cells were stained for intracellular IFN-γ. Lymphocytes were gated on CD3(-) events to exclude NKT cells

3.5   *Stimulation of NKT Cells by* α-*GalCer*

Induction of hepatic NKT cells in vivo can be achieved through injection of α-GalCer (Axxora cat. no. ALX-306-027) either intravenously or intraperitoneally. One can then perform intracellular cytokine staining, within 2–3 h, for IFN-γ to identify activated hepatic NKT cells. A host of cytokines are also readily detectable in the serum of treated animals *(29)*. The effects of NKT activation include secondary stimulation (by virtue of IFN-γ secretion) of NK lytic activity which is detectable within 12–18 h following α-GalCer injection *(30)*. α-GalCer-reactive NKT cells increase as a function of age and systemic treatment of older mice with this compound can result in significant liver pathology and death *(31,32)*. Within 24 h after the administration of α-GalCer, NKT cells may no longer be detectable in the liver, a phenomenon originally attributed to activation-induced cell death. It has been suggested, however, that the inability to detect these cells is the result of internalization of the TCR/CD3 complex as well as down-modulation of NK1.1 *(33)*. Additionally, it has been suggested that subsequent to a single dose of α-GalCer, NKT cells proliferate rapidly then enter a state of long-term hyporesponsiveness with many of the hallmarks of anergy *(34,35)*.

Stimulation of NKT cells by α-GalCer can also be done in vitro. Such stimulation enhances NKT lytic activity and induces them to secrete several cytokines. As shown in Table 1, stimulation of splenic and hepatic mononuclear cells by overnight culture with α-GalCer induces the secretion of significant amounts of IFN-γ, MIP-1α, and IP-10. Interestingly, hepatic, but not splenic, NKT cells also secrete significant amounts of IL-4 and IL-6. In addition to cytokine secretion, α-GalCer enhances the cytotoxic activity of NKT cells. NKT cells do not kill YAC-1 (the classical NK-susceptible target cell) or P815 (an NK-resistant mastocytoma). However, NKT can mediate R-ADCC by using anti-CD3 mAb (which binds to CD3 on the surface of NKT cells) and an Fc-γ-expressing target cells (P815 works well for this assay). As shown in Fig. 7, overnight stimulation of hepatic mononuclear cells with α-GalCer markedly enhances the R-ADCC activity of NK cells.

Table 1 Cytokine concentration (pg/mL)

Culture conditions	IL-4	IL-6	IFN-γ	MIP-1α	IP-10
Splenocytes					
Vehicle	0.6	18.0	9.2	188.3	28.2
α-GalCer	35.2	18.0	19,284.0	5,271.6	412.8
Hepatocytes					
Vehicle	3.3	157.4	172.2	37.1	113.5
α-GalCer	3,818.2	1,560.6	22,978.6	9,940.6	616.6

Splenocytes and hepatic mononuclear cells were isolated from naïve C57BL/6 mice and cultured with vehicle (DMSO) or α-GalCer (100 ng/mL) overnight. Supernatants were harvested and analyzed by multiplex cytokine bead array (Linco Research, St. Charles, MO). Data are expressed as pg/mL.

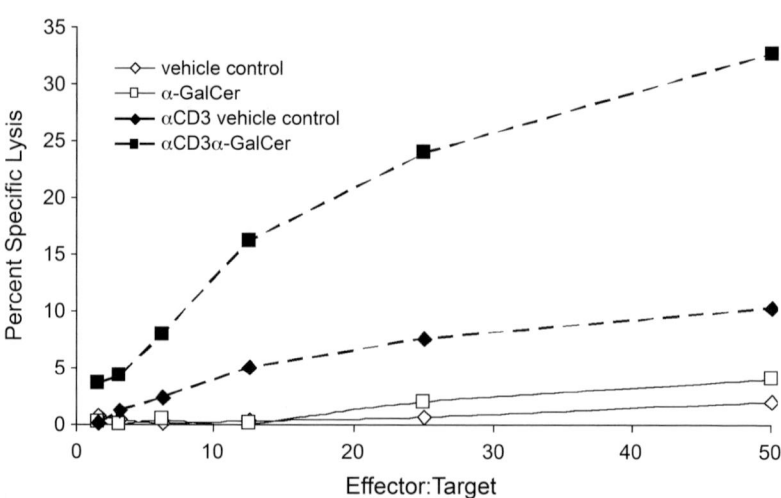

Overnight Culture of Hepatic MNC

Fig. 7 R-ADCC activity of NKT cells stimulated, in vitro, with α-GalCer. Hepatic mononuclear cells were isolated from naïve C57BL/6 mice and cultured with vehicle (DMSO) or α-GalCer [100 ng/mL] overnight. The cultured cells were harvested, counted and used as effectors in a 4-h ^{51}Cr-release assay. The targets were the NK-resistant P815 cells. To determine R-ADCC activity, some wells received 2 μg/well of the α-CD3 mAb 145-2C11 (represented by the dashed lines). There is little killing of P815 targets without the addition of α-CD3. In vitro culture with α-GalCer greatly enhances R-ADCC activity

Acknowledgments   This work was supported by VA Merit Review and by Grant AA014418 from the National Institutes of Health.

4   Notes

1. In general, the methodology for the examination of splenic NK and NKT cell is straightforward. Hepatic mononuclear cells, on other hand, may pose a challenge as to the technique and recovery. It is imperative that the liver circulatory system be flushed in order to get rid of circulating lymphocytes that might otherwise dilute hepatic lymphocytes and NKT cells. We found that flushing through the portal vein works best, although it takes some practice. We found it helpful, when teaching the technique, to search for images of the murine hepatic circulation on the Internet so that one can have a schematic as to where the portal vein is. However, even with the best technique, recovery of hepatic lymphocytes is meager. If deemed acceptable by an experimental protocol, one may have to pool the lymphocytes from several livers for adequate numbers.

2. NK cells are part of the innate immune system and are dependent on several cytokines, with IL-15 playing a central role, for their maturation and proliferation. However, mice reared in gnotobiotic environments may have NK cells as determined by surface markers but these cells have little or no cytotoxicity. Upon moving mice to a less-sterile environment, NK cells assert their presence rather quickly. Indeed, a very high spontaneous NK-cell cytotoxic activity should prompt one to worry about a possible infection in the mouse colony. Although this creates a problem when one wants to examine the cytotoxic activity of freshly obtained NK cells, one can work around it by depleting B cells (either by plate adherence or using paramagnetic particles coupled with an anti-mouse immunoglobulin reagent). If that is not enough, one can add IL-2 in the wells of the microtiter plates used for the ^{51}Cr-release assay. This short-term stimulation augments the killing activity per NK cell without changing the spectrum of susceptible target cells. This is different from the induction of LAK which require longer stimulation with IL-2 and which are able to kill NK-resistant target cells. Addition of type 1 IFN can also augment NK cell activity without changing the spectrum of susceptible target cells.

3. Enrichment of various lymphocyte subsets by paramagnetic particles is a fast and convenient method. We favor the use of negative selection unless one can identify a non-activating surface marker for positive selection (e.g., use CD5 instead of CD3 for enrichment of NKT cells). One should avoid using NK1.1 for positive selection as it is an activating marker. Indeed, one of the approaches for the activation of NK cells for examination of intracellular IFN-γ, is to stimulate the cells with plate-bound PK136 (anti-NK1.1) mAb. A word of caution about using paramagnetic particles is that one loses a large percentage of the starting population. A useful approach to estimate as to how many cells to use in the starting population is to start by estimating how many of the purified subsets one

needs and use a starting number that is ten times as much (this ratio applies for NK and NKT not necessarily for T or B cells which exist in much larger percentages). Another important issue, which was alluded to already, arises when one uses a mouse strain whose NK cells do not express NK1.1. Although NK from all mouse strains express DX5 (CD49b), when one examines NK1.1 vs DX5 in an NK1.1+ strain, one is struck by a small but consistent subset which is NK1.1+DX5-. Moreover, there is another rare subset that is NK1.1-DX5+. The exact characteristics and functions of these two minor subsets are not well delineated.

4. Coupling surface and cytoplasmic staining is a powerful and convenient tool for clearly identifying which cell is the major source of a given cytokine. Although the methodology is relatively straightforward, credible results may be difficult to obtain unless one selects the brightest antibodies possible. The brightness of most surface markers dims considerably after fixation and permeabilization. A useful approach is to first titrate available antibodies using surface staining only. Then, titrate all available antibodies for the cytoplasmic marker using cytoplasmic staining alone. One can then select the brightest antibody for each marker and do the combined surface/cytoplasmic staining. However, even with this approach, certain cytokines are very difficult to identify. In general, commercially available antibodies for IFN-γ give bright staining. Most of the available antibodies for IL-4, on the other hand, are extremely dim and one cannot easily make any conclusions with the staining obtained. Nevertheless, this approach is much more informative than measuring cytokines in the supernatants unless one starts with a population that is >99% pure. ELISPOT circumvents the purity problem but is semi-quantitative at best. This is especially important if one is examining an NKT cell population that has some NK cells. Stimulation of NKT cells will induce them to secrete IFN-γ, which will then activate NK cells to induce them to secrete more IFN-γ as well as other cytokines.

References

1. MacGregor, R. R., and Louria, D. B. (1997) Alcohol and infection. *Curr. Clin. Top. Infect. Dis.* **17**, 291–315.
2. Brown, L. M. (2005) Epidemiology of alcohol-associated cancers. *Alcohol.* **35**, 161–168.
3. Boffeta, P., and Hashibe, M. (2006) Alcohol and cancer. *Lancet Oncol.* **7**, 149–156.
4. Orange, J. S., and Ballas, Z. K. (2006) Natural killer cells in human health and disease. *Clin. Immunol.* **118**, 1–10.
5. Heusel, J. W., and Ballas, Z. K. (2003) Natural killer cells: emerging concepts in immunity to infection and implications for assessment of immunodeficiency. *Curr. Opin. Pediatr.* **15**, 586–593.
6. Yokoyama, W. M., and Kim, S. (2006) How do natural killer cells find self to achieve tolerance? *Immunity.* **24**, 249–257.
7. Lodoen, M. B., and Lanier, L. L. (2006) Natural killer cells as an initial defense against pathogens. *Curr. Opin. Immunol.* **18**, 391–398.
8. Kronenberg, M. (2005) Towards an understanding of NKT cell biology: progress and paradoxes. *Annu. Rev. Immunol.* **23**, 877–900.

9. Cook, R. T., Li, F., Vandersteen, D., Ballas, Z. K., Cook, B. L., and LaBrecque, D. R. (1997) Ethanol and natural killer cells. I. Activity and immunophenotype in alcoholic humans. *Alcohol Clin. Exp. Res.* **21**, 974–980.

10. Meadows, G. G., Blank, S. E., and Duncan, D. D. (1989) Influence of ethanol consumption on natural killer cell activity in mice. *Alcohol Clin. Exp. Res.* **13**, 476–479.

11. Saxena, Q. B., Saxena, R. K., and Adler, W. H. (1981) Regulation of natural killer activity in vivo: part IV–high natural killer activity in alcohol drinking mice. *Indian J. Exp. Biol.* **19**, 1001–1006.

12. Abdallah, R. M., Starkey, J. R., and Meadows, G. G. (1988) Toxicity of chronic high alcohol intake on mouse natural killer cell activity. *Res. Commun. Chem. Pathol. Pharmacol.* **59**, 245–258.

13. Kronenberg, M., and Gapin, L. (2002) The unconventional lifestyle of NKT cells. *Nat. Rev. Immunol.* **2**, 557–568.

14. Chan, C. W., Crafton, E., Fan, H. N., Flook, J., Yoshimura, K., Skarica, M., Brockstedt, D., Dubensky, T. W., Stins, M. F., Lanier, L. L., et al. (2006) Interferon-producing killer dendritic cells provide a link between innate and adaptive immunity. *Nat. Med.* **12**, 207–213.

15. Ballas, Z. K., Rasmussen, W., and van Otegham, J. K. (1987) Lymphokine-activated killer (LAK) cells. II. Delineation of distinct murine LAK-precursor subpopulations. *J. Immunol.* **138**, 1647–1652.

16. Watanabe, H., Ohtsuka, K., Kimura, M., Ikarashi, Y., Ohmori, K., Kusumi, A., Ohteki, T., Seki, S., and Abo, T. (1992) Details of an isolation method for hepatic lymphocytes in mice. *J. Immunol. Methods.* **146**, 145–154.

17. Arase, H., Saito, T., Phillips, J. H., and Lanier, L. L. (2001) Cutting edge: the mouse NK cell-associated antigen recognized by DX5 monoclonal antibody is CD49b (alpha 2 integrin, very late antigen-2). *J. Immunol.* **167**, 1141–1144.

18. Kobayaski, E., Motoki, K., Uchida, T., Fukushima, H., and Koezuka, Y. (1995) KRN7000, a novel immunomodulator, and its antitumor activities. *Oncol. Res.* **7**, 529–534.

19. Hameg, A., Apostolou, I., Leite-De-Moraes, M., Gombert, J. M., Garcia, C., Koezuka, Y., Bach, J. F., and Herbelin, A. (2000) A subset of NKT cells that lacks the NK1.1 marker, expresses CD1d molecules, and autopresents the alpha-galactosylceramide antigen. *J. Immunol.* **165**, 4917–4926.

20. Ballas, Z. K., and Rasmussen,W. (1990) Lymphokine-activated killer (LAK) cells. IV. Characterization of murine LAK effector subpopulations. *J. Immunol.* **144**, 386–395.

21. Ballas Z. K., Rasmussen, W. L., and Krieg, A. M. (1996) Induction of NK activity in murine and human cells by CpG motifs in oligodeoxynucleotides and bacterial DNA. *J Immunol.* **157**, 1840–1845.

22. Rothenfusser, S., Tuma, E., Wagner, M., Endres, S., and Hartmann, G. (2003) Recent advances in immunostimulatory CpG oligonucleotides. *Curr. Opin. Mol. Ther.* **5**, 98–106.

23. Daeron, M. (1997) Fc receptor biology. *Annu. Rev. Immunol.* **15**, 203–234.

24. Emoto M., Emoto, Y., and Kaufmann, S. H. (1997) TCR-mediated target cell lysis by CD4+NK1+ liver T lymphocytes. *Int. Immunol.* **9**, 563–571.

25. Ballas, Z. K., and Rasmussen, W. (1990) NK1.1+ thymocytes. Adult murine CD4-, CD8- thymocytes contain an NK1.1+, CD3+, CD5hi, CD44hi, TCR-V beta 8+ subset. *J. Immunol.* **145**, 1039–1045.

26. Bartizal, K. F., Salkowski, C., Pleasants, J. R., and Balish, E. (1984) The effect of microbial flora, diet, and age on the tumoricidal activity of natural killer cells. *J. Leukoc. Biol.* **36**, 739–750.

27. Godfrey, D. I., Hammond, K. J., Poulton, L. D., Smyth, M. J., and Baxter, A. G. (2000) NKT cells: facts, functions and fallacies. *Immunol Today.* **21**, 573–583.

28. O'Connor, K. A., Holguin, A., Hansen, M. K., Maier, S. F., and Watkins, L. R. (2004) A method for measuring multiple cytokines from small samples. *Brain Behav. Immun.* **18**, 274–280.

29. Parekh, V. V., Singh, A. K., Wilson, M. T., Olivares-Villagomez, D., Bezbradica, J. S., Inazawa, H., Ehara, H., Sakai, T., Serizawa, I., Wu, L., et al. (2004) Quantitative and qualitative differences in the in vivo response of NKT cells to distinct alpha- and beta-anomeric glycolipids. *J. Immunol.* **173**, 3693–3706.

30. Carnaud, C., Lee, D., Donnars, O., Park, S. H., Beavis, A., Koezuka, Y., and Bendelac, A. (1999) Cutting edge: cross-talk between cells of the innate immune system: NKT cells rapidly activate NK cells. *J. Immunol.* **163**, 4647–4650.

31. Inui, T., Nakagawa, R., Ohkura, S., Habu, Y., Koike Y, Motoki, K., Kuranaga, N., Fukasawa, M., Shinomiya, N., and Seki, S. (2002) Age-associated augmentation of the synthetic ligand-mediated function of mouse NK1.1 ag(+) T cells: their cytokine production and hepatotoxicity in vivo and in vitro. *J. Immunol.* **169**, 6127–6132.

32. Sagiyama, K., Tsuchida, M., Kawamura, H.,Wang, S., Li, C., Bai, X., Nagura, T., Nozoe, S., and Abo, T. (2004) Age-related bias in function of natural killer T cells and granulocytes after stress: reciprocal association of steroid hormones and sympathetic nerves. *Clin Ex. Immunol* **135**, 56–63.

33. Wilson, M. T., Johansson, C., Olivares-Villagomez, D., Singh, A. K., Stanic, A. K., Wang, C. R., Joyce, S., Wick, M. J., and Van Kaer, L. (2003) The response of natural killer T cells to glycolipid antigens is characterized by surface receptor down-modulation and expansion. *Proc. Natl. Acad. Sci. USA.* **100**, 10913–10918.

34. Crowe, N. Y., Uldrich, A. P., Kyparissoudis, K., Hammond, K. J., Hayakawa, Y., Sidobre, S., Keating, R., Kronenberg, M., Smyth, M. J., and Godfrey, D. I. (2003) Glycolipid antigen drives rapid expansion and sustained cytokine production by NK T cells. *J. Immunol.* **171**, 4020–4027.

35. Parekh, V. V., Wilson, M. T., Olivares-Villagomez, D., Singh, A. K., Eu, L., Wang, C. R., Joyce, S., and Van Kaer, L. (2005) Glycolipid antigen induces long-term natural killer T cell anergy in mice. *J. Clin. Invest.* **115**, 2328–2329.

19
Polyclonal and Antigen-specific Responses of T Cells and T Cell Subsets

Betty M. Young, Susan Wiechert, Ruth A. Coleman, Prajwal Gurung, and Robert T. Cook

Summary Evaluation of the functional responses of T cells is of importance in determining the mechanism(s) of immunodeficiency resulting from chronic alcohol abuse and other conditions that lead to immune dysfunction. Mice that are chronically exposed to 20% (w/v) ethanol in water develop immunodeficiency and have T cells with abnormal activation profiles, reduced total numbers, increased CD4/CD8 ratios, and an increased memory/naïve phenotype ratio. These cells also have abnormal antigen-specific responses after inoculation of the ethanol mice with model infectious organisms. Study of the functional abnormalities of these cells requires a reliable system that can present appropriate activation stimuli in vitro for the generation of polyclonal or antigen-specific responses in enriched or purified T cells, free of the influence of previously ethanol exposed accessory cells. In this chapter, we describe protocols to assess the T cell response to polyclonal stimulation through the T cell receptor and the use of a model infectious disease bacterium,

Listeria monocytogenes, that allows evaluation of the T-cell response to specific peptide epitopes of the bacterium after previous inoculation.

Keywords T cells; chronic alcohol abuse; mice; *Listeria monocytogenes*; antigen-specific T cells.

1   Introduction

Considerable efforts have been made to determine the mechanistic changes that are responsible for immune deficiency and autoimmunity in the chronic alcoholic. Alterations that could reflect regulatory disturbances involving some aspect of T-cell function include: 1) increased serum levels of IgA, IgG, and IgM *(1)*; 2) reduced cell-mediated immunity such as a loss of DTH reactions *(2,3)*; 3) reduced lymphocyte numbers, altered subset distributions, persistent T-cell activation, and increased CD4/CD8 ratio *(1)*; 4) changes in cytokine levels *(1)*, which can be acute or chronic. In work to characterize the T cells of chronic alcoholics, it was established that alcoholics have an increased percentage of activated CD8+ T cells, measured as HLADR surface expression *(4)*. Subsequent work by three- and four-color flow cytometry showed that these patients also have a significant shift from a naïve cell expression of the leukocyte common antigen (CD45RA+) toward a memory cell phenotype (CD45RO+) *(5)*. This change is present in both CD8+ and CD4+ T cells and is accompanied by a significant reduction of L-selectin and an increase in CD11b on the surface of CD8+ cells, and by a loss of L-selectin in some but not all patients' CD4+ T cells. Most alcoholics also have stably increased expression of the carbohydrate-rich marker, CD57 *(6)*. The CD57+ T cell subsets of both patients and controls respond to polyclonal stimulation through the TCR with a rapid burst of IFN-γ and TNF-α production, but not IL-4. In addition, the CD57+ subset does not require a second signal for this production, whereas the CD57- subset does. These findings are consistent with the concept that this subset is a differentiated effector cell with cytotoxic potential and a TH1 immediate response cytokine profile. It has been demonstrated also that there is a substantial increase in rapid cytoplasmic IFNγ production in the T cells of chronic ethanol consuming mice *(7)*, along with an increase in the memory phenotype in the activated T cells of these mice. Because there are also increases in IL-4 production, the findings do not exclude TH2 excess in the chronic ethanol mice, but it has not been possible to demonstrate any TH1/TH2 skewing in antigen-specific T cells of chronic ethanol mice. Recently, it has been observed in our laboratory that in spite of the activated phenotype of the CD8+ T cells in these mice, the production of antigen-specific CD8+ cells during the primary response to the intracellular bacterium *Listeria monocytogenes*, is significantly reduced (Gurung et al., submitted 2007). The adequacy of both polyclonal and antigen-specific T-cell responses is of increasing interest as we attempt to understand the mechanisms of the immunodeficiency of chronic alcohol abuse. This chapter describes some of the basic techniques available for characterization of these responses.

2 Materials

2.1 Isolation of T cells From Spleen, Thymus, or Lymph Node

1. Dulbecco's phosphate buffered saline (DPBS, Invitrogen, Carlsbad, CA).
2. RPMI-1640 media (Invitrogen, Carlsbad, CA).
3. Lysis buffer (lyses red blood cells in cell suspension): $155\,mM$ NH_4Cl, $1\,mM$ Tris-HCl, pH 7.20–7.5. Mix 4.15 g of NH_4Cl (Fisher Scientific, Pittsburgh, PA) with 0.061 g of Trizma® Base (Sigma, St. Louis, MO.) to make 1 L with reagent grade water. Adjust pH to 7.2–7.5. Filter to sterilize and store at room temperature (RT).
4. Trypan blue stain 0.4% (Invitrogen, Carlsbad, CA). Stain used to discriminate live cells from dead cells when counting cells in a hemocytometer.

2.2 Column Fractionation of Spleen Cell Suspension

1. Magnetic Cell Sorting (MACS) Kit (Miltenyi Biotech, Auburn, CA). Kits contain biotinylated-antibody cocktail and streptavidin microbeads for cell separation. Negative-selection kits are available for: CD4+ cells (cat. no. 130–090–860), CD8+ cells (cat. no. 130–090–859), and B cells (cat. no. 130–090–862).
2. Column buffer (for the MACS separation system): $2\,mM$ ethylene diamine tetraacetic acid (EDTA), 0.5% bovine serum albumin (BSA) in DPBS. Place 1 liter of DPBS on stir plate with stir bar. Add 0.75 g of EDTA (Sigma, St Louis, MO) and stir until dissolved. Then add 5 g of biotechnology grade BSA (Amresco, Solon, OH) and stir until dissolved (see **Note 1**). Store at 4°C.
3. Columns and magnetic holders for cell separation (Miltenyi Biotech, Auburn, CA).
4. Biotinylated-anti CD45R/B220 antibody (clone # RA3–6B2) for B-cell depletion of splenocytes.
5. Streptavidin microbeads (Miltenyi Biotech, Auburn, CA) for B-cell depletion of splenocytes.
6. Lipopolysaccharide (LPS, Sigma, St Louis, MO). Purified from bacteria, lyophilized and sterilized by γ-irradiation (see **Note 2**).

2.3 Flow Cytometry Staining

1. Flow cytometry wash buffer: PBS-0.1% azide (used for washing flow cytometry tubes). Add 1.0 g of sodium azide (Fisher Scientific, Pittsburgh, PA) to 1 liter of DPBS. Store at RT.

2. Flow cytometry fixative (used to fix cells before flow cytometry): Mix 1 part 10% buffered formalin with nine parts flow cytometry wash buffer. Make fresh each week. Store at RT.

2.4   Stimulation and Cytokine Response of Cells

1. Conjugation buffer (for immobilizing antibody on culture plates): Mix 6.055 g of Tris-HCl (Sigma, St. Louis, MO) with 0.9 mL of $2N$ HCl (Sigma, St. Louis, MO) in 1 liter of reagent grade water. Adjust pH to 9.25–9.50. Filter to sterilize and store at RT.
2. Purified anti-CD3 antibody (clone #1445–2C11), (Becton Dickinson, Sparks, MD). Dilute antibody in conjugation buffer (1 µg/mL). Dilute antibody just before use. Allow antibody to conjugate to plate for 4 h at RT or overnight at 4°C (*see* **Note 3**).
3. Purified, azide-free, low endotoxin, anti-CD28 antibody (clone # 37.51; Becton Dickinson, Sparks, MD). Add 1 µg/mL to media in anti-CD3 positive control well after washing and before adding cells.
4. Culture medium – RPMI-10, Mix 500 mL of RPMI with L-glutamine, 50 mL of heat-inactivated fetal calf serum (HI-FCS, Invitrogen, Carlsbad, CA), 100 µL of 50 mM beta-mercaptoethanol (Sigma, St. Louis, MO), 100 µL of 50 mg/ml Gentamicin (Invitrogen, Carlsbad, CA). Filter to sterilize and store at 4°C. Use 1.0 mL/well for a 24-well tissue culture plate (*see* **Note 4**).
5. BD Cytofix/Cytoperm™ Plus Kit (with BD GolgiPlug™ Protein Transport Inhibitor; Becton Dickinson, Sparks, MD).
6. Micro-porous tape (3 M, Minneapolis, MN).

2.5   Antigen-Specific T Cells, Using Attenuated, L. Monocytogenes *as a Model Organism*

1. ACT A- *Listeria monocytogenes*, DPL 1942 (*8*). The stock we used was kindly provided by John Harty, University of Iowa. *See* **Note 5** for production of frozen stocks of attenuated Listeria.
2. Tryptic Soy Broth (TSB) Media for culture of, *L. monocytogenes*: Mix 30.0 g of BD Bacto™ Tryptic Soy Broth powder (Becton Dickinson, Sparks, MD) in 1 liter of reagent grade water. Warm to dissolve powder and autoclave. For plates, add 15.0 g of Bacto™ Agar (Becton Dickinson, Sparks, MD) per liter. Add streptomycin sulfate (Sigma, St. Louis, MO) to a final concentration of 50 µg/mL.
3. *Listeria monocytogenes* Listeriolysin O molecule (LLO)-specific peptide (Bio-Synthesis, Inc., Lewisville, TX). The peptide sequence for the C57Bl/6 CD4+ epitope

used in our lab is NEKYAQAYPNVS (LLO AA# 190–201). For Balb/c the CD8+ epitope is GYKDGNEYI (LLO AA# 91–99) *(9)*.

4. Intracellular staining buffer for cytoplasmic staining. Flow cytometry wash buffer (*see* Subheading 2.3) with 1% HI-FCS.

5. Sterile saline solution for injection of Listeria. 0.9% Sodium chloride, inj., USP, (Hospira, Inc., Lake Forest, IL).

6. Igepal for homogenization of Listeria-containing mouse organs. Igepal C630 (o ctylphenoxy)polyethoxyethanol (Sigma, St. Louis, MO). Igepal is a surfactant to help release Listeria from spleen/liver cells.

7. Variable speed tissue homogenizer to homogenize organs. Tissue Tearor Model 985–370 (Biospec Products, Inc., Bartlesville, OK).

8. P815 cell line – (ATCC # TIB-64™; cultured in RPMI-10).

2.6 Adoptive Transfer

1. Pentamer wash buffer: 0.1% sodium azide, 0.1% biotechnology grade BSA in DPBS.

2. Pentamer fixation buffer: 1% HI-FCS, 2.5% formaldehyde (Fisher Scientific, Pittsburgh, PA) in DPBS.

3 Methods

3.1 Isolation of T Cells From Spleen, Thymus, or Lymph Node

1. Organ harvest: Place animals in a CO_2 chamber until respiration ceases. Remove animals and perform cervical dislocation. Excise desired organs and place them in 5 mL of either buffer (DPBS) or culture medium (RPMI-10).

2. Mechanical disruption: several alternative methods can be used:

   a. Transfer organs and buffer to a 60 × 15-mm Petri dish. Take two sterilized glass microscope slides with frosted ends. Place the organ on the frosted side of one slide and crush the tissue with the frosted end of the other. Spleens can be cut into two pieces to facilitate crushing.

   b. Place a cell dissociation sieve with a 60-mesh screen in a 100 × 15-mm Petri dish and transfer the organs and buffer to the sieve. Press the tissue through the screen using either a tissue grinder pestle or the plunger of a 10-mL syringe. The screen helps to sift out pieces of connective tissue. This method is especially effective when a number of organs are being pooled.

   c. Place a small volume of buffer in a glass tissue homogenizer. Grind tissue with the pestle to make a single-cell suspension.

2. Removal of red blood cells: Transfer the cell suspension to a 15-mL culture tube, rinsing with enough buffer to fill the tube. Pellet the cells for 5 min at 250 RCF.

Aspirate the media from the pellet. Resuspend the cell pellet in lysis buffer, using 1 ml of lysis buffer/organ and incubate at RT for 1–2 min. Rinse away the lysis buffer by adding approx 10 volumes of buffer or media.

3. Quick Spin - Pull the chunks of connective tissue to the bottom of the tubes by centrifuging them until the force just reaches 70 RCF. Remove the tubes from the centrifuge and decant the supernatant into clean tubes. Pellet the cells in the new tube as before. Alternatively, cell suspensions may be filtered through nylon mesh cell strainers (Becton Dickinson, Sparks, MD).

3. Count: Resuspend the cells in a small volume (~1 mL) of buffer or media. Determine the cell density by counting an aliquot of the cells diluted 1:100 with Trypan blue stain in a hemocytometer. The cells are now ready for culture as whole splenocytes or for further fractionation of the population. Save cells for lineage marker analysis if using flow cytometry (10^6 cells per tube).

3.2 Column Fractionation of Spleen Cell Suspension

Cell suspensions may be fractionated to allow studies of purified populations or to enrich for rare cell types. The refinement of magnetic bead separations allows easy fractionation of T and B cells with 90%+ purity. Cells are isolated by either positive or negative selection. The positive selection methods can result in activation of the selected cells, which is often an undesirable effect in culture.

Briefly, cells are incubated with a biotinylated antibody cocktail followed by anti-biotin microbeads that react with the antibodies attached to the cells. Elution of the cell/bead suspension over a column in a magnetic field retains the labeled cells on the column allowing the un-labeled (negatively selected) cells to pass unimpeded. The column is then removed from the magnetic field and the labeled (positively selected) cells are eluted, if desired.

3.2.1 Column Purification of CD4+ T Cells by Negative Selection

1. Antibody incubation: Prepare single-cell suspension as discussed in Subheading 3.1. Suspend the cells in column buffer with 40 µL of buffer for every 10^7 total cells. Add 10 µL of biotinylated-antibody cocktail from kit (Miltenyi cat. no. 130–090–860) for every 10^7 total cells and incubate at 4°C for 10 min.

2. Microbead incubation: Add an additional 30 µL of column buffer and 20 µL of anti-biotin microbeads (provided in kit) for every 10^7 cells. Mix and incubate 15 min at 4°C.

3. Wash cells in approx 10x volume of column buffer and pellet by centrifuging for 5 min at 250 RCF. Resuspend them in de-gassed column buffer (*see* **Note 1**) with 500 µL of column buffer for every 10^8 cells.

4. Magnetic column elution: Use one Miltenyi LS size column for every 10^8 cells. Place the columns in the proper magnetic holder and pre-wet columns with 3 mL of de-gassed column buffer, discarding the eluate. Load 10^8 cells (0.5 mL) on each

column and let them run through, collecting the eluate. Wash the column four times using 3 mL of de-gassed column buffer in each wash. Pellet and resuspend the cells in the desired buffer, pooling tubes as needed. The negatively selected cells are then ready for culture or further treatment. Take aliquots of 10^6 cells for lineage marker analysis if using flow cytometry to determine the purity of the cell fractionation. If positively selected cells are to be collected, remove the column from the magnet holder and elute the retained cells into a clean tube with four 3 mL washes of column buffer. Otherwise, discard the used column.

3.2.2   Column Purification of CD8+ T Cells by Negative Selection

Follow the same protocol as the CD4+ purification given above (Miltenyi Biotech Inc., cat. no. 130–090–859).

3.2.3   Column Purification of B Cells with Miltenyi B-Cell Kit; Overnight Activation

1. Antibody incubation: prepare single-cell suspensions of splenocytes from two or three normal mice. Resuspend cells in column buffer with 40 μL of buffer and 10 μL of antibody cocktail from kit (Miltenyi Biotech Inc., cat. no. 130–090–862) for every 10^7 cells. Incubate at 4°C for 10 min.
2. Microbead incubation: Add an additional 30 μL of column buffer and 20 μL of anti-biotin microbeads for every 10^7 cells. Incubate 15 min at 4°C. Wash once with approx 10x volumes of column buffer. Resuspend cells in de-gassed column buffer at 500 μL/10^8 cells.
3. Magnetic column elution: prepare column and elute cells as in Subheading 3.2.1., step 4.
4. Resuspend purified cells in RPMI-10 at 2×10^6 cells/mL and add 20 μg/mL LPS. Incubate overnight at 37°C, 5% CO_2 (*see* Note 2).
5. The next day, wash cells twice with DPBS or RPMI to remove LPS. Resuspend in RPMI-10. Cells are now ready for peptide loading as described in Subheading 3.5.4., step 1.

3.2.4   B-Cell Depletion of Splenocytes

1. Antibody incubation: Prepare a single-cell splenocyte suspension as discussed in Subheading 3.1. Resuspend the cells in RPMI or DPBS without serum. Add biotinylated-anti-CD45R/B220 antibody at a concentration of 1 μg antibody/106 spleen cells. Incubate 20 minutes at 4°C.
2. Microbead incubation: Wash away excess antibody with approx 10x volumes of RPMI or DPBS by pelleting cells once. Add streptavidin microbeads (Miltenyi Biotech Inc., cat. no. 130–048–101) with 10 μL of beads for every 107 cells and incubate 15 min at 4°C to allow the attachment of labeled cells to the beads.

Wash once with approx 10x volumes of column buffer. Resuspend cells in de-gassed column buffer at 500 μL/108 cells.
3. Magnetic column elution: Prepare column and elute cells as in Subheading 3.2.1., step 4.

3.3. Flow Cytometry Staining

Staining lineage markers on the cells is used to demonstrate both that the starting cell population was as expected and that any fractionation of the cells was successful. The markers used may vary from experiment to experiment. Markers to document the makeup of the donor animals would include markers for the various cell lineages: T cells (e.g., CD3, CD4, CD8), B cells (e.g., CD19), macrophages and DCs (e.g., CD11b, CD11c), etc., as determined by the aim of the experiment. The same panel of stains is used on any purified or enriched cell suspension to demonstrate that the fractionation was successful.

1. Stain cells: Place 10^6 cells in a 12 × 75-mm polystyrene tube in a volume of approx 100 μL. Add the antibodies to the cells. Antibodies either can be from commercial sources or can be locally produced from hybridoma lines. Commercially produced antibodies may have a recommended working concentration; however, the optimum concentration of antibody should be determined by titrating the response of the cells used in any particular laboratory.
2. Incubate stain: incubate cells with antibodies in the dark at either RT or 4°C for 15–20 min.
3. Wash: wash to remove excess antibody with 1 ml of flow cytometry wash buffer. Pellet cells for 5 min at 250 RCF.
4. Secondary antibody: if a biotin-labeled primary antibody is used, add the fluorescent-labeled-streptavidin conjugate in wash buffer at 1 μg/10^6 cells and incubate as in Section 3.3 step 3 for 20 min. Wash as above.
5. Fix: Resuspend cells in 0.3 mL of flow fixative. Store at 4°C in dark until ready to analyze (see **Note 6**).

3.4 Stimulation and Cytokine Response of Cells After Polyclonal Stimulation

The response of cells is measured by exposing a controlled population of cells to defined stimuli in culture. Positive and negative control conditions are included to aid in determination of the cells' response to the test stimuli. The volume of cells to be used in culture is determined by the parameters being measured and by the number of cells available. Cytokine production provides one means to quantify the functional response of the cells being tested. Cytokine production can be quantified by measuring either the total (soluble) cytokines secreted or the intracellular cytokines. Intracellular staining allows the identification of the cell type producing the cytokines within a given 4–6 hour period of stimulation (*see* **Note 7**). Brefeldin A prevents the release of cytokines from cells by blocking protein transport from the endoplasmic reticulum (ER) to the Golgi apparatus. The number and subset of cells producing cytokines can be identified using combined cell surface marker and cytoplasmic staining.

1. Preparation of the culture plate: Volumes (1.0 mL/well) and concentrations given are for a 24-well culture plate (*see* **Note 4**). Prepare the plates by adding a poly-clonal stimulator, such as anti-CD3 antibody, to the positive-control wells. Immobilized monoclonal anti-CD3 antibody (clone #145–2C11) activates T cells in culture in a polyclonal fashion providing a good positive control. Dilute the antibody to a working concentration of 1 μg/mL with conjugation buffer and pipet into (positive control) culture wells (1.0 mL/well). Seal with Parafilm® (Pechiney Plastic Packaging, Menasha, WI) and place plate at 4°C overnight or keep at RT for 4 h. Wash the wells as discussed in ***Note 3***. Do not allow wells to dry out. After the final wash of the anti-CD3 wells, add 0.5 mL of culture medium to *all* wells that will have cells. If intracellular cytokine levels will be measured, use media containing 2 μl/mL BD GolgiPlug™ (BD Cytofix/Cytoperm™ Plus kit, Becton Dickinson, Sparks, MD) and place 0.5 mL in the plates/wells for those samples. The final Brefeldin A concentration in the wells after the addition of cells will be 1 μg/mL. The cells being cultured can be either from whole organ preps (e.g., whole splenocytes), B-depleted or purified T cells (CD4+ or CD8+). The prepara-tion of single-cell suspensions is discussed in Subheading 3.1.

2. Adjust the concentration of cells to be cultured to 4.0×10^6/mL and add 0.5 mL of the cell suspension to the wells (final concentration of 2×10^6 cells/mL). Cover and seal lid to plate with micro-porous tape (3 M, Minneapolis, MN.). Incubate for the desired time period (*see* **Note 7**). Cells and/or culture superna-tants may be harvested for assay at the end of the incubation period.

3. For supernatant collection, pipet the liquid contents of each well into a centrifuge tube. Centrifuge the tubes to pellet any cells and then transfer the supernatants to tubes. Freeze at −80°C for cytokine ELISA or Cytokine Bead Array (CBA) analysis.

4. To collect cells resuspend them with a Pasteur pipet. Transfer the suspended cells to a centrifuge tube. Wash the well twice with the same volume of media to increase cell recovery. Centrifuge to pellet the cells and discard the superna-tant. The cells are then ready for cell surface marker staining, cytoplasmic stain-ing, or RNA/DNA isolation, etc.

3.5 Antigen-Specific T Cells, Using Attenuated, L. Monocytogenes *as a Model Organism*

Antigen-specific T cells can be induced in the experimental animal model for study of adaptive immunity in the experimental animal encompassing primary, secondary and memory responses. Inoculation of the experimental animal with an infectious organism (either virulent or attenuated) or with an isolated immunogenic molecule produces an antigen-specific response that can be measured later by stimulating cells in culture with the infectious organism, a specific molecule or a peptide epitope.

Any cell preparation method discussed above (Subheadings 3.1. and 3.2.) may be used, depending on the aim of the experiment. The cell preparation may be from either individual mice or pooled populations. Both the economics and logistics of cell population purification discourage processing a large number of samples.

Listeriolysin O (LLO) T-cell peptide epitopes have been identified *(9)* that can be used to measure T-cell specific responses. The peptides used in our lab are targeted to CD4+ or CD8+ T cells in C57Bl/6 or Balb/c mice, respectively. The specific peptide chosen must be titered with the cell system being studied to determine the optimum concentration for culture.

3.5.1   Inoculation

ACT A - *Listeria monocytogenes* (DPL 1942), may be used to infect mice and elicit a measurable immune response to known specific peptide epitopes *(see* **Note 5**). ACT A - Listeria has been engineered so that the intracellular bacterium is unable to nucleate host cellular actin and therefore is incapable of moving from cell to cell to amplify the infection; it is thus less likely to kill the host *(8)*.

1. Preheat a shaking incubator to 37°C. Prepare a 50-mL culture tube with 10 mL of TSB media containing 50 µg/mL streptomycin. Remove 1 mL of media and save to use as the blank in a spectrophotometer. Thaw a 1-mL aliquot of ACT A- Listeria and add it to the 9 mL of media in the culture tube. Incubate in the pre-warmed shaking incubator for approx 1 h. After 30 min, the OD_{600} should be at least 0.01. Continue monitoring OD_{600} until it is 0.06–0.1 (log phase). Remove the tube from the incubator and hold it at RT. An OD_{600} of 1.0 = 10^9 CFU/mL.

2. Calculate the amount of inoculum to be prepared. Example: Inoculation of 10 mice with 0.3-mL doses of 5×10^5 CFU Listeria requires 3.0 mL, but extra should be prepared. Prepare 6.0 mL (20 doses):
   20 doses $\times$ (5×10^5 CFU/dose) = 1×10^7 CFU needed.
   A culture with an OD_{600} = 0.06 contains 6×10^7 CFU/mL.
   (1×10^7 CFU needed) / (6×10^7 CFU/mL) = 0.167 mL culture needed. Mix the 0.167 mL of Listeria culture with 5.833 mL of sterile saline and inject mice intraperitoneally. For intravenous infection, the number of CFU's of Listeria given the mice is reduced to 10^5 CFU and the injected volume is decreased to 0.1 mL.

3. Prepare four or five 10-fold serial dilutions of the inoculum. Plate 20 µL of each dilution on separate TSB-strep agar plates to determine the actual dose of bacteria that was given to the experimental animals. Incubate plates overnight at 37°C and count the colonies the next day. (# colonies) $\times$ (dilution factor) $\times$ 50 = # CFU/mL.

3.5.2   Cell Preparation

Harvest organs from Listeria-infected mice (*see* **Note 8**). Prepare single-cell suspensions for culture from mice as discussed above (Subheading 3.1). Prepare positive and negative control wells, as well as the LLO-specific peptide wells, and add cells as in Subheading 3.4. Positive control wells contain anti-CD3 +/− anti-CD28 as needed. If measuring intracellular cytokines, add Brefeldin A (GolgiPlug™ in kit) to the wells 4–6h before harvesting the cells (Subheading 3.4., step 1).

3.5.3 Culture

Culture the cells in a 37°C, 5% CO_2 incubator for the time period determined to provide the desired response for the cells being tested (usually 6h; *see* **Note 7**). A complete set of test/control wells is required for each time period used. After the time has elapsed, supernatants can be collected to measure soluble cytokines (Subheading 3.4) or cells may be collected from Brefeldin A-treated wells (Subheading 3.4) for measurement of intracellular cytokines. Freeze supernatants for soluble cytokine measurement, at −80°C until assayed. Stain cells for intracellular cytokine expression as discussed in Subheading 3.5.5. below.

3.5.4 Antigen Presenting Cells (APCs)

When culturing purified T cells (CD4+, CD8+ cells), APCs may be pre-loaded with specific peptide antigen. Either activated purified B cells or cell lines (e.g., P815 cells) may be used.

1. Whole splenocytes: In a whole spleen preparation, the native APCs are present and functional and therefore no additional APCs are required. This is the starting point for determining optimum antigen concentrations for culture. Prepare whole spleen, single-cell suspensions from either pooled or individual animals. Add prepared cells to the test wells (negative control, positive control and peptide) with a final concentration of 2×10^6 cells/mL. The concentration of peptide must be titrated in the cell system being studied. Start with a range of serial dilutions (e.g. 10^{-4} M, 10^{-5} M, etc) and determine the concentration that induces the optimum response. Culture the cells for the specified time and collect cells (Subheading 3.5.5., step 5) and/or supernatants for analysis.
2. Loaded B cells: Prepare activated B cells, beginning the day before, as discussed in Subheading 3.2.3., culturing (activating) them overnight with LPS. Wash two or three times with approx 10x volumes of DPBS or RPMI to remove LPS. Adjust B-cell concentration to 2×10^6 cells/mL in RPMI-10 and add the specific peptide at the concentration found to be optimum when culturing whole splenocytes with the peptide (*see* Subheading 3.4., step 1). Incubate for 1h at 37°C. Wash twice with 10x volumes of RPMI. Add loaded B cells to culture wells along with the purified T cells in a range of several T cell:B cell ratios such as 100:1 down to 1:1. No additional peptide is added to the wells.
3. P815 cell line as APCs: P815 (ATCC # TIB-64™) is a mouse lymphoblast-like mastocytoma cell line that can be used as antigen presentation cells. They can be loaded with specific peptides in the same manner as loading purified

B cells (Subheading 3.5.1). P815 cells, grown in suspension, are washed once with RPMI-10. The cell concentration is adjusted to 5×10^6 cells/mL and peptide added to the optimal concentration as determined by titration in the cell system being used. Incubate for 1 h at 37°C. Wash once or twice with RPMI-10. The cells are then ready to add to the prepared culture plate. It is best to initially test a range of cell ratios as with the B cells (Subheading 3.5.4., step 2). A negative control well is required in which nonpeptide-loaded P815 cells are added to the purified T cells. This is used to determine non-specific binding during analysis of the flow cytometry data.

3.5.5 Intracellular Staining for Cytokines after Brefeldin A Treatment

Harvest cells as described in Subheading 3.3., step 5. This staining is a multistep process. The cell surface markers (CD4, CD8, etc.) must be stained first, then the cells are permeabilized to allow the antibodies against the cytokines to reach the cytoplasmic constituents of the cells. Figure 1 shows a plot of purified CD4+ cells stained for cytoplasmic interferon-γ (IFN-γ) and CD44+ expression. CD4+ T cells were isolated from C57Bl/6 mice 7 d after inoculation with ACT A- Listeria, and incubated with LLO peptide for 6 h.

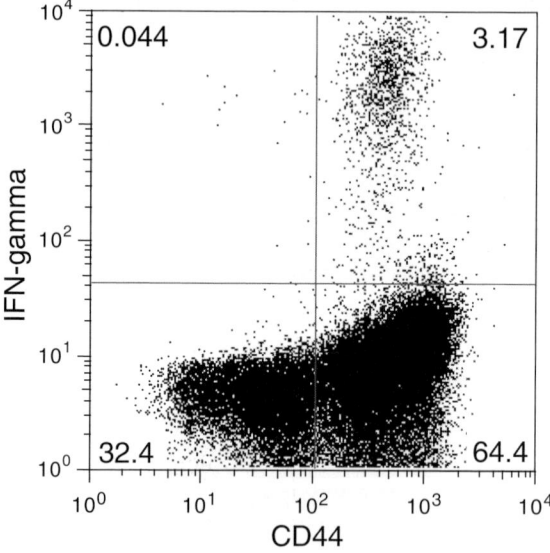

Fig. 1 Pattern of cytoplasmic IFN-γ and surface CD44+ expression in a culture of purified CD4+ cells. CD4+ cells were isolated from C57Bl/6 mice 7 d after inoculation with ACT A- *Listeria monocytogenes* and cultured for 6 h with B cells loaded with listeriolysin peptide (AA# 190–201) at a ratio of 1:1 with added GolgiPlug™, then permeabilized, stained and analyzed on a BD™ FACSCalibur. The IFN-γ+ cells represent those responding specifically to the peptide during the peak of the primary response. The numbers represent the percentage of CD4+ cells in each quadrant

1. Stain the surface markers as described in Subheading 3.3. After the final wash step, omit the fixation buffer. Instead, the cells are washed twice with 1.0 mL of intracellular staining buffer.
2. Resuspend the cell pellet in 250 μL of the Cytofix/Cytoperm solution provided in the BD Cytofix/Cytoperm™ Plus kit (Becton Dickinson, Sparks, MD.). Vortex the cells as the buffer is added to decrease clumping of the cells. Incubate 20 min at 4°C.
3. Prepare the Perm/Wash buffer (approx 4.2 mL/sample will be needed) by diluting the 10x stock provided in the kit with reagent grade water. Wash the cells twice with 1.0 mL of diluted Perm/Wash buffer, vortexing the tube as the buffer is added.
4. During the second wash of the cells, prepare fluorescent-labeled anti-cytokine solution by diluting the antibody in Perm/Wash buffer (50 μL of diluted antibody/sample). Vortex the cell pellet while adding the antibody and incubate in the dark at 4°C for 30 min (*see* **Note 9**).
5. Wash twice with 1.0 mL of Perm/Wash buffer. Vortex while adding buffer.
6. Resuspend in 250 μL of staining buffer and vortex well. Keep at 4°C in the dark until the samples can be run on a flow cytometer. They must be run within 24 h or signal could be decreased.

3.6 Time Course of Response

Establish a response curve by measuring the cytokine response over time. A typical response curve for *Listeria monocytogenes* infection is shown in Fig. 2. Culture cells from Listeria-infected mice at various times after inoculation (e.g. daily,

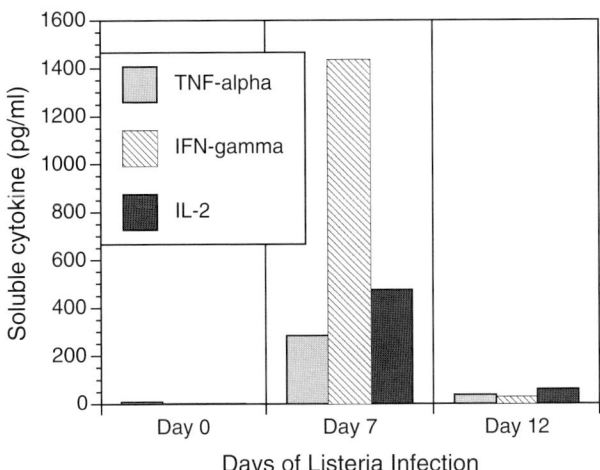

Fig. 2 Time course of soluble cytokine production after ACT A- *Listeria monocytogenes* inoculation. Unfractionated spleen cells cultured for 8 h with listeriolysin peptide (190–201) at a concentration of 10^{-4} M. Time course with uninfected (Day 0), seven and twelve day Listeria inoculated C57Bl/6 mice

weekly, monthly) to measure immune response over time depending on the parameters being studied. Soluble (total) cytokines may be measured by either cytokine ELISA or by CBA assay (e.g. Becton Dickinson's BD™ CBA or Bio-Rad's Bio-Plex™).

3.7 Adoptive Transfer

Adoptive transfer of immunity can be used to study primary and secondary immune responses. Congenic strains of mice (e.g., C57 Bl/6J [CD45.2+] and B6.SJL-Ptprc^a Pep^b/BoyJ [CD45.1+]) allow one to discriminate between the responding cell types. Monoclonal antibodies are available to separate CD45.1+ cells (clone # A20) and CD45.2+ cells (clone # 104). An immune response is induced in the donor mice. Then T cells are transferred to congenic recipient mice. Inoculation of the recipient animals produces cytokine responses from both donor (secondary response) and recipient (primary response) T cells.

Donor mice are inoculated with Listeria. T cells are isolated, usually after 28+ days (see **Note 8**), and transferred intravenously to congenic recipients. The recipient mice are then inoculated immediately with Listeria. At the peak primary response time for Listeria infection, 6–7 d later, spleen cells are cultured with and without Listeria-specific peptide and then the cytokine response is measured.

Figure 3A illustrates the resolution of congenic cells in a whole spleen culture one day after adoptive transfer of CD4+ T cells. Staining for intracellular cytokine expression and gating on CD4+ cells allows the discrimination of cytokine-producing donor cells from cytokine-producing recipient cells as shown in Fig. 3B.

3.7.1 Adoptive Transfer Utilizing Congenic Strains of Mice

1. Inoculate the donor mice (e.g., the CD45.1+ strain, B6.SJL-Ptprc^a Pep^b/BoyJ) with Listeria as in Subheading 3.5.1.
2. Isolate T cells from donor mice at the desired interval by column purification or B220-depletion as in Subheading 3.2. Resuspend cells in DPBS and inject intravenously or via retro-orbital sinus into recipient mice (e.g., the CD45.2+ strain, C57Bl/6J). Adoptive transfers have been performed in our lab using up to a 1:1 ratio of donor to recipient splenic lymphocytes as starting material. Average cell recoveries after purification of donor cells allow the routine transfer of 1×10^7 CD4+ cells and 8×10^6 CD8+ cells. Fewer transferred cells may be preferred in some experimental protocols.
3. Infect recipient mice as in Subheading 3.5.1. This can either be done the same day or the day following the cell transfer.
4. After 6 d (peak memory response time), harvest organs and culture cells, either as whole spleen or purified T-cell fractions as described previously. Intracellular staining for cytokines combined with haplotype staining discriminate the response of donor cells from those of the recipient (Fig. 3B). Analysis requires

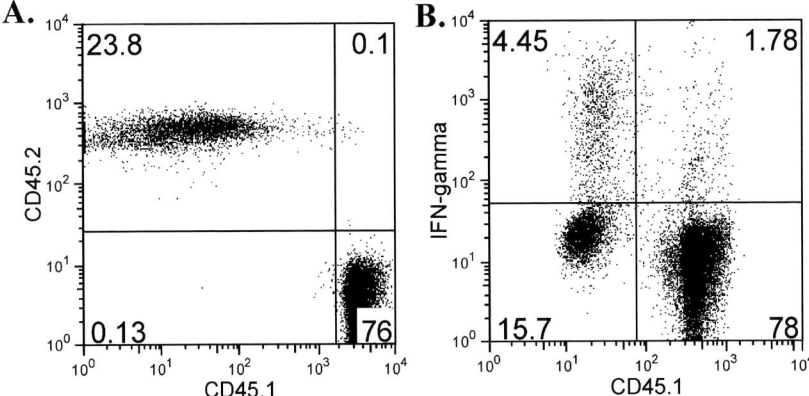

Fig. 3 Patterns of congenic CD4+ cells after adoptive transfer. Donor (CD45.2+) CD4+ cells were transferred to recipient (CD45.1+) mice. Recipients were inoculated with Listeria 1 day later and organs harvested 6 d after Listeria inoculation. Analyzed on BD™ FACSCalibur. (**A**) Whole splenocytes gated for total CD4+ cells. (**B**) Whole splenocytes cultured 6 hours with CD4+ LLO-specific peptide and GolgiPlug™. Stained for cytoplasmic IFN-γ. In this case the donor cells were obtained 58 d after Listeria inoculation. Thus, the donor haplotype (CD45.1⁻) response is the response of memory cells at the peak time of memory response (6 d), whereas the recipient response (CD45.1+) is a primary response, and is one day before the usual peak primary response at 7 d after inoculation

gating for the transferred T cells (CD4+ or CD8+) followed by displaying anti-cytokine-labeled cells against either the donor or recipient haplotype.

3.7.2 Fluorescent-labeled MHC Class I Pentamers and MHC Class II Tetramers

These are commercially available to identify antigen-specific T cells (CD8+ and CD4+ respectively) without the need for congenic strains of mice. Antigen-specific T cells are generated and cultured as discussed previously.

1. Collect 10^6 cells to stain for flow cytometry. Wash with 1.0 mL and resuspend in 50 μL of pentamer wash buffer.
2. Add 10 μL of fluorescent-labeled pentamer and incubate 10–15 min at RT in the dark.
3. Wash with 1.0 mL of pentamer wash buffer and add 50 μL of pentamer wash buffer containing fluorescent-labeled anti-T-cell antibody.
4. Incubate on ice 20–30 min in the dark.
5. Wash twice with 1.0 mL of pentamer wash buffer and resuspend in 0.3 mL of pentamer fix buffer. Store the samples in the dark at 4°C until they can be analyzed by flow cytometry.

Acknowledgments This work was supported by NIH AA 014405 and the Department of Pathology, University of Iowa.

4 Notes

1. Column fractionation buffer: The EDTA must be completely dissolved in the DPBS before the BSA is added or the EDTA will not dissolve well. This buffer must be de-gassed at RT for at least 1 h before applying it to a column. If the buffer is not de-gassed, bubbles may form inside the column, greatly slowing the flow rate and decreasing the cell recovery. De-gassing is easily accomplished: place column buffer in a sterile Erlenmeyer flask and apply a vacuum for 1 h (ordinary house vacuum, if available, is usually sufficient).
2. LPS preparations from several different bacterial strains are available. It is necessary to screen a selection of preparations with any cell system to find the optimal combination of LPS source and concentration. All LPS preparations used should be purified to eliminate cell stimulation by TLR ligands other than LPS.
3. Wash the anti-CD3-conjugated wells of the culture plate before adding cells to remove all traces of the conjugation buffer. Wash twice with 2 mL of DPBS or RPMI. Wash once more with 1 mL of culture media. Do not allow wells to dry out between wash steps. Purified antibody that is suitable for use in co-stimulation, immunoprecipitation, flow cytometry, and western blotting is desired. The concentration used here (1 µg/mL) has been titered in our lab and has proven to provide maximum response in T-cell cultures. If a less intense response is desired, the antibody concentration may be diluted. Soluble anti-CD3 antibody may be used in some culture applications in place of immobilized antibody and is simply added to the culture well. The antibody preparation used for this must be azide-free and low endotoxin (NA/LE).
4. Consistency in the plasticware (culture plates and flasks) used for tissue culture is important. Slight differences in manufacturing processes can have an effect on how cells will fare in culture. Different sources of plasticware should be tested with the cell system under investigation. Cell volumes in culture can be scaled proportionately up or down (from the 1.0 mL/well used for a 24-well tissue culture plate) if a different plate configuration is preferred.
5. Production of frozen stock of attenuated *Listeria monocytogenes*. Listeria must be cycled through mice periodically to maintain the ability to enter cells.

   a. Inoculate a 10-mL TSB-strep broth culture with Listeria and grow to log phase ($OD_{600} = 0.06$). *See* Subheading 3.5.1.
   b. Inject 10^7 CFU (in 0.3 mL of sterile saline) into two or three Balb/c mice (intraperitoneally). Do this as late in the day as possible.
   c. As early as possible the next day, harvest and weigh livers and spleens from all infected mice. Attenuated Listeria is cleared from the body quickly so keep the in vivo time to no more than 16–18 h.

d. Homogenize individual organs in 0.2% Igepal in reagent grade water, 10 mL for liver and 5 mL for spleen, using the high speed on a Tissue Tearor (Biospec Products, Inc., Bartlesville, OK.). Homogenize 10–15 s until tissue is liquefied. Prepare three 10-fold serial dilutions of the organ homogenates with 0.2% Igepal and plate 20 μL of each sample, including the whole organ homogenates, on individual TSB-strep agar plates and incubate overnight at 37°C. Listeria infection is typically reported as CFU/gram in liver and total CFU in spleen.

e. Late the next day, pick one, isolated, colony to inoculate a 10 mL TSB-strep overnight culture. Incubate in a shaking incubator at 37°C.

f. Use the entire 10 mL overnight culture to inoculate 0.5 – 1.0 liter of TSB-strep broth. Incubate in a shaking incubator at 37°C until the $OD_{600} = 0.1$ (log phase).

g. Freeze 1.0 mL aliquots of bacteria in TSB-strep broth and store at −80°C. No glycerol is required.

6. Cells that have been stained for flow cytometry should be analyzed as quickly as possible. Storing them for more than 1–2 d can result in auto-fluorescence that may interfere with the analysis.

7. The time period for stimulation depends on the functional response being measured. Memory cells (CD44+) will respond rapidly and are typically measured after 6 hours of stimulation. If the responsiveness of the naïve subset (CD44−) is sought, soluble cytokine is measured after 24–72 hours, in the presence of anti-CD3 antibody with anti-CD28 antibody for the second signal.

8. For the primary Listeria response, seven days after inoculation is typically used. A secondary exposure (boost) is best measured at 6 d after the booster inoculation. The contraction phase of the T-cell response is estimated at 11 or 12 ds after inoculation *(10)* and memory levels after 28+ days.

9. Phycoerythrin works quite well as the fluor for intracellular staining. Use other fluorescent tags for the surface markers.

References

1. Cook, R. T. (1998) Alcohol abuse, alcoholism, and damage to the immune system—a review. *Alcohol Clin. Exp. Res.* **22**, 1927–1942.
2. Paronetto, F. (1993) Immunologic reactions in alcoholic liver disease. *Semin. Liver Disease* **13**, 183–195.
3. MacGregor, R. R., and Louria, D. B. (1997) Alcohol and infection. *Curr. Clin. Topics Infect. Dis.* **17**, 291–315.
4. Cook, R. T., Garvey, M. J., Booth, B. M., Goeken, J. A., Stewart, B., and Noel, M. (1991) Activated CD-8 cells and HLA DR expression in alcoholics without overt liver disease. *J. Clin. Immunol.* **11**, 246–253.
5. Cook, R. T. Waldschmidt, T. J. Ballas, Z. K., et al. (1994) Fine T-cell subsets in alcoholics as determined by the expression of L-selectin, leukocyte common antigen, and beta-integrin. *Alcohol Clin. Exp. Res.* **18**, 71–80.
6. Song, K., Coleman, R. A., Alber, C., Ballas, Z. K., Waldschmidt, T. J., Mortari, F., LaBrecque, D. R., and Cook, R. T.(2001) TH1 cytokine response of CD57+ T-cell subsets in healthy controls and patients with alcoholic liver disease. *Alcohol* **24**, 155–167.

7. Song, K., Coleman, R. A. Zhu, X., et al. (2002) Chronic ethanol consumption by mice results in activated splenic T cells. *J. Leuk. Biol.* **72,** 1109–1116.

8. Brundage, R. A. Smith, G. A. Camilli, A., Theriot, J. A. and Portnoy, D. A. (1993) Expression and phosphorylation of the Listeria monocytogenes ActA protein in mammmalian cells. *Proc. Natl. Acad. Sci. USA* **90,** 11890–11894.

9. Geginat, G., Schenk, S., Skoberne, M., and Goebel, W. H. H. (2001) A novel approach of direct ex vivo epitope mapping identifies dominant and subdominant CD4 and CD8 T cell epitopes from *Listeria monocytogenes. J. Immunol.* **166,** 1877–1884.

10. Badovinac, V. P. Porter, B. B., and Harty, J. T. (2002) Programmed contraction of CD8+ T cells after infection. *Nat. Immunol.* **3,** 619–626.

20
B-Cell Studies in Chronic Ethanol Mice

Shilpi Verma, Carla-Maria A. Alexander, Michael J. Carlson,
Lorraine T. Tygrett, and Thomas J. Waldschmidt

Summary Chronic alcohol abuse leads to multiple defects in the immune system,
leading to an increased risk of infectious disease and malignancy. Immune lesions
encompass both the innate and adaptive arms and include deficiencies in the B-cell
compartment. Long-term alcoholics exhibit loss of B cells in the periphery and
diminished ability to generate protective antibodies. To better mimic the chronic
alcoholic patient, our group has used an ethanol-in-drinking-water mouse model.
Mice consuming alcohol in this manner progressively develop a range of immune
abnormalities, including defects in humoral immunity. To document and explore
B-cell lesions in ethanol-consuming mice, our laboratory has used a broad panel of
technologies. These include protocols to define the physical state of B cells in the
bone marrow and periphery, in vitro approaches to test B-cell activation potential
and in vivo experiments to document the humoral competence of the host. These
key techniques are detailed in the present chapter.

Keywords B lymphocytes; proliferation; differentiation; antibody production;
flow cytometry; immunohistology.

L. E. Nagy (ed.), *Alcohol: Methods and Protocols*
© Humana Press 2008

1 Introduction

It is well understood that long-term alcohol consumption compromises the human immune system. This fact is best underscored when examining the susceptibility of chronic alcoholics to infectious disease. Alcoholic patients have a greatly increased risk of infection with extracellular bacteria, intracellular bacteria, and viruses *(1–9)*. Numerous reports have documented the alcoholic population to exhibit greater rates of bacterial pneumonia, sepsis, meningitis, and peritonitis *(1–9)*. Tuberculosis and disease caused by other intracellular agents are associated with long-term alcohol abuse, as are elevated rates of hepatitis B, hepatitis C and HIV infection *(1,2,4-7)*. Collectively, these findings point to multiple lesions in both innate and adaptive immunity.

A large body of work had indeed documented lesions in the innate immune system, with a range of defects in granulocytes, monocytes/macrophages and natural killer (NK) cells *(2–7, 10,11)*. The human T-cell compartment likewise shows abnormalities after chronic alcohol consumption, with evidence of both anergy and hyperactivity *(2–5,12–15)*. Of interest to our laboratory are reports demonstrating B-cell dysfunction in chronic alcoholics. A number of investigators have shown diminished numbers of peripheral blood B cells (relative and absolute) after long-term alcohol consumption *(15–18)*. Alcoholics also exhibit poor Ab levels upon immunization with the hepatitis B vaccine, a response that requires the participation of T helper cells *(19–22)*. These findings suggest not only attrition of B cells, but diminished ability to generate protective Ab. In contrast, it has been commonly reported that alcoholic patients have elevated concentrations of serum IgA and IgG *(2–5,16,17,23–26)*. A number of investigators have further demonstrated the presence of autoantibodies reactive against various self-tissues *(4,23)*. Finally, in vitro studies have shown peripheral blood B cells of alcoholics to secrete higher levels of Ig compared with normal controls after incubation in medium alone, or after stimulation with mitogens or cytokines *(25,26)*.

As detailed in the chapter by Cook and colleagues, our group has used an ethanol-in-drinking-water mouse model to assess the B-cell compartment in long-term alcohol consuming mice. Consistent with the field of B-cell immunology, we have used a range of standard in vivo and in vitro techniques to evaluate humoral competence. This chapter highlights key protocols used in these studies to evaluate the physical state of the B-cell population, the ability to proliferate and differentiate in vitro and, importantly, the capability to generate antibody responses in vivo. As such, the chapter is divided into three sections that reflect these different technological approaches.

2 Materials

2.1 Physical Assessment of the B-Cell Compartment in Mice

2.1.1 Flow Cytometric Analysis

1. Balanced Salt Solution (BSS): Recipe from R.I. Mishell and R. W. Dutton *(27)*.
2. Fico/Lite-LM from Atlanta Biologicals, Norcross, GA. This product is made specifically for separation of murine leukocytes (density = 1.086).
3. Staining buffer: 1X BSS, 5% bovine calf serum, 0.1% (w/v) NaN$_3$. The buffer can be stored at 4°C for 2–3 wk.
4. Mouse FACS fixative: 1% formaldehyde in 1.25X PBS, 12.5 mL of 10X PBS/ NaN$_3$, 2.7 mL of 37% formaldehyde solution. Bring volume to 100 mL with deionized H$_2$O. Store in brown bottle at 4°C.
5. Staining tubes: Fisher 1.5-mL conical tubes (cat. no. 04–978–145).
6. Desktop micro-centrifuge (e.g. Fisher micro-centrifuge).
7. 12 × 75-mm tubes from BD Falcon (cat. no. 352052).

2.1.2 Histological Analysis

1. Optimal Cutting Temperature compound (OCT) from Tissue-Tek (Sakura Finetek, Torrance, CA; cat. no. 4583).
2. Vectashield mounting medium from Vector Laboratories, Burlingame, CA (at. no. H-1000).
3. 100 mM Tris-buffered saline solution: 100 mM Tris-HCl, pH 7.3, 0.9% (w/v) NaCl. The buffer can be stored up to several months at 4°C.
4. 100 mM Tris-HCl solution, pH 7.3: Dissolve 12.1 g of Tris Base in 800 mL of water. Adjust to pH 7.3 with concentrated HCl. Adjust volume to 1 liter with water. The buffer can be stored up to several months at 4°C.
5. Blocking solution: 1 mL 2.4G2 (anti-murine CD16/32 mAb, 1 mg/mL), 1 mL of goat serum, 8 mL of 100 mM Tris-buffered saline, 0.2 g of bovine serum albumin, 5 µL of Tween 20.
6. Tissue Path base mold (15×15×5 mm) from Fisher Scientific (cat. no. 22–038217).

2.2 In Vitro Functional Assessment of B-Cell Competence

2.2.1 MACS Bead Enrichment

1. CD43 Microbeads from Miltenyi Biotec, Auburn, CA (cat. no. 498–01).
2. LS separation columns from Miltenyi Biotec (cat. no. 130–042–401).
3. Midi MACS magnet and magnet stand.

4. Reaction buffer: 1X PBS with 0.5% BSA and 2 mM ethylene diamine tetraacetic acid (EDTA). To make a 500-mL reaction buffer, add 0.372 g of EDTA, 2.5 g of BSA to 50 mL of 10X PBS and bring up the volume to 500 mL with deionized H$_2$O (*see* **Note 1**).

2.2.2 Culturing of B Cells

1. Complete medium: RPMI 1640, 10% low endotoxin fetal calf serum, 5×10^{-5} M 2-mercaptoethanol, 1% v/v penicillin-streptomycin, and 1% v/v L-glutamine. Complete medium should be made fresh for each experiment.
2. 96-well flat-bottom plates.

2.3.3 B-Cell Proliferation

1. CFSE: 5-(and-6-)-carboxyfluorescein diacetate succinimidyl ester from Invitrogen-Molecular Probes, Eugene, OR (cat. no. C-1157). Prepare a 5 mM stock solution in DMSO, aliquot and store desiccated at −20°C.
2. Tritiated (6-^{3}H) thymidine 1 mCi/mL (37 Mbq/mL) in aqueous solution from PerkinElmer, Wellesley, MA (cat. no. NET-355). Prepare a stock solution of 40 μCi solution by diluting 40 μL into 960 μL of sterile BSS.

2.2.4 B-Cell Apoptosis

1. Merocyanine 540 (MC540) from Sigma-Aldrich, St. Louis, MO (cat. no. 323756). Prepare a 1 mg/mL stock solution in deionized H$_2$O. Remove clumps by passage through a 0.22-μm disk filter. Store at 4°C protected from light. Replace after 1 mo.
2. Propidium Iodide (PI) from Sigma-Aldrich (cat. no. P4170). Prepare a 25 μg/mL stock solution in PBS. Store at 4°C protected from light.

2.2.5 B-Cell Activating Agents

1. Anti-IgM: Both monoclonal and polyclonal anti-IgM Ab can be agonistic and provide a strong signal 1 (B cell receptor crosslinking by antigen) mimic. It is important to note that only a subset of rat anti-mouse IgM mAb are agonistic (e.g., b7–6 (*28*)). If goat or rabbit polyclonal anti-IgM is used, it is best to employ a Fab'$_2$ preparation to avoid suppression of activating signals mediated by IgM and FcγRII crosslinking. Recombinant IL-4 can be added to boost the response. Anti-IgM Ab stimulate well in fluid phase.
2. Anti-CD40: A number of hamster or rat (e.g., 1C10 (*29*)) anti-CD40 monoclonal Ab serve as mimics of CD40 ligand (CD154) and can provide a strong signal 2 to B cells. Recombinant IL-4 can be added to boost the response. Anti-CD40 Ab stimulate well in fluid phase.

3. TLR ligands: Murine B cells contain a range of TLR that activate most B-cell subsets when engaged. TLR ligands can be purchased from a number of companies (e.g. InvivoGen, San Diego, CA). Recommended ligands are Pam3CSK4 (TLR2), LPS (TLR4), R848 (TLR7/8) and CpG containing oligonucleotides (TLR9). The CpG recommended for mouse B cells is 1826 (TCCATGA<u>CG</u>TTCCTGA<u>CG</u>TT).

2.3 In Vivo Functional Assessment of B-Cell Competence

2.3.1 Antigens

1. Trinitrophenol- (TNP) Ficoll from Biosearch Technologies, Novato, CA (cat. no. F-1300). TNP-Ficoll is a prototypic T cell independent type 2 (TI-2) antigen that is highly immunogenic when administered to most strains of mice. The main utility of a hapten-carrier based immunogen is the ability to measure the induced Ab response using an enzyme-linked immunosorbent assay (ELISA) directed against the hapten portion of the conjugate. TNP-Ficoll is typically given at a dose of 5–20 µg intraperitoneally (i.p.) for optimal responses.
2. TNP-Keyhole limpet hemocyanin (KLH) from Biosearch Technologies (cat. no. F-1300). TNP-KLH is a prototypic T cell-dependent antigen that is highly immunogenic because of the large size and complex nature of the KLH carrier. TNP-KLH is typically given at a dose of 50–100 µg i.p. for optimal responses.
3. NP-KLH from Biosearch Technologies (cat. no. N-5060). NP-KLH is typically given at a dose of i.p. 50–100 µg for optimal responses.
4. Sheep red blood cells (SRBCs) from Colorado Serum Company, Denver, CO (cat. no. CS1112U). 0.2 mL of 10% v/v SRBC solution (equivalent to $1-5 \times 10^8$ SRBC) is typically given i.p. for optimal responses.
5. Phycoerythrin (R-PE, MW 240,000 g/mol) solution from Chromaprobe, Maryland Heights, MO (cat. no. R-201). PE is typically given at a dose of 100–200 µg i.p. for optimal responses.

2.3.2 Adjuvants

1. Alum solution: $AlK(SO_4)_2$ from Sigma-Aldrich (cat. no. 237086). Precipitating antigens with alum creates an immunogen that induces a TH2-directed Ab response. Add 7.36 g of dodecahydrous $AlK(SO_4)_2$ to 40 mL of deionized H_2O.
2. CpG oligonucleotide 1826. Mixing antigen with CpG oligonucleotides will not only lead to potent activation of dendritic cells, but also promote a TH1-directed Ab response.
3. Ribi adjuvant, Ribi ImmunoChem Research, Hamilton, MT. Ribi is an immunostimulant based on muramyl peptides and tends to invoke a TH "neutral" response that includes both TH1 and TH2 isotypes.

2.3.3 General ELISA Materials

1. Coupling Buffer: $0.05\,M$ Tris, pH 9.5 with 0.02% NaN_3. Add 3.03 g of Tris base and 0.1 g of NaN_3 to deionized H_2O and bring up the volume to 500 mL.
2. Reaction Buffer: 5% (w/v) powdered skim milk in 1X PBS with 0.02% NaN_3.
3. Substrate Buffer: $0.05\,M$ sodium carbonate, pH 9.8. Add 1.16 g of Na_2CO_3 and 0.1 g of $MgCl_2$ (hexahydrate) to deionized H_2O and bring up the volume to 500 mL.
4. Wash Buffer: 0.9% NaCl with 0.05% Tween 20, pH 7.0. Add 81 g of NaCl and 4.5 mL of Tween 20 to deionized H_2O and bring up the volume to 9 liters.
5. Alkaline phosphatase (AP) streptavidin from Zymed Laboratories (Invitrogen), South San Francisco, CA (cat. no. 43–4322)
6. p-Nitrophenyl phosphate [(pNPP) phosphatase substrate] from Sigma-Aldrich (cat. no. S0942, 5-mg tablets)
7. Immulon 2 HB high binding immunoassay microplates from Thermo Fisher Scientific, Waltham, MA (cat. no. 3455).

2.3.4 TNP-Specific ELISA

1. Capture antibodies:

   Goat anti-mouse IgM, Southern Biotech, Birmingham, AL (cat. no. 1021–01)
   Goat anti-mouse IgG1, Southern Biotech (cat. no. 1070–01)
   Goat anti-mouse IgG2a, Southern Biotech (cat. no. 1080–01)
   Goat anti-mouse IgG2b, Southern Biotech (cat. no. 1090–01)
   Goat anti-mouse IgG3, Southern Biotech (cat. no. 1100–01)
   Goat anti-mouse IgA, Southern Biotech (cat. no. 11165–01)
   Rat IgG anti-mouse IgE mAb. Our laboratory uses the EM95 hybridoma; other rat anti-mouse IgE mAb are commercially available (*see* **Note 2**).

2. Mouse anti-TNP mAb for establishment of standard curves: For each isotype-specific anti-TNP assay performed, a matching murine anti-TNP mAb is needed. Most of these can be purchased from commercial sources (e.g., BD Pharmingen). Alternatively, our laboratory has collected IgM, IgG1, IgG2a, IgG2b, IgG3, IgA and IgE anti-TNP hybridoma cell lines. Anti-TNP mAb are purified by passing hybridoma culture supernatants over a TNP-HGG-Sepharose column and eluting with DNP-glycine.
3. Human gamma globulins (HGG) from Sigma-Aldrich (cat. no. G4386 Cohn fraction II, III).
4. Picrylsulfonic acid solution (2,4,6-Trinitrobenzenesulfonic acid solution) from Sigma-Aldrich (cat. no. P2297). Picrylsulfonic acid solution comes as 5% (w/v) or 50 mg/mL solution. Picrylsulfonic acid is very soluble in water-based buffers. Avoid contact of the solution with skin.
5. (+)-Biotin N-succinimidyl ester (biotin) from Sigma-Aldrich (cat. no. H1759)

6. Dimethyl sulfoxide [(DMSO) anhydrous, ≥99.9%] from Sigma-Aldrich (cat. no. 276855).

2.3.5 NP-Specific ELISA

1. Capture reagents: NP_{30}-BSA and NP_3-BSA from Biosearch Technologies (cat. no. N-5050).
2. Hyper-immune NP-specific sera from C57BL/6 mice.
3. Biotin-conjugated goat anti-mouse IgG1 from Southern Biotech.

2.3.6 Flow Cytometric Analysis for GC and Memory B Cells

1. FITC-peanut agglutinin (PNA) from Vector Laboratories (cat. no. FL-1071).
2. Phycoerythrin (R-PE, MW 240,000 g/mol) solution from Chromaprobe.

3 Methods

3.1 Physical Assessment of the B-Cell Compartment in Mice

As a first step in evaluating the B-cell compartment of mice, it is of value to examine whether B-cell maturation is occurring normally in the bone marrow, and a full complement of subsets exists in peripheral sites. As documented by a number of investigators, murine B cells undergo a defined developmental program with each stage delineated by surface markers *(30)*. Once at the immature stage, B cells exit the bone marrow and home to the spleen. In this organ, B cells continue their maturational sequence as transitional B cells until they become fully mature follicular or marginal zone B cells *(31)*. Accordingly, the spleen is a rich site of B-cell subsets. Lymph nodes and mucosal aggregates contain primarily follicular B cells. In the mouse, the peritoneal cavity also harbors two populations of B cells, the B1 and B2 subsets. B2 B cells are very similar to follicular B cells found in spleen and lymph nodes, and B1 B cells compose the B1a and B1b populations *(30)*. Importantly, B-cell subsets in the periphery are also defined by differential display of surface markers. As such, their detection and enumeration can be easily accomplished by flow cytometry. Tables 1–3 summarize the subsets in the bone marrow and peripheral sites, and key markers that enable their definition. Figure 1 provides examples of multicolor flow cytometric analyses of B-cell populations residing in bone marrow, spleen, lymph node, and peritoneal cavity.

In addition to flow cytometric assessment of subsets, the physical state of the B-cell compartment can be evaluated by *in situ* analysis. Specifically, multicolor

Table 1 Maturational Stages of Bone Marrow B Cells

	Stage A	Stage B/C	Stage D	Stage E
Marker	Pre-pro-B cell	Pro-B cell	Pre-B cell	Immature
B220	++	++	++	+++
CD19	–	+	+	+
HSA[a]	++	+++	+++	+++
CD43[b]	++	++	–	–
BP-1	–	–/+	++	–
IgM	–	–	cytoplasmic	+++

[a] Heat-stable antigen or CD24.
[b] CD43 isoform recognized by the S7 mAb.

Table 2 Splenic B-Cell Subsets

Marker	T1[a]	CD21[lo] T2	CD21[hi] T2	Follicular	MZ
IgM	+++	+++	+++	++	+++
IgD	–	+	+++	+++	+/–
HSA[b]	++++	++++	+++	++	+++
CD21/35	–	+	+++	++	+++
CD23	–	+++	+++	+++	+/–

[a] T = transitional
[b] Heat-stable antigen or CD24.

Table 3 Peritoneal B Cell Subsets

Marker	B1a	B1b	B2
IgM	+++	+++	++
IgD	+/–	+/–	+++
CD5	+	–	–
CD11b	+	+	–
CD23	+/–	+/–	+++

frozen-section immunofluoresence can be valuable in detecting morphologic abnormalities in B cell and T cells zones within spleen and lymph nodes *(32)*.

3.1.1 Flow Cytometric Analysis of B Cells From Bone Marrow, Spleen, Lymph Nodes, and Peritoneal Cavity

1. Remove spleen or lymph nodes and mince tissue between frosted ends of microscope slides in 1X BSS to obtain a single cell suspension.
2. Bone marrow cells can be obtained from femur and tibia of mice. Bones are removed and single cell suspension made using 3-mL syringe and 23-gage needle (filled with 1X BSS) to expel marrow.
3. Single-cell suspension from peritoneal cavity can be obtained by performing peritoneal lavage using 1X BSS.

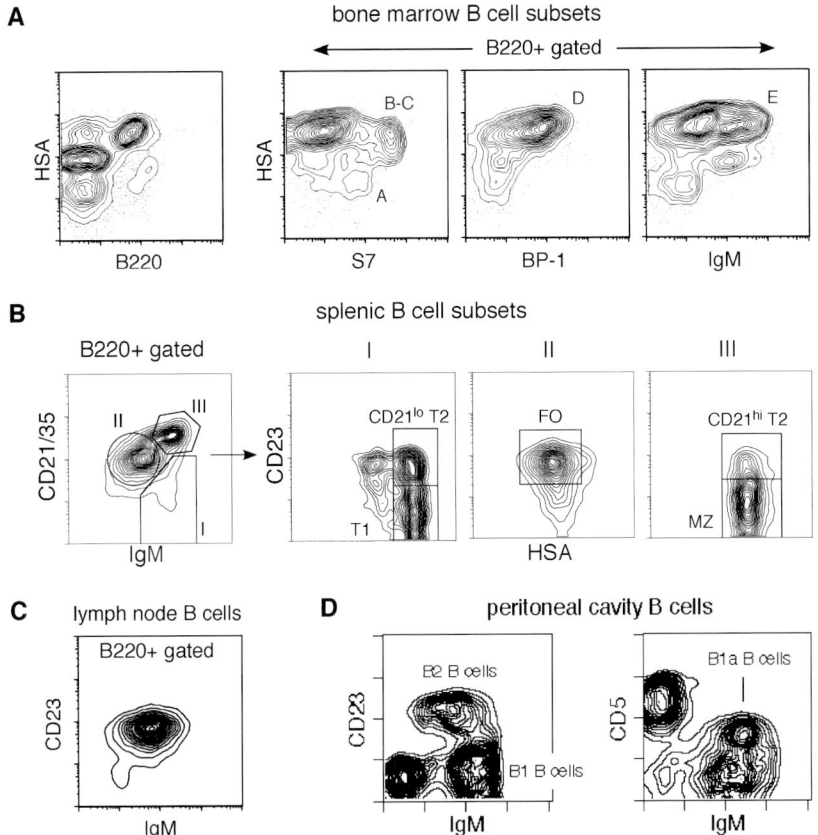

Fig. 1 Flow cytometric definition of murine B cell subsets. (**A**) Early B-cell subsets in bone marrow (BM). BM was harvested from adult BALB/c mice and stained with B220, HSA and either S7, BP-1, or IgM. The plots show early B cells (fractions A-E) identified using these markers. (**B**) Splenic B-cell subsets. Splenocytes from C3H mice were stained with anti-CD21/35, IgM, CD23, and HSA to distinguish among the transitional and mature subsets. (**C**) Lymph node (LN) B cells, and (**D**) Peritoneal cavity (PC) B cells. LN and PC lymphocytes were harvested from BALB/c mice and stained with markers listed in the plots. LNs contain primarily recirculating, mature follicular B cells while the PC harbors a mix of B1 (B1a and B1b) and B2 B cells

4. Place cell suspension in a 15-mL conical tube and bring up to 15 mL final volume with 1X BSS.
5. Wash in a refrigerated centrifuge for 7 mins at 400 g.
6. Remove supernatant from pellet and resuspend cells in approx 2 mL of 1X BSS.
7. Underlay cell suspension with approx 2 mL Fico/Lite-LM and spin at room temperature for 20 min at 950 g.
8. Harvest viable mononuclear cell interface with glass pipet and wash in staining buffer for 7 min at 400 g in a 15-mL conical tube.

9. Remove supernatant and resuspend pellet in staining buffer for cell count. Add sufficient volume that will result in a $2–5 \times 10^7$ cells/mL in order to facilitate cell counting (don't dilute too much).

10. Pipet 10^6 cells in 10–15 μL of volume into each staining tube (Fisher 1.5-mL conical tube; *see* **Note 3**).

11. Add blocking solution to each tube: 10 μL of 2.4 G2 (10–20 μg/tube) and 10 μL of normal rat serum.

12. After the cells have been mixed with the blocking solution, add primary Ab at the proper dilution to each tube. Final staining volume should be no more that 75 μL.

13. Vortex tubes briefly to mix.

14. Incubate 20 min on ice protected from light. Vortex briefly after 10 min of incubation. Cover and incubate for an additional 10 min.

15. Wash tubes 2X in staining buffer. For washing, add 1 mL of staining buffer and spin tubes for 1 min in a desktop micro-centrifuge that will accommodate the 1.5-mL conical tubes. Aspirate supernatant. Vortex pellet to resuspend the cells, add 1 mL of staining buffer, spin and aspirate again.

16. Vortex pellet to resuspend cells. If secondary reagents are being used, add proper dilutions of the reagent to the tubes.

17. Incubate 20 min on ice protected from light. Vortex briefly after 10 min of incubation. Cover and incubate an additional 10 min.

18. Wash tubes 2X in staining buffer.

19. After the last wash, resuspend cells in 200 μL of mouse FACS fixative and transfer to 12×75-mm tubes for flow cytometric analysis.

20. Store cells at 4°C protected from light. Most stained and fixed murine cell preparations can be stored for up to 2 wk before analysis.

3.1.2 Histological Analysis of B Cells From Secondary Lymphoid Tissues

3.1.2.1 Tissue Freezing

1. Presoak the harvested/dissected tissue block in 20% sucrose/1X PBS solution for 20–30 min at 4°C. Do not exceed this time limit.

2. Soak up the extra sucrose by placing the tissue on a KimWipe and gently patting with a forceps.

3. Place the OCT bottle in liquid nitrogen until 70–80% percent of the OCT is frozen (do not freeze the entire bottle contents). Pour cold OCT in the Tissue Path base mold and place the tissue block in the OCT containing mold. Make sure the tissue block is completely submerged in OCT. Cooling the OCT before freezing the tissue in liquid nitrogen prevents cracking of the specimen during the snap-freezing process.

4. Slowly place the base mold containing the tissue block into liquid nitrogen. Initially, place the bottom of the base mold on liquid nitrogen for approximately 10 s. As the OCT starts to freeze, submerge the entire block/base mold into liquid nitrogen.

5. Wrap the frozen block/base mold in aluminum foil and label it. Store the frozen tissue at ⁻80°C until ready to section.

3.1.2.2 Tissue Sectioning

1. Retrieve the frozen tissue block/base mold and transfer on ice to the cryostat. Do not let any part of the block thaw.
2. Place a small amount of OCT on the chuck. Turn the block/base mold up-side-down and pop the frozen tissue out by applying pressure to the bottom of the mold. Place the bottom of the tissue block (side that was next to the base form) on the OCT and set it on the rack to allow freezing of the chuck in the cryostat. This allows the tissue block to firmly adhere to the chuck. (Do not use too much OCT on the chuck or uneven placement of the tissue block will occur).
3. Place the chuck on the cryostat and start sectioning. You can remove the extra OCT by cutting 50-µm sections until frozen tissue is plainly seen.
4. Move the blade setting back to 8 µm (or less) and continue sectioning. We have found −15°C to work well for murine tissue sectioning.
5. When making the initial cuts, pay attention to the smoothness of the cut and any possible holes in the sections. If you see holes or tears in serial sections, it is likely that the portion of the tissue block being cut is damaged. Try to remove the damaged portion using a razor or scalpel blade.
6. After every tissue cut, pause momentarily prior to the next cut so that you can gently grab the end of the frozen section with a brush. Carefully continue the cut while simultaneously pulling the sample with the brush. This is essential for quality sections. Failure to do this will result in torn sections.
7. Place the tissue section onto the adjacent cryostat stage. Position a glass slide over the section in the following manner: place the short (nonfrosted) edge on the stage while holding the remaining slide at a ~30 degree angle over the section. Carefully drop the slide onto the tissue section allowing it to firmly and evenly adhere. Turn the slide over and let it air dry for 30 mins to 1 h at room temperature (*see* **Note 4**).
8. Place the air-dried samples in ice cold acetone for 10 min. After the acetone bath, remove as much fluid as possible by gently tapping the slide and blotting excess fluid at the edge of the slide with a KimWipe. Be careful not to touch the tissue section. Store the slides in an airtight container at −80°C until ready to stain.

3.1.2.3 Tissue Staining

1. Remove slides from −80°C and rehydrate in 1X PBS for 5 min.
2. Remove the PBS by gently tapping the slide and blotting excess fluid at the edge of the slide with a KimWipe. Be careful not to touch the tissue section (*see* **Note 5**).
3. Make a circle around the section using a hydrophobic (PAP) pen and immediately add 200 µL of blocking solution. (Never let the tissue section dry). Incubate the section in blocking solution for 1 hour at room temperature.

4. Remove excess blocking solution by gently tapping the slides and blotting remaining fluid as in step 2. Add 200 μL of the primary Ab diluted to the appropriate concentration in blocking solution that does not contain 2.4G2 mAb. Incubate for 1 hour at room temperature (*see* **Note 6**).

5. Wash slides three to four times in 1X PBS for 5 min each time.

6. Following the last wash, remove as much PBS as possible by tapping and blotting as in step 2. Incubate the sections with secondary Ab diluted to the appropriate concentration in blocking solution that does not contain 2.4G2 mAb.

7. Wash slides three to four times in 1X PBS for 5 min each time.

8. Following the last wash, remove as much PBS as possible by tapping and blotting as in step 2. Carefully add approx 10–12 μL of Vectashield mounting medium onto the sections. Delicately place a cover slip over the section and avoid getting any bubbles trapped on the specimen (*see* **Note 7**).

9. Seal the sides of the cover slip with nail polish and visualize your slides using fluorescence or confocal microscopy. The slides can be stored for several months at 4°C with the exception of sections stained with Cy3-conjugated Ab. Cy3 will lose its intensity after 2 wk in Vectashield.

3.2 In Vitro Functional Assessment of B-Cell Competence

Flow cytometric and histologic analyses provide information as to the in vivo physical status of the B cell compartment. To assess the functional competence of B cells, one can perform in vivo challenges with antigen (Subheading 4) or alternatively, examine purified B cells in vitro. Testing the activation potential of B cells in vitro has the advantage of exploring the capacity of B cells to respond without the need for other cellular interactions. In vitro experiments are thus an excellent means by which cell autonomous lesions can be detected. Importantly, a range of reagents exist that can mimic in vivo activating signals. These include anti-IgM, anti-CD40 and Toll-like receptor (TLR) ligands. In addition, assessment of cell death, cell expansion and differentiation are relatively straightforward.

3.2.1 MACS Bead Enrichment of B Cells

1. Harvest spleen and prepare a single cell suspension in 1X BSS. Wash by centrifuging for 7 min at 400 g. Resuspend in 2 mL of complete medium. For an unimmunized spleen, it is recommended that a Percoll separation be performed to enrich for resting lymphocytes. For an immunized spleen, a Ficoll separation should be performed to enrich for total mononuclear cells.

2. Harvest the desired interface and wash with 1X BSS. Resuspend the cell pellet in 500 μl reaction buffer and perform a cell count.

3. The ratio of beads to cells you wish to achieve is 10 μL of beads to 1×10^7 cells (as per the Miltenyi protocol) in 90 μL of reaction buffer. Adjust buffer volume and add beads. If resulting purity is not >95%, one can perform a bead to cell

titration to optimize conditions.Incubate for 15 min at 4°C, (in the refrigerator: do NOT put on ice), mixing at the half-way point. Wash with 10X volume of reaction buffer, centrifuge at 200 g for 10 min.

4. Resuspend cells in 500 µL of reaction buffer. You can resuspend up to 10^8 cells in this volume. DO NOT vary volume.

5. Prepare LS column for NEGATIVE SELECTION. The B cells of interest will be in the fall-through while all CD43$^+$ cells (T cells, NK cells, myeloid cells) will adhere to the magnetic matrix. Place LS column in the Midi MACS magnet apparatus. Wash LS column with 3 mL of reaction buffer.

6. Place cell suspension on the column and collect fall-through in a clean tube. Wash column 3X with 3 mL of reaction buffer. Pool all of the fall-through and wash the cells as quickly as possible. Resuspend in the appropriate solution (buffer or medium) and do a cell count. The typical purity observed with the negatively selected fall-through fraction is depicted in Fig. 2.

7. Remove the LS column from the magnet and place in a clean tube. Place 5 mL of reaction buffer on the column and with the plunger, rapidly push the buffer through the column. This fall-through is the POSITIVE SELECTION fraction.

8. The negatively selected B-cell fraction can be further subdivided by using either negative or positive selection protocols.

3.2.2 Culturing of B Cells

1. After enrichment, cell counting and suspension in complete medium, cells are dispensed in 96 well flat bottom plates at $5–10 \times 10^4$ per well in 200 µL of volume. Higher concentrations or round bottom wells are not recommended for 96-well

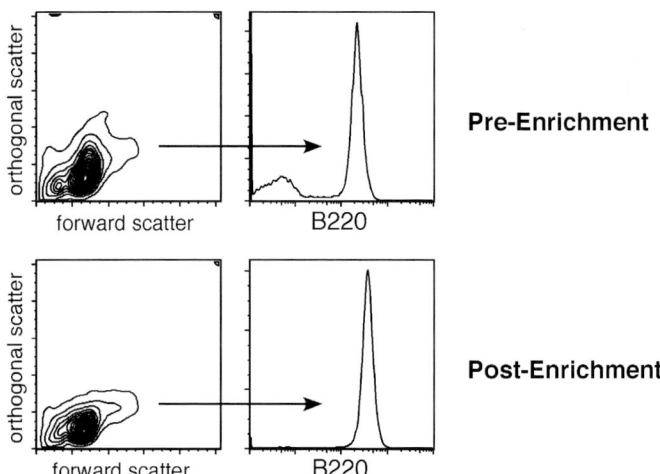

Fig. 2 MACS bead enrichment of murine B cells. Splenocytes were labeled with anti-CD43 magnetic microbeads and passed over an LS column attached to the MACS magnet apparatus. The fall-through was pooled, washed and an aliquot stained with anti-B220 (pan-B cell marker). The plots illustrate pre- and post-enrichment preparations

plates as murine B cells do not perform well when overly crowded. For some activation experiments, cell-cell contact is important. If one needs to dispense cells at less than 5×10^4 per well, it is advised that one edge of the 96-well plate be slightly tilted (with one or two glass slides) to facilitate cell clustering on the opposite side of the well.

2. Murine B cells are highly dependent upon the reducing capacity of 2-mercaptoethanol in order to undergo and sustain their activation sequence. Since this agent is highly volatile, it is recommended that the outside wells of 96-well plates not be used to culture B cells, but instead be filled with complete medium (containing 2-mercaptoethanol) alone.

3.2.3   Proliferation of B Cells

3.2.3.1   By Thymidine Uptake

1. During the last 4–8 h of culture, dispense 25 µL of the ^{3}H-thymidine stock solution into each well. This volume provides 1.0 µCi per well.
2. Harvest well contents onto glass filters to determine ^{3}H-thymidine uptake. Cells cultured in complete medium alone (without any stimulus) serve as background controls.

3.2.3.2   By CFSE Labeling

1. Harvest cells from wells and wash 2X in 1X PBS/5% serum.
2. Resuspend cells in 1X PBS at $1–5 \times 10^7$ cells/mL (*see* **Note 8**).
3. Prepare a 50 µM solution of CSFE in PBS (1:100 dilution of the stock).
4. Incubate cells for 10 min at 37°C with CFSE at a final concentration of 5–10 µM (*33,34*).
5. Stop the reaction with cold complete medium. Wash cells 3X with 1X BSS (*see* **Note 9**).
6. After the last wash culture cells with activating stimuli.
7. After the culture period, stain cells for surface markers (if desired) and analyze by flow cytometry. CFSE is a fluorescein-based dye with excitation and emission spectra identical to FITC-conjugated antibodies. The CFSE profile of B cells cultured with different agonists is shown in Fig. 3.

3.2.4.   Apoptosis of B Cells

1. After culture period, harvest cells, wash in cold BSS and resuspend in staining buffer at approximately 2×10^6 cells/400 µL. Take the cell samples to the flow cytometer prior to adding stains.
2. Add 2 µL of the MC540 stock solution (1 mg/mL) to the cell suspension, mix gently and incubate for 3 min at room temperature.

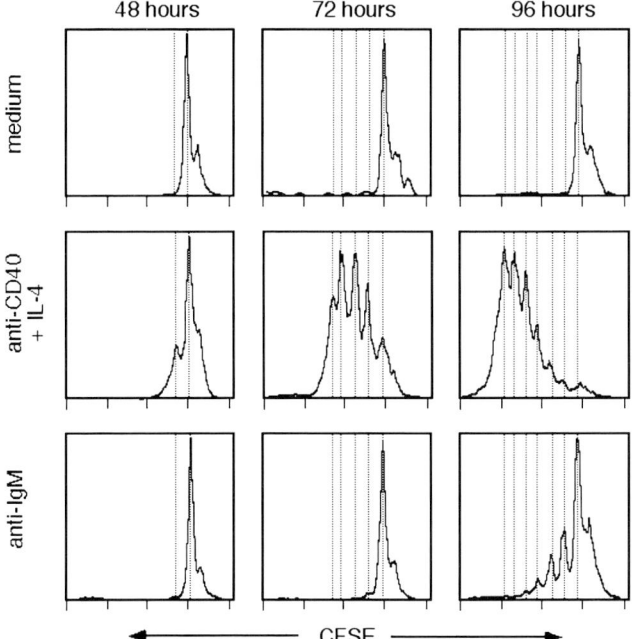

Fig. 3 Detection of B-cell cycling by CFSE fluorescence. Enriched splenic B cells were labeled with CFSE and incubated for the designated time periods in either complete medium alone, anti-CD40 (1C10; 20 μg/mL) plus 100 units/mL of recombinant IL-4, or with anti-IgM (b7–6; 10 μg/mL). After the culture period, cells were harvested washed and analyzed by flow cytometry. Vertical lines delineate increasing numbers of divisions

3. Add 10 μL of the PI stock solution (25 μg/mL) to the cell suspension and mix gently.
4. After addition of PI, analyze the cells immediately by flow cytometry *(35)*. If the experiment consists of a large number of samples, add MC540 and PI in a staggered manner to avoid incubation of the cells with MC540/PI over a prolonged period.
5. To score for all events (viable, apoptotic and dead) the forward and orthogonal gates should be left OPEN. Conventional scatter exclusion thresholds (for viable lymphocyte analysis) will lead to an underestimate of late stage apoptotic and dead cells.
6. Collect MC540 (in the PE channel) with logarithmic amplification; PI can be collected in either linear of logarithmic mode, depending upon the desired degree of separation. Collect at least 50,000 events per sample. An example of MC540 staining without or with PI staining can be seen in Fig. 4.

3.2.5 Differentiation of B Cells

1. Purified B cells can also be induced to undergo differentiation (IgM secretion) and class switch recombination (IgG and IgE secretion) when activated with

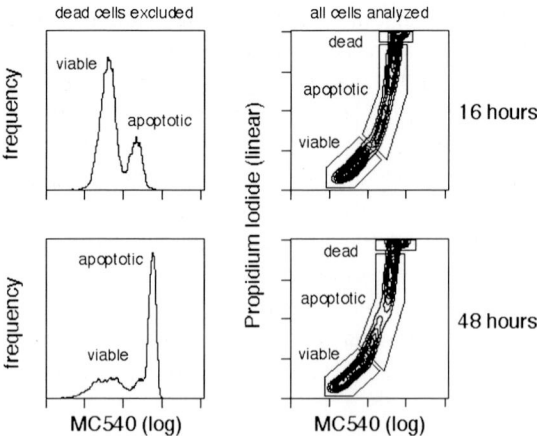

Fig. 4 Detection of B cell apoptosis with MC540 and PI staining. Purified murine splenic marginal zone B cells were stimulated in 96-well plates with anti-IgM for either 16 or 48 h, followed by harvest, washing, staining and flow cytometric analysis. Marginal zone B cells undergo rapid apoptosis after BCR ligation. The left panels (MC540 histograms) are obtained from the PI negative gate

appropriate stimuli. Both CD40 agonists and TLR ligands induce IgM secretion when cultured with B cells. To induce switching, key cytokines must be added. One can simulate TH2 conditions (leading to IgG1 and IgE production) by adding IL-4 along with CD40 agonists or TLR ligands (*36*). Similarly, one can add IFN-γ to mimic TH1 conditions (leading to IgG1 and IgG2 production (*37*)).

2. B cells, CD40 agonists, or TLR ligands ± cytokines are incubated for 6 days, after which culture supernatants are harvested and total IgM, IgG1, IgG2 and IgE assayed by isotype-specific ELISA (*see* **Note 10**).

3.3 *In Vivo Functional Assessment of B-Cell Competence*

In vitro experiments are useful for mapping functional responses of purified B cells, but they cannot provide insights into the overall in vivo state of humoral competence. Because a range of cellular interactions are required for successful T cell-independent (TI) and T cell-dependent (TD) Ab responses, challenge of the intact host is required. Immunization with both TI and TD antigens can be utilized, with readouts including antigen- and isotype-specific ELISA, germinal center formation and generation of memory cells.

3.4 *Immunization Procedures*

In general, mice are immunized by two different routes in order to induce B cell responses. An i.p. challenge will result in the concentration of antigen in the spleen, and strong Ab levels in serum. Subcutaneous immunization will engage the draining lymph nodes and also lead to measurable Ab levels in the serum.

A range of antigens can be used to assess in vivo competence of the humoral response. TI immunogens are typically carbohydrate in nature, and haptenated-Ficoll is a commonly used prototypic TI-2 antigen. TD antigens consist largely of protein, and are often haptenated. Although a range of proteins can be used to generate hapten-carrier conjugates, KLH is often the carrier of choice given its size and immunogenicity. Importantly, haptenated antigens (TI and TD) are very useful when assessing in vivo B cell responses given the ease by which hapten-specific ELISA can be designed. In addition to hapten-carrier conjugates, most of which are soluble molecules, one can immunize with large particulate antigens. The latter are highly immunogenic and quite useful when examining germinal center reactions.

3.4.1 Adjuvant-Based Challenge

Adjuvants have been an extremely useful tool to both the basic researcher and clinician in boosting Ab titers after immunization. It is now understood that adjuvants activate the innate immune system thereby creating an environment highly conducive to dendritic cell, T-cell, and B-cell activation. As such, adjuvants are commonly used with soluble protein antigens, which by their nature are not strong immunogens. Additionally, the use of adjuvants allows one to direct T cell-driven B cell responses along a TH1, TH2, or even TH "neutral" path. As discussed previously, CpG-containing oligonucleotides drive TH1 Ab profiles, alum precipitation results in a TH2 Ab pattern and Ribi generates a mixed TH1/TH2 response.

When challenging with an antigen/adjuvant mix, serum Ab will be measurable by the end of the first week and will increase during the second and third week (especially the switched isotypes). As such, it is important to determine serum titers at multiple time points to fully analyze the Ab producing capacity of the host. It is further advised to administer a secondary challenge with the same antigen/adjuvant mix. If given within 2–3 wk of the first immunization, the second challenge will test the ability of the animal to generate a "boosted" response (i.e., the ability to produce additional Ab producing cells during an ongoing response). If the secondary immunization is given more than a month after the primary injection, one can then test the capacity to generate a "recall" response (i.e. the ability to recruit and activate memory T cells and B cells). If secondary challenges are given, Ab titers should be measured one week post-injection.

3.4.1.1   Preparation of Alum [Al(OH)$_3$] Slurry

1. Pipet 2 mL of AlK(SO$_4$)$_2$ solution into a 15 mL polycarbonate tube.
2. Add 1 drop of 0.2% phenol red solution.
3. While stirring, add approximately 2 mL of 0.5 N NaOH dropwise. A heavy precipitate will form. Keep adding the NaOH until an orange-pink color appears.

4. Leave the mixture set for 10 min at room temperature.
5. Centrifuge at 200 g for 3 min at room temperature.
6. Discard supernatant.
7. Add 2 mL of deionized H_2O, shake the tube, and centrifuge again at 200 g for 3 min.
8. Discard supernatant.
9. Suspend the precipitate in 2 mL of 1X PBS and transfer into a bottle. Store at room temperature. The adjuvant contains approximately 10–12 mg/mL dry matter (*38–40*).

3.4.1.2 Adsorption of Protein Antigen onto Alum

1. Prepare a 1 mg/mL solution of protein (or hapten-protein) antigen in 1X PBS.
2. Add equal volumes of antigen solution and alum slurry into a small polypropylene tube and incubate for 30 min at 37°C.
3. After incubation, mix thoroughly and draw up into a syringe. Inject mice with necessary volume to obtain total desired amount of protein (e.g., 100 µg = 200 µL; *see* **Note 11**).

3.4.2 Nonadjuvant Particulate Antigens

In addition to soluble antigens administered with adjuvants, particulate antigens can also be used to generate powerful B-cell responses. Given the immunogenicity of large, complex particulate antigens, co-administration of an adjuvant is unnecessary. A commonly used particulate antigen for mouse studies is SRBC. This antigen is inexpensive and induces a marked response when given either i.p. or subcutaneously. Although not as useful for examining serum Ab titers (given the inherent difficulty in designing quantitative assays for Ab directed against complex particles with multiple epitopes), SRBC induce large germinal center responses. The same is true for other large complex particulate antigens, such as inactivated or attenuated viruses.

3.4.2.1 ELISA for TNP-Based Antigens

Hapten-protein conjugates offer the advantages of an immunogenic carrier coupled with a defined chemical moiety against which Ab can be formed. As such, one can induce a strong B cell response, and measure this response by designing an ELISA specific for the hapten. TNP is a commonly used hapten and can be conjugated (as 2,4,6-trinitrobenzenesulfonic acid) to a range of carrier proteins.

When designing an ELISA to measure TNP-specific Ab, one can take a conventional approach whereby TNP conjugated to a protein unrelated to the immunogen carrier is used as the "capture" substrate. Although this approach can work well, it may lead to an underestimate of low-titer anti-TNP isotypes (e.g., IgE anti-TNP Ab) as the result of substantial excess of other Ab classes (particularly IgM anti-TNP Ab) and competition for available epitopes. To avoid this potential pitfall, we have designed an ELISA whereby the capture reagent is specific for each mouse isotype (anti-IgM, IgG1, IgG2a, IgG2b, IgG3, IgA, IgE). Serum diluted appropriately (depending upon the anticipated concentration of each isotype) is added to the wells for capture, followed by washing. The detection agent is HGG coupled with both TNP and biotin. The TNP hapten will be bound by anti-TNP specific Ab captured on the plate, and the biotin portion will allow for binding of AP-streptavidin. The latter reagent catalyzes the final chromogenic reaction of the ELISA. In order to establish standard curves (and hence mass unit measurements), we have further obtained hybridomas secreting either IgM, IgG1, IgG2a, IgG2b, IgG3, IgA, or IgE anti-TNP mAb. In this manner, one can specifically quantitate the anti-TNP binding Ab activity for each isotype.

3.4.2.1.1   *Preparation of TNP-HGG-Biotin*

TNP Haptenation of HGG

1. Carefully dissolve 5 mg of HGG in a small volume of in 4% (w/v) $NaHCO_3$ and dialyze against the same buffer (at least 1 liter). Change the dialysis buffer once. Place the HGG solution in a small snap-top 1.5-mL volume polypropylene tube.
2. Add sufficient volume of the picrylsulfonic acid solution to the HGG to achieve a 30:1 (picrylsulfonic acid:HGG) molar ratio. This should result in an approximate final haptenation ratio of 20:1. (If starting with 5 mg of HGG, add 5.2 µL of picrylsulfonic acid solution). Cover the tube with foil and place in a 40°C water bath for 2–4 h.
3. Dialyze the TNP-HGG against PBS at 4°C protected from light (2 liter volume), with at *least* two buffer changes to remove unconjugated TNP.
4. Spin the TNP-HGG preparation at 9000 g in a high-speed microcentrifuge for 10 minutes to remove precipitate. (A small amount of denatured material is to be expected.)

Determination of TNP-HGG Concentration

1. Prepare a dilution of the TNP-HGG conjugate (1:20 to 1:40 range) in PBS buffer.
2. Determine absorption at 280 nm and 350 nm. Zero the spectrophotometer at both 280 and 350 with the PBS buffer (*see* **Note 12**).
3. Absorption of protein can be calculated as follows:

   Abs280 − (Abs350)(0.373) = corrected absorbance of protein (*see* **Note 13**).
4. Concentration of protein (*see* **Note 14**):

   corrected Absorbance of protein / extinction coefficient = concentration of HGG solution (mg/mL).
5. To determine the TNP-HGG concentration, multiply the value obtained in step 4 with the dilution factor (from step 1).

Biotin Conjugation of TNP-HGG

1. Dialyze TNP-HGG against 0.1 M NaHCO$_3$ changing the dialysis buffer once. Place the TNP-HGG in a small glass vial.
2. Weigh out a small amount of biotin and dissolve in DMSO to achieve a concentration of 1–2 mg/mL. This solution must be made fresh.
3. Add 120 μg of biotin per mg of HGG protein. Cover vial with foil and place on a rotator. Incubate for 4 h at room temperature.
4. Dialyze TNP-HGG-biotin conjugate against PBS/NaN$_3$ at 4°C protected from light (2 liter volume), with at least two buffer changes. Centrifuge if necessary to remove any precipitate.

3.4.2.2   ELISA for TNP-Specific IgM, IgG1, IgG2a, IgG3, and IgE

1. Coat Immulon 2 HB plates with 100 μL of the capture Ab diluted in coupling buffer. Leave one or two wells uncoated; these will be used as blank wells. Coat overnight at 4°C or for 2 h at 37°C. The capture Ab for each isotype with their recommended final concentration are listed below:

Goat anti-mouse IgM: recommended final concentration = 10 μg/mL
Goat anti-mouse IgG1: recommended final concentration = 5 μg/mL
Goat anti-mouse IgG2a: recommended final concentration = 5 μg/mL
Goat anti-mouse IgG2b: recommended final concentration = 10 μg/mL
Goat anti-mouse IgG3: recommended final concentration = 5 μg/mL
Goat anti-mouse IgA: recommended final concentration = 1 μg/mL
Rat anti-mouse IgE mAb (EM95): recommended final concentration = 10 μg/mL

2. Wash plates 3X in wash buffer.
3. Block plate with 200 μL of reaction buffer. Incubate 1 h at 37°C.
4. Wash plates 3X in wash buffer.
5. Add samples to the wells. The final volume for each serum sample should be 100 μL. Dilutions are done in the reaction buffer. The recommended final serum dilutions for each isotype specific anti-TNP antibodies are listed below:

IgM = 1:400–1:800
IgG1 = 1:200–1:400
IgG2a = 1:200–1:400
IgG2b = 1:200–1:400
IgG3 = 1:400–1:800
IgA = 1:200–1:400
IgE = 1:10

6. The standard curve for each isotype is established using anti-TNP specific mAb. Start the standard curve at recommended concentration and perform 1:2 serial dilutions until 10 concentration points are achieved. The final volume for each standard dilution should be 100 μL.

7. Dilutions are done in the reaction buffer. Incubate 1 h at 37°C. The recommended starting concentrations for the isotype-specific anti-TNP mAb used in our laboratory are as follows:

mouse IgM anti-TNP mAb - 4 µg/mL
mouse IgG1 anti-TNP mAb - 1 µg/mL
mouse IgG2a anti-TNP mAb - 1 µg/mL
mouse IgG2b anti-TNP mAb - 1 µg/mL
mouse IgG3 anti-TNP mAb - 4 µg/mL
mouse IgA anti-TNP mAb - 2 µg/mL
mouse IgE anti-TNP mAb - 2 µg/mL

8. Wash plates 3X in wash buffer.
9. Bound TNP-specific Ab are determined using the TNP-HGG-biotin detection reagent. The recommended final concentration for this reagent is 5–10 µg/mL. Dilute detection reagent in reaction buffer and add 100 µL to each well. Incubate 1 h at 37°C.
10. Wash plates 3X in wash buffer.
11. Dilute AP-streptavidin 1:500 in reaction buffer (final concentration is 3 µg/mL) and add 100 µL to each well. Incubate 1 h at 37°C.
12. Wash plates 3X in wash buffer.
13. Weigh out phosphatase substrate to achieve a concentration of 2 mg/mL in substrate buffer. Add 100 µL per well.
14. Observe plates for development of yellow color. When color becomes visible, read on ELISA plate reader at 405 and 540 nm. Do several readings so that an optimal standard curve is obtained. This is typically achieved when the highest concentration in the standard curve attains an optical density of at least 2.0.

3.4.2.3   ELISA for NP-Specific IgG1

When conjugated to protein carriers, the hapten (4-hydroxy-3-nitrophenyl) acetyl (NP) induces a genetically restricted response in C57Bl/6 (or other Igh^b) mice *(41)*. The responding B cells are drawn from a limited number of founding clones, most of which express an antigen receptor consisting of a V_H186.2 rearranged heavy chain and a λ light chain. With a narrow range of B cells comprising the response, one can reliably measure the affinity of Ab generated during the reaction. Importantly, this allows for determination of affinity maturation within the anti-NP response due to somatic hypermutation (e.g., Trp → Leu mutation at position 33 in the V_H186.2 gene) and selection. By measuring affinity increases of serum Ab with an ELISA over time, one can obtain an accurate measure of germinal center "health" because B-cell differentiation, somatic mutation and affinity-based selection must occur normally with the GC for the NP response to mature. As detailed below, the ELISA entails measurement of IgG1 anti-NP Ab against a high and low valency NP-conjugated carrier (capture reagents), and representing the data as a ratio of low affinity:high affinity binding Ab. Early in the response, the ratio is heavily weighted in favor of low affinity Ab (binding well to high valency but poor to low valency capture substrates). As somatic mutation and selection ensues, and high affinity anti-NP Ab dominate, the ratio normalizes (1:1) with Ab now capable of binding both low and high valency substrates equally well.

3.4.2.3.1 Generation of Hyperimmune Sera for NP ELISA Standard

1. Immunize 3–4 C57BL/6 mice i.p. with $100\,\mu g$ of NP-KLH precipitated with alum. Immunize with a high epitope density NP-KLH (i.e. NP_{25}-KLH or greater). When using alum as adjuvant, the majority of switched Ab will be of the IgG1 isotype.
2. After approx 6wk (42–45d) bleed out the mice and pool the sera. Freeze the pooled sera in small volume $(6–10\,\mu L)$ samples and store at $-80°C$.
3. Determine the level of total IgG1 present in the pooled serum standard using a murine IgG1-specific ELISA. When planning serum dilutions for this assay, anticipate a high titer of serum IgG1.
4. The total amount of IgG1 is then assigned an arbitrary unit number per mL to be used for subsequent quantitation in the standard curve. For example, if the total amount of IgG1 is found to be 8mg/mL, the unit value of 8000 Units/mL can be assigned.
5. The hyper-immune serum pool is used as standard for both the NP_{30}-BSA and NP_3-BSA ELISA. Each time a new standard is generated, it is important to test in the NP ELISA in order to determine the proper initial dilution. The success of the assay rests with the experimental values being contained within the linear portion of the standard curve.

3.4.2.3.2 NP ELISA

1. Coat two sets of Immulon 2HB plates, one with NP_{30}-BSA and the other with NP_3-BSA. Coat with $100\,\mu L$ of the capture reagents diluted in coupling buffer. The recommended final concentration for the NP-BSA conjugates is $50\,\mu g/mL$. Leave one or two wells uncoated; these will be used as blank wells. Coat overnight at 4°C or for 2h at 37°C.
2. Wash plates 3X in wash buffer.
3. Block plate with $200\,\mu L$ of reaction buffer. Incubate 1h at 37°C.
4. Wash plates 3X in wash buffer.
5. Add samples to the wells. The final volume for each serum sample should be $100\,\mu l$. Dilutions are done in the reaction buffer. When scoring for IgG1 anti-NP Ab, the recommended final serum dilution is 1:400 (this will depend upon the anticipated strength of the response). The standard curve is established using hyper-immune sera from C57BL/6 mice. The recommended initial dilution is 1:1000 (or if the starting activity is 8000U/mL, start the standard curve at 8U/mL). Perform 1:2 serial dilutions until 10 concentration points are achieved (or 0.016U/mL). Incubate 1h at 37°C.
6. Wash plates 3X in wash buffer.

7. Bound NP-specific IgG1 is detected using goat-anti-mouse IgG1 biotin. The recommended final concentration for this reagent is 5 μg/mL. Dilute the detection reagent in reaction buffer and add 100 μl to each well. Incubate 1 h at 37°C.

8. Wash plates 3X in wash buffer.

9. Dilute AP-streptavidin 1:500 in reaction buffer (final concentration is 3 μg/mL) and add 100 μL to each well. Incubate 1 h at 37°C.

10. Wash plates 3X in wash buffer.

11. Weigh out phosphatase substrate to achieve a concentration of 2 mg/mL in substrate buffer. Add 100 μL per well.

12. Observe plates for development of yellow color. When color becomes visible, read on ELISA plate reader at 405 and 540 nm. Do several readings so that an optimal standard curve is obtained. This is typically achieved when the highest concentration in the standard curve attains an O.D. of at least 2.0 (*see* **Note 15**).

13. Based on the standard curve, experimental samples are assigned a U/mL value. Importantly, the value obtained from the high valency capture reagent (NP_{30}-BSA) and low valency capture reagent (NP_3-BSA), can be represented as a ratio to provide a means by which affinity maturation can be followed over time.

3.4.3. Flow Cytometric Assays for GC and Memory B Cells

Determination of the quantity and quality of induced Ab after antigen challenge is an important means by which humoral competence can be measured. In addition, one can examine the in vivo events that lead to Ab formation after immunization with TD antigens. Two key processes that contribute to Ab producing B cells are GC induction and memory cell formation. Any TD antigen will generate both a GC reaction and population of memory cells in competent hosts. As discussed above, particulate antigens are preferred when studying GC as they are capable of invoking a marked GC B-cell population when injected either i.p. (to measure GC in the spleen) or subcutaneously (to examine GC in lymph nodes). In this regard, SRBC have proven to be a potent and reliable antigen when inducing and studying GC *(42)*. In some strains, up to 10% of a splenic suspension will consist of GC B cells after SRBC challenge (Fig. 5A). This becomes valuable when investigating the cellular and molecular events that occur within GC.

Although SRBC and other particulate Ag are potent immunogens, they do not provide an easy means by which Ag-reactive B cells can be identified and tracked. In general, it is difficult to follow the fate of Ag-specific B cells after activation in non-BCR transgenic strains. One approach to accomplish this is the use of a protein that is both TD antigen and fluorochrome. Based on the report of Hardy and co-workers utilizing Phycoerythrin (PE) as an immunogen and detection agent *(43)*, our laboratory has demonstrated PE to not only induce GC and memory B cell responses when administered to mice, but also to accurately enumerate antigen-specific B cells by flow cytometry. This technique works in any normal or diseased strain that can undergo a TH cell-driven B cell response, and eliminates the need to engineer BCR-transgenic mice. Towards this end, the use of PE has been of value

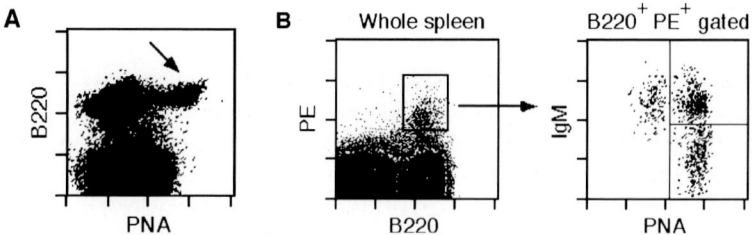

Fig. 5 Flow cytometric detection of germinal center (GC) B cells. Adult BALB/c mice were immunized with SRBC (**A**) and PE in alum (**B**) and germinal center responses were analyzed flow cytometrically. In A, the arrow demarcates the B220^hiPNA^hi GC population 8 d after SRBC challenge. In (**B**), the left panel demonstrates the PE-binding B cell population (B220^hiPE^+) 12 d after challenge. The right panel shows the IgM and PNA staining of the B220^hiPE^+ population. Note that most of the induced B220^hiPE^+ cells are PNA^hi and hence GC B cells. The plot further shows the distribution of IgM+ and IgM^- subsets within the antigen-specific GC population

to our laboratory examining the dysregulated B cell response in lupus-prone mice. Of note, immunization with PE leads to generation of a PE-specific GC response that constitutes approx 0.5%, and a memory cell response that totals ~0.05% of a total splenocyte suspension. As such, it is important to collect a high number of events ($>1 \times 10^6$) when performing flow cytometric analysis.

3.4.3.1   Flow Cytometric Analysis of the CG Response

3.4.3.1.1   *Preparation of SRBC for Immunization*

1. Add 1 mL of SRBC suspension to 9 mL 1X PBS and spin at 900 g for 10 min (*see* **Note 16**).
2. Aspirate PBS, vortex and wash again with 10 mL of PBS.
3. Bring up the volume to 2 mL with sterile 1X PBS.
4. Inject 200 μL of i.p. for studying splenic GC, or 50 μL in the foot-pad to examine the response in popliteal lymph nodes.

3.4.3.1.2   *Preparation of PE for Immunization*

1. Dialyze 0.5- to 1.0-mL PE solution (Chromaprobe) against 1X PBS (2 liter volume) changing the dialysis buffer once.
2. Remove dialyzed PE and measure its exact volume. This step is performed to calculate the concentration of PE. This is the PE stock solution.
3. Make a working solution of PE (2 mg/mL) in 1X PBS.
4. Store both the stock and working solution at 4°C. The working solution is used for immunizing mice (*see* **Note 17**).
5. Inject mice i.p with 200 μg of PE precipitated in alum. Using alum as adjuvant will result in a TH2-biased Ab response, with IgG1 being the most predominant switched isotype.

3.4.3.1.3   Detection of Murine GC B Cells by Flow Cytometry

1. After challenge with either SRBC or PE (for PE immunization, see section 4.2.4.2.1), spleens or popliteal lymph nodes are harvested at days 4, 8, 12, and 18 after immunization and stained to study the germinal center response.
2. To delineate the entire GC B cell population, stain the cell suspension with a pan-B cell reagent (anti-B220 or anti-CD19 mAb) and PNA (peanut agglutinin). GC B cells are B220^{hi} or CD19^{hi} and PNA^{hi} as shown in Fig. 5A. Other mAbs can be used to further subset the GC B cell compartment (reviewed in *(42)*; *see* **Note 18**).
3. When using PE as immunogen, the ensuing antigen-specific B cells can be detected with PE as described above. Cell suspensions are stained with a pan-B cell reagent (anti-B220 or anti-CD19 mAb), PNA and PE. GC B cells specific for PE are B220^{hi} or CD19^{hi}, PNA^{hi} and PE-binding, as shown in Fig. 5B. Additional mAbs can be used to further subset the PE-specific GC B cells. In Fig. 5B, an anti-IgM mAb was added to illustrate the ability to detect IgM positive and negative PE-binding GC B cell populations.

3.4.3.2   Enumeration of Murine Memory B Cells by Flow Cytometry

In the mouse, there is no distinct marker for memory B cells. Whereas a range of surface molecules have been defined that demarcate murine memory T cells, a clear marker or panel of markers for memory B cells does not exist. In order to track memory B cells, the PE system has again proven useful. Upon immunization of mice with PE, a GC response follows that will result, after several weeks, in a population of PE-specific memory B cells. These B cells can again be detected by PE binding, and hence their persistence in secondary lymphoid organs can be documented. In our laboratory, we are able to detect PE-specific memory B cells months after primary challenge.

3.4.3.2.1   Detection of PE-Specific Murine Memory B Cells With Flow Cytometry

1. Sacrifice mice immunized with PE at different time points after challenge and harvest splenocytes or lymph nodes.
2. To detect antigen specific memory B cells, cell suspensions are stained with a pan-B cell reagent (anti-B220 or anti-CD19 mAb) and PE. GC B cells specific for PE are B220^{hi} or CD19^{hi} and PE-binding, as illustrated in Fig. 6. Because memory B cells are present at very low frequency (~0.05%), it is essential to collect at least 1×10^6 events at the flow cytometer. To reliably collect this number of events, it will be important to scale up the staining procedure (e.g. to collect 2–3×10^6 events, start with 6×10^6 cells) (*see* **Note 19**).

Acknowledgments The authors wish to thank Ms Teresa Duling for expert help with flow cytometry. This work was supported by RO1 AA014400.

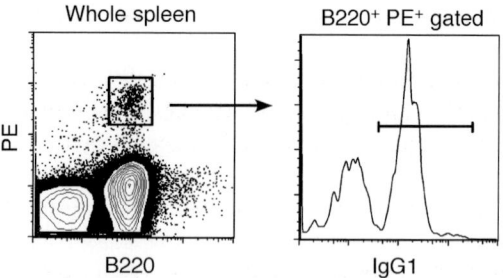

Fig. 6 Flow cytometric detection of memory B cells. Adult BALB/c mice were immunized with PE in alum and splenocytes harvested 24 wk after challenge. The left panel shows the PE-binding memory B cell population (B220^{hi}PE$^+$) still present in the spleen after 24 wk. The right panel illustrates the proportion of IgG1$^+$ and IgG1$^-$ cells within the memory B-cell subset

4   Notes

1. The use of EDTA helps prevent clumping of cells on the column; however, it will decrease cell viability if cells are left in the buffer too long. Always wash cells out of this buffer as quickly as possible. If cell viability is an issue, BSA can be substituted with 0.5% (v/v) bovine calf serum.

2. If examining IgG2 Ab responses in C57BL/6 mice (or other Igh^b strains) polyclonal anti-mouse IgG2c preparations are now commercially available.

3. Our laboratory stains cell samples for flow cytometric analysis using individual 1.5-mL tubes. The use of 96-well plates works equally well.

4. It is important that the slides thoroughly dry to prevent sections from detaching during the staining process.

5. It is critical not to contact the tissue section at any point during the staining process.

6. It is not necessary to wash the slides after the blocking step. The primary Ab solution can be added immediately after tapping and blotting.

7. Do not add too much Vectashield as this may result in a loose coverslip and tissue shearing. This will lead to a blurred image at the microscope.

8. If cells will be labeled at concentrations less than 10^7 cells/mL, the suspension buffer should include 5% serum.

9. If cells will be labeled at concentrations less than 1 × 10^7 cells/mL, wash cells 2X with 1X PBS/5% serum and perform the last wash in 1X PBS.

10. If B cells are obtained from C57BL/6 mice, it is important to note that this strain does not have the IgG2a constant region gene, but rather the murine IgG2c gene. When scoring for IgG2; therefore, ELISA must be designed to measure either IgG2b or IgG2c.

11. Storage: It is best to make up a fresh protein/alum immunogen mix for each series of immunizations.

12. NaN$_3$ will change the baseline absorption of buffers if included as preservative.

13. This correction is necessary because 37.3% of the 280 nm absorption will be due to TNP
14. The extinction coefficient value of IgG is typically 1.4.
15. With high titer, high affinity anti-NP samples, the chromogenic reaction can occur rapidly. Be prepared to read the ELISA plates quickly after addition of substrate.
16. Always use SRBCs within 4 wk of purchase. SRBC greater than 1 mo of age will lose potency.
17. The PBS buffers must not contain any preservative (e.g., NaN$_3$) if the PE is used for in vivo challenge.
18. PNA is a lectin specific for terminal galactosyl (β-1,3) N-acetylgalactosamine residues. These residues are upregulated on GC B cells.
19. When performing memory cell experiments, it is common to find Fc bound Ab on the surface of harvested B cells. To avoid a high background due to cytophilic Ab, it may be important to perform a brief low acid treatment with the cell suspension before staining.

References

1. Sternbach, G. L. (1990) Infections in alcoholic patients. *Emerg. Med. Clin. North Am.* **8**, 793–803.
2. MacGregor, R. R., and Louria, D. B. (1997) Alcohol and infection. *Curr. Clin. Top. Infect. Dis.* **17**, 291–315.
3. Baker, R. C., and Jerrells, T. R. (1993) Recent developments in alcoholism: immunological aspects. *Recent Dev. Alcohol.* **11**, 249–271.
4. Cook, R. T. (1998) Alcohol abuse, alcoholism, and damage to the immune system—a review. *Alcohol. Clin. Exp. Res.* **22**, 1927–1942.
5. Szabo, G. (1999) Consequences of alcohol consumption on host defense. *Alcohol Alcohol.* **34**, 830–841.
6. Nelson, S., and Kolls, J. K. (2002) Alcohol, host defense and society. *Nat. Rev. Immunol.* **2**, 205–209.
7. Happel, K. I., and Nelson, S. (2005) Alcohol, immunosuppression, and the lung. *Proc. Am. Thorac. Soc.* **2**, 428–432.
8. Ruiz, M., Ewig, S., Torres, A., Arancibia, F., Marco, F., Mensa, J., Sanchez, M., and Martinez, J. A. (1999) Severe community-acquired pneumonia. Risk factors and follow-up epidemiology. *Am. J. Respir. Crit. Care. Med.* **160**, 923–929.
9. de Roux, A., Cavalcanti, M., Marcos, M. A., Garcia, E., Ewig, S., Mensa, J., and Torres, A. (2006) Impact of alcohol abuse in the etiology and severity of community-acquired pneumonia. *Chest.* **129**, 1219–1225.
10. MacGregor, R. R. (1986) Alcohol and immune defense. *JAMA.* **256**, 1474–1479.
11. Messingham, K. A., Faunce, D. E., and Kovacs, E. J. (2002) Alcohol, injury, and cellular immunity. *Alcohol.* **28**, 137–149.
12. Cook, R. T., Waldschmidt, T. J., Ballas, Z. K., Cook, B. L., Booth, B. M., Stewart, B. C., and Garvey, M. J. (1994) Fine T-cell subsets in alcoholics as determined by the expression of L-selectin, leukocyte common antigen, and beta-integrin. *Alcohol. Clin. Exp. Res.* **18**, 71–80.
13. Cook, R. T., Ballas, Z. K., Waldschmidt, T. J., Vandersteen, D., LaBrecque, D. R., and Cook, B. L. (1995) Modulation of T-cell adhesion markers, and the CD45R and CD57 antigens in human alcoholics. *Alcohol. Clin. Exp. Res.* **19**, 555–563.

14. Song, K., Coleman, R. A., Alber, C., Ballas, Z. K., Waldschmidt, T. J., Mortari, F., LaBrecque, D. R., and Cook, R. T. (2001) TH1 cytokine response of CD57+ T-cell subsets in healthy controls and patients with alcoholic liver disease. *Alcohol.* **24**, 155–167.

15. Laso, F. J., Madruga, J. I., Lopez, A., Ciudad, J., Alvarez-Mon, M., San Miguel, J., and Orfao, A. (1996) Distribution of peripheral blood lymphoid subsets in alcoholic liver cirrhosis: influence of ethanol intake. *Alcohol. Clin. Exp. Res.* **20**, 1564–1568.

16. Mili, F., Flanders, W. D., Boring, J. R., Annest, J. L., and DeStefano, F. (1992) The associations of alcohol drinking and drinking cessation to measures of the immune system in middle-aged men. *Alcohol. Clin. Exp. Res.* **16**, 688–694.

17. Cook, R. T., Waldschmidt, T. J., Cook, B. L., Labrecque, D. R., and McLatchie, K. (1996) Loss of the CD5+ and CD45RA[hi] B cell subsets in alcoholics. *Clin. Exp. Immunol.* **103**, 304–310.

18. acanella, E., Estruch, R., Gaya, A., Fernandez-Sola, J., Antunez, E., and Urbano-Marquez, A. (1998) Activated lymphocytes (CD25+ CD69+ cells) and decreased CD19+ cells in well-nourished chronic alcoholics without ethanol-related diseases. *Alcohol. Clin. Exp. Res.* **22**, 897–901.

19. Degos, F., Duhamel, G., Brechot, C., Nalpas, B., Courouce, A. M., Tron, F., and Berthelot, P. (1986) Hepatitis B vaccination in chronic alcoholics. *J. Hepatol.* **2**, 402–409.

20. Mendenhall, C., Roselle, G. A., Lybecker, L. A., Marshall, L. E., Grossman, C. J., Myre, S. A., Weesner, R. E., and Morgan, D. D. (1988) Hepatitis B vaccination. Response of alcoholic with and without liver injury. *Dig. Dis. Sci.* **33**, 263–269.

21. Nalpas, B., Thepot, V., Driss, F., Pol, S., Courouce, A. M., Saliou, P., and Berthelot, P. (1993) Secondary immune response to hepatitis B virus vaccine in alcoholics. *Alcohol. Clin. Exp. Res.* **17**, 295–298.

22. Pillot, J., Poynard, T., Elias, A., Maillard, J., Lazizi, Y., Brancer, M., Dubreuil, P., Budkowska, A., and Chaput, J. C. (1995) Weak immunogenicity of the preS2 sequence and lack of circumventing effect on the unresponsiveness to the hepatitis B virus vaccine. *Vaccine.* **13**, 289–294.

23. Johnson, R. D., and Williams, R. (1986) Immune responses in alcoholic liver disease. *Alcohol. Clin. Exp. Res.* **10**, 471–486.

24. Smith, W. I., Jr., Van Thiel, D. H., Whiteside, T., Janoson, B., Magovern, J., Puet, T., and Rabin, B. S. (1980) Altered immunity in male patients with alcoholic liver disease: evidence for defective immune regulation. *Alcohol. Clin. Exp. Res.* **4**, 199–206.

25. Drew, P. A., Clifton, P. M., LaBrooy, J. T., and Shearman, D. J. (1984) Polyclonal B cell activation in alcoholic patients with no evidence of liver dysfunction. *Clin. Exp. Immunol.* **57**, 479–486.

26. Giron, J. A., Alvarez-Mon, M., Menendez-Caro, J. L., Abreu, L., Albillos, A., Manzano, L., and Durantez, A. (1992) Increased spontaneous and lymphokine-conditioned IgA and IgG synthesis by B cells from alcoholic cirrhotic patients. *Hepatology.* **16**, 664–670.

27. Mishell, R. I., and Dutton, R. W. (1967) Immunization of dissociated spleen cell cultures from normal mice. *J. Exp. Med.* **126**, 423–442.

28. Leptin, M., Potash, M. J., Grutzmann, R., Heusser, C., Shulman, M., Kohler, G., and Melchers, F. (1984) Monoclonal antibodies specific for murine IgM I. Characterization of antigenic determinants on the four constant domains of the mu heavy chain. *Eur. J. Immunol.* **14**, 534–542.

29. Heath, A. W., Wu, W. W., and Howard, M. C. (1994) Monoclonal antibodies to murine CD40 define two distinct functional epitopes. *Eur. J. Immunol.* **24**, 1828–1834.

30. Hardy, R. R., and Hayakawa, K. (2001) B cell development pathways. *Annu. Rev. Immunol.* **19**, 595–621.

31. Thomas, M. D., Srivastava, B., and Allman, D. (2006) Regulation of peripheral B cell maturation. *Cell Immunol.* **239**, 92–102.

32. Mebius, R. E., and Kraal, G. (2005) Structure and function of the spleen. *Nat. Rev. Immunol.* **5**, 606–616.

33. Lyons, A. B., and Parish, C. R. (1994) Determination of lymphocyte division by flow cytometry. *J. Immunol. Methods.* **171**, 131–137.

34. Hodgkin, P. D., Lee, J. H., and Lyons, A. B. (1996) B cell differentiation and isotype switching is related to division cycle number. *J. Exp. Med.* **184**, 277–281.

35. Mower, D. A., Jr., Peckham, D. W., Illera, V. A., Fishbaugh, J. K., Stunz, L. L., and Ashman, R. F. (1994) Decreased membrane phospholipid packing and decreased cell size precede DNA cleavage in mature mouse B cell apoptosis. *J. Immunol.* **152**, 4832–4842.

36. Erickson, L. D., Foy, T. M., and Waldschmidt, T. J. (2001) Murine B1 B cells require IL-5 for optimal T cell-dependent activation. *J. Immunol.* **166**, 1531–1539.

37. Davis, H. L., Weeratna, R., Waldschmidt, T. J., Tygrett, L., Schorr, J., and Krieg, A. M. (1998) CpG DNA is a potent enhancer of specific immunity in mice immunized with recombinant hepatitis B surface antigen. *J. Immunol.* **160**, 870–876.

38. Gupta, R. K., Rost, B. E., Relyveld, E., and Siber, G. R. (1995) Adjuvant properties of aluminum and calcium compounds, in *Vaccine Design: The Subunit and Adjuvant Approach* (Powell, M. F., and Newman, M. J., eds.), Plenum Press, NY, pp. 229–248.

39. Rinella, J. V., Jr., White, J. L., and Hem, S. L. (1996) Treatment of aluminum hydroxide adjuvant to optimize the adsorption of basic proteins. *Vaccine.* **14**, 298–300.

40. Weissburg, R. P., Berman, P. W., Cleland, J. L., Eastman, D., Farina, F., Frie, S., Lim, A., Mordenti, J., Nguyen, T. T., and Peterson, M. R. (1995) Characterization of the MN gp120 HIV-1 vaccine: antigen binding to alum. *Pharm. Res.* **12**, 1439–1446.

41. Kelsoe, G. (1995) In situ studies of the germinal center reaction. *Adv. Immunol.* **60**, 267–288.

42. Wolniak, K. L., Shinall, S. M., and Waldschmidt, T. J. (2004). The germinal center response. *Crit. Rev. Immunol.* **24**:39–65.

43. Hayakawa, K., Ishii, R., Yamasaki, K., Kishimoto, T., and Hardy, R. R. (1987) Isolation of high-affinity memory B cells: Phycoerythrin as a probe for antigen-binding cells. *Proc. Natl. Acad. Sci. USA.* **84,** 1379–1383.

21
Histological Analysis of Bone

Urszula T. Iwaniec, Thomas J. Wronski, and Russell T. Turner

Summary Bone is an important target tissue for alcohol. Moderate alcohol consumption may slow bone loss during aging, but alcohol consumption inhibits bone growth during adolescence, and alcohol abuse in adults is an important risk factor for osteoporosis. Various techniques have been applied for evaluating the impact of alcohol on bone, including densitometry for assessment of bone mass and density, computed tomography for evaluation of bone microarchitecture, serum biochemistry for measurement of markers of global bone resorption and formation, and histomorphometry for assessment of cellular activity. Of these methods, histomorphometry is the gold standard for assessing bone because it is the only method for the direct *in situ* analysis of bone cells and their activities. The procedures described in this chapter provide tools for the histomorphometric characterization of the effects of alcohol on cancellous and cortical bone growth and turnover.

Specifically detailed are processes for embedding, cutting, staining, and evaluating histological bone specimens with a focus on rodent models.

Keywords Bone; skeleton; histomorphometry; osteoblasts; osteoclasts.

1   Introduction

The skeleton is a target organ for alcohol, and alcohol abuse is considered to be one of the most important lifestyle risk factors for osteoporosis. Because of the many limitations associated with performing studies in humans, animal models are a mainstay of research directed toward understanding the physiological effects of alcohol consumption on bone metabolism *(1)*. Many of the detrimental effects of alcohol on bone were first identified in animal models *(1)*. Alcohol consumption is a potent inhibitor of bone growth in animals and reduces peak bone mass (Fig. 1). In laboratory animals, as in humans, alcohol consumption also inhibits bone remodeling *(2–4)*. Although progress is continually being made, many unanswered questions remain regarding the effects of alcohol on bone. For example, the etiology of alcohol-induced bone loss is poorly understood. Specifically, comorbidity factors such as

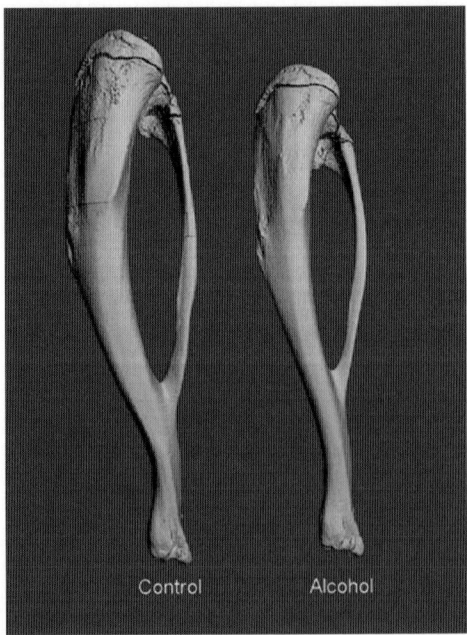

Fig. 1 Alcohol consumption is a potent inhibitor of bone growth as illustrated by the tibias removed from rats fed control and alcohol diets during growth

nutrition and pattern of alcohol consumption are thought to be important in the etiology of osteoporosis, but the roles of these variables have not been explored in depth. The interaction of alcohol and pharmaceuticals intended to treat osteoporosis is also largely unexplored, as is the mechanism for the putative beneficial effects of moderate alcohol consumption on bone *(5)*. Many studies in humans and animal models have focused on changes in biochemical markers and bone mass and density in response to alcohol. However, to fully understand the effects of alcohol on the skeleton, it is necessary to also measure the responses of bone cells to alcohol consumption. Bone histomorphometry is the gold standard for evaluating changes in bone cell number and activity, and the methods that are used in our laboratories to prepare and measure bone specimens for routine quantitative histomorphometric evaluation are the focus of this chapter. For techniques used by other laboratories, please see the *Handbook of Histology Methods for Bone and Cartilage (6)*.

2   Materials

2.1   Fluorochrome Labeling

1. Calcein injection solution (*see* **Note 1**): For 10 mL of injection solution, at a calcein concentration of 10 mg/mL, 0.1 g of calcein (Sigma, cat. no. C0875), 0.2 g of sodium bicarbonate (Sigma, cat. no. S8875), and 10 mL of 0.9% sterile saline are combined and allowed to stir for a few minutes until completely dissolved. The solution is then transferred from the beaker to a rubber top injection vial covered with aluminum foil.
2. Declomycin injection solution: The declomycin injection solution is prepared by combining 0.1 g of declomycin (Sigma, cat. no. D6140) in 10 mL of 0.9% sterile saline. The solution is allowed to stir vigorously for at least 2 h until completely dissolved.

2.2   Tissue Collection

1. 20-mL glass scintillation vials (Fisher, cat. no. 03-337-5).
2. specimen containers (Fisher, cat. no. 14-375-148).
3. ethanol.

2.3   Fixation

Phosphate buffered formalin solution: To prepare 100 mL of 10% phosphate buffered formalin, combine 10 mL of 37% formaldehyde (Fisher, cat. no. F79-500), 90 mL dH$_2$O, 1.86 g of sodium phosphate monobasic (Fisher, cat. no. S369-500), and 0.42 g of sodium hydroxide (Fisher, cat. no. S392-212). Adjust the pH to 7.0-7.4.

2.4 Embedding

Preparation of the solutions for infiltration and embedding of bone tissue is conducted under a fume hood. Water is removed from the methyl methacrylate using calcium chloride. We combine calcium chloride (Fisher, cat. no. C614-10) pellets (100-mL volume beaker) with methyl methacrylate (Fisher, cat. no. O3629-4) (200 mL) in a Nalgene container with a tight-fitting lid. The mixture is shaken vigorously for about 15 s, allowed to settle for approx 30 s, and filtered (Fisher, cat. no. 5802-240). Desired amounts of dibutyl phthalate (Fisher, cat. no. D-30) and benzoyl peroxide (Aldrich Chemical, cat. no. 179981) are then added to the filtered methyl methacrylate. The combined reagents are stirred for 2–5 h as described to follow:

Solution 1: methyl methacrylate: 8.5 mL per bone, dibutyl phthalate: 1.5 mL per bone; solution is stirred for approximately 2 h under the hood.

Solution 2: methyl methacrylate: 8.5 mL per bone, dibutyl phthalate: 1.5 mL per bone, benzoyl peroxide: 0.1 g per bone; solution is stirred for approximately 4 h under the hood.

Solution 3: methyl methacrylate: 8.5 mL per bone, dibutyl phthalate: 1.5 mL per bone, benzoyl peroxide: 0.25 g per bone; solution is stirred for a minimum of 4 h under the hood.

Solution 4: methyl methacrylate: 15.3 mL per bone, dibutyl phthalate: 2.7 mL per bone, benzoyl peroxide: 0.45 g per bone; solution is stirred for a minimum of 5 h under the hood.

2.5 Sectioning Undecalcified Bone

1. Glass slides (Fisher, cat. no. 12-550-34).
2. Staining trays (Electron Microscopy Sciences, cat. no. 70312-24).
3. Gelatin solution: To make sufficient gelatin for 100 slides, 600 mL of dH$_2$O is mixed with 10.5 g of 300 bloom gelatin (Sigma, cat. no. G2500) in a flask with a stir bar. While stirring, the solution is heated to 60°C, left at 60°C for 5 min, followed by addition of 30 mL glycerin (Fisher, cat. no. G33-500).

2.6 Staining

1. Permount (Fisher, cat. no. SP15-100).
2. Glass coverslips (Fisher, cat. no. 12-548-5A).

All staining solutions are prepared on the day of staining. The quantities described to follow are sufficient to stain two jars of slides (i.e., 16 slides).

Solution 1, silver nitrate: 3 g of silver nitrate (Fisher, cat. no. S181-25) is added to 60 mL of dH$_2$O. The solution is stirred in an aluminum foil-covered beaker (silver nitrate is light sensitive) and transferred into a foil-covered staining jar when completely dissolved.

Solution 2, sodium carbonate-formaldehyde: 5 g of anhydrous sodium carbonate (EMD, cat. no. SX0395-1), 25 mL of formaldehyde (Fisher, cat. no. F79-500), and 75 mL dH$_2$O are combined and stirred until completely dissolved.

Solutions 3A and 3B, Farmer's diminisher: For solution 3A, 10 g of sodium thiosulfate (Fisher, cat. no. S446-500) is added to 100 mL of dH$_2$O. For solution 3B, 1 g of potassium ferricyanide (Fisher, cat. no. P232-500) is added to 10 mL of dH$_2$O. Both solutions are stirred until completely dissolved. 50 mL of solution 3A is combined with 2.5 mL of solution 3B before use (solution stable for less than 30 min).

Solution 4, tetrachrome (*see* **Note 2**): 1.8 g of tetrachrome (Polysciences, cat. no. 02783) is added to 60 mL of dH$_2$O and vigorously stirred (in foil-covered beaker) for at least 3 h. The solution is heated on a low setting for the last hour of stirring, and subsequently filtered into a foil-covered staining jar.

2.7   *Ground Cortical Sections*

1. Catalyst (Tap Plastics; MEKP catalyst, cat. no. 08478).
2. Plastic (Tap Plastics; tap casting resin, cat. no. 00191).

3   Methods

3.1   *Fluorochrome Labeling*

Fluorochrome labeling is used to identify sites undergoing bone mineralization and for measurement of mean osteoblast activity and overall bone formation rate (*7*). We use the fluorochrome calcein for labeling cancellous bone. Young rats (<6 mo old) are typically injected subcutaneously at 10 and 2 d before necropsy wheras older rats (>6 mo old) are injected subcutaneously at 14 and 4 d before necropsy (*see* **Note 3**). The longer labeling interval in the older rats is to compensate for the age-related decrease in matrix deposition and mineralization. This fluorochrome regimen results in deposition of double labels at bone surfaces that are actively mineralizing at the time of label injection (*see* **Note 4**). For a dose of 15 mg/kg, 0.15 mL per 100 g body weight is injected subcutaneously in the dorsal scapular region of the rat.

Many investigators limit their analysis to cancellous bone. However, alcohol has important effects on cortical bone. For biological (slow bone mineralization rate) and technical (thicker tissue sections) reasons, labeling intervals optimized for cancellous bone measurements are too closely spaced for their analysis in cortical bone. This problem can be generally circumvented by administering three fluorochrome labels, the first of which is injected 2 or more weeks before necropsy. The interval between labels 1 and 3 is optimized for cortical bone and the interval between labels 2 and 3 is optimized for cancellous bone. Label number 1 must be different than labels 2 and 3 to allow its unambiguous identification. We often use declomycin followed by two calcein labels when labeling both cortical and cancellous bone in the same animal. Using the triple-label protocol, cortical bone measurements are performed using the first and third labels, whereas cancellous bone measurements are performed using the second and third labels. Figure 2 shows examples of fluorochrome labeling of cancellous and cortical bone.

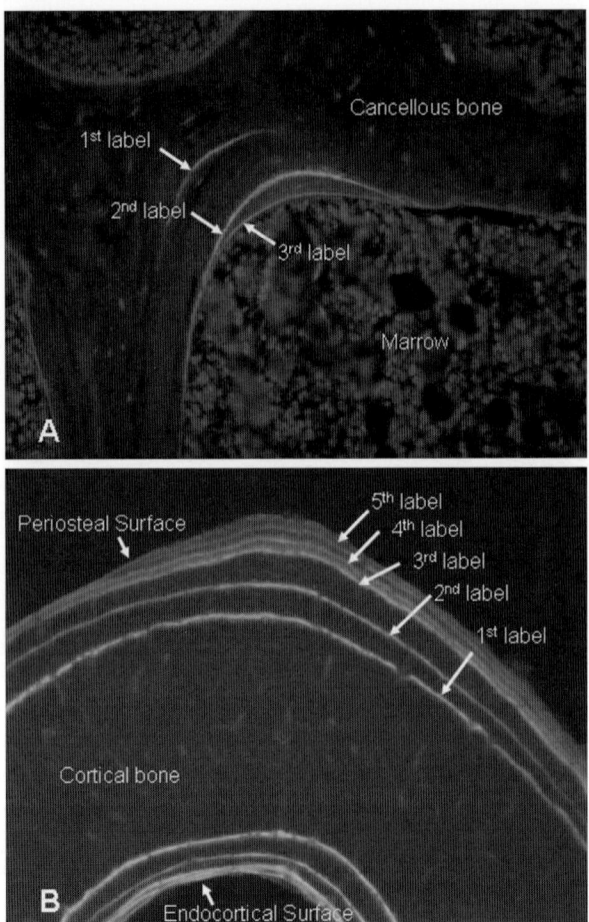

Fig. 2 Fluorochrome labels in cancellous (**A**) and cortical (**B**) bone. (**A**) 5-μm thick, unstained, plastic-embedded section viewed under ultraviolet light. The faint first label is declomycin. Labels 2 and 3 (calcein) can be used to measure double label surface and interlabel width. These two endpoints are then used to calculate bone formation rate. (**B**) A 20-μm thick, unstained, unembedded ground section viewed under ultraviolet light. Calcein was administered every 2 wk to a rapidly growing rat. Note that the labels get closer together, indicating a decrease in osteoblast activity, as the animal aged

3.2   Tissue Collection

The data obtained from histomorphometry are usually limited to relatively small amounts of tissue. There is no single location that is representative of the entire skeleton. Observations confined to a single portion of the skeleton should not be generalized to the entire skeleton. It is, therefore, recommended that studies be performed at sampling sites representative of the axial and appendicular skeleton and of cortical and cancellous bone.

Lumbar vertebrae are representative of the axial skeleton. There is no great advantage as to which vertebra is used. To minimize damage, it is easier to excise several vertebrae and remove adhering soft tissue ex vivo. Because it is easy to confuse vertebra, especially after excision, we recommend simultaneous removal of lumbar vertebrae and the last thoracic vertebra with rib intact to facilitate orientation.

The tibia is commonly used for assessment by bone histomorphometry in rat studies. The proximal tibia (rich in cancellous bone) and tibia-fibula synostosis (exclusively cortical bone) are representative of the appendicular skeleton. The distal femur is an acceptable alternative to the proximal tibia as a cancellous sampling site. Some investigators use a mid-shaft site in the femur for assessment of cortical bone. However, because alcohol decreases longitudinal bone growth and the proximal and distal ends of long bones grow at different rates, it is better to standardize skeletal sampling sites based on anatomical location. This is more easily accomplished in the tibia because the tibia and fibula are fused throughout post-natal life. Other investigators chose the tibia for histomorphometry because they prefer to perform mechanical testing on the femur. The tibia is excised easily using a scalpel. Cutting tendons adhering to the distal portion of the bone allows for easy muscle removal.

3.3   Fixation

We use numerous fixation techniques. Our choice depends upon how the tissue will be processed and analyzed. For routine studies, we use 70% alcohol. Technically, alcohol is not a fixative because it does not cross link macromolecules. However, alcohol is adequate for preservation of cellular detail (Fig. 3). The principal advantages of using alcohol are lower toxicity compared with fixatives, availability and low cost, ability to ship samples from remote locations in 70% alcohol, and superior retention of activity of the enzymes alkaline phosphatase (osteoblasts) and acid phosphatase (osteoclasts) in alcohol-preserved tissue compared with fixed tissues. A good alternative to 70% alcohol is phosphate-buffered formalin. Formalin-fixation of tissues may improve cellular detail. If the researcher is interested in in-situ enzyme localization, it is important to perform the formalin fixation at 4°C and remove the fixative solution and replace it with 70% alcohol within 24 h. Other fixatives are used for specialized applications such as immunohistochemistry and electron microscopy *(8,9)*.

Fix tissue capped in vials in 10% phosphate-buffered formalin for only 24 h and transfer to 70% ethanol for storage until processing. Long-term storage of bone samples in this formalin solution should be avoided as partial decalcification of bone will eventually occur. Use a volume of fixative sufficient to completely cover the tissue. We typically fix bones in 20-mL glass scintillation vials or, for larger volumes of fixative and multiple bone samples, in specimen containers.

3.4 Embedding

To achieve proper sectioning for analysis of cancellous bone, the substrate that the bone is embedded in must match the hardness of the bone tissue. The method described in this section retains fluorochrome labels for dynamic histomorphometry and is suitable for measurements of bone cells following staining. However, the method is not suitable for immunohistochemistry. Most methods for facilitating antigen retrieval use demineralized paraffin-embedded tissue (8). Also, investigators who intend to perform enzyme localization may want to consider alternative low temperature embedding procedures (10).

We use methyl methacrylate for embedding undecalcified bone in 20 mL glass vials. Prior to embedding (at necropsy), the bones are faced off (exposure of marrow cavity by removal of the cranial surface of long bones and ventral surface of vertebrae) with a razor blade to allow for penetration of the plastic and easy positioning. The bones are dehydrated in increasing concentrations of ethanol and cleared in xylenes (Fisher, cat. no. X5–4) according to the schedule outlined to follow:

Fresh 70% ethanol	1 d in solution (start on Monday)
95% ethanol	2 d in solution (Tuesday and Wednesday)
100% ethanol	1 d in solution (Thursday)
100% ethanol	3 d in solution (Friday)
Xylenes	1 d in solution (Monday)

All solutions, except xylene, are chilled to 4°C, and all bones in solution are stored at 4°C. Bones can remain in all solutions, except xylene, for 2–3 extra days if necessary. Bones should not stay in xylene for more than 24 h. Bones are switched to new 20-mL glass vials following clearing in xylene. After dehydration, the bones are infiltrated in solutions composed of methyl methacrylate, dibutyl (n-butyl) phthalate (softener) and benzoyl peroxide (catalyst). The bones go through a series of four different solution changes.

After a solution has been mixed, it is poured into the glass vials containing the bone and, with the exception of solution 4, placed at 4°C for 2 or more days. Vials with solution 4 remain at room temperature overnight in a hood. The following morning, the vials (uncapped) are placed in a dessicator attached to a vacuum line. The vacuum is initiated and released at thirty minute intervals over 6-8 hours to ensure infiltration of plastic into bone. At the end of the day, the bones are positioned, using a dissection probe, at the center of the bottom of the vial with the "faced-off" surface facing the bottom. The vials (with lids unscrewed or holes drilled in lids) are then placed in a water bath (43°C) overnight to accelerate polymerization (*see* **Note 5**). The follow-ing morning, the plastic is checked for stage of solidification. Vials that are fully polymerized are removed from the water bath, left in the fume hood overnight, and transferred to 4°C (to facilitate cracking of glass). If there is a liquid plastic layer thicker than 1 mm on the top, the water temperature is increased (between 44 and 47°C) to accelerate the solidification of the remaining unpolymerized plastic. To obtain the specimen block, the chilled vials are placed in a towel (to avoid injury), cracked with a hammer, and the broken glass removed. Any sticky, partially polym-erized plastic is removed with a razor blade. Prior to sectioning, the blocks are placed in a vice clamp and trimmed with a saw (Fig. 4).

Care should be taken when disposing of methyl methacrylate solutions 1–3. The methyl methacrylate waste should be kept at 4°C or hydroquinone (Fisher, cat. no. H329-500) should be added to the waste bottles to prevent accidental uncontrolled polymerization capable of rupturing the waste container.

3.5   Sectioning Undecalcified Bone

The goal of sectioning is to obtain defect-free histological specimens. The tissue sections used for quantitative histomorphometry must be consistent in thickness and anatomical location (Fig. 5). Dynamic histomorphometry must be performed on undecalcified sections in order to retain fluorochromes that had been deposited into the mineralizing bone matrix. Fluorochrome labels appear sharper in thinner sections but are brighter in thicker sections. Thus, we recommend using the thinnest unstained sections that can be routinely produced by your microtome in which the labels can be adequately visualized. This thickness generally ranges from 4 to 12 μm. Cell measurements can be performed by staining the plastic embedded undecalcified sections as described below. To optimize cellular detail, use the thin-nest sections that can be routinely prepared.

We normally cut 4-μm-thick longitudinal sections, which are suitable for both staining and visualization of fluorochromes labels. Sectioning undecalcified bone requires a specialized heavy duty microtome (e.g., Leica 2265 microtome). The blocks are placed in the microtome using specialized holders and trimmed to a desired anatomical position within the bone. The sections are lifted off the knife with a pair of fine-tipped tweezers and placed on gelatinized slides. Gelatinization of slides (see below) allows for superior adherence of sections. After placing the section on a gelatinized slide, the section is flattened using a fine paintbrush dipped in 40% alcohol (*see* **Note 6**). The section is then covered with a 1 cm × 1 cm piece of plastic (Glad brand sandwich bags work well) and air dried for approximately 15–30 minutes. Typically 6–10 sections/bone are cut. Excess gelatin is scraped off from both the front and back of the slide with a razor blade and multiple slides (e.g., 6–10) are placed in between two blank slides and clamped with two small c-clamps. The clamped slides are then placed in a drying oven at 53°C overnight. The next morning, the slides are unclamped, separated, and the plastic covers removed from the sections. The best sections with minimal artifacts are then chosen for analysis.

Gelatinized slides are prepared prior to sectioning to facilitate section adhesion. The slides (Fisher, cat. no. 12-550-34) are precleaned by placing in staining trays (Electron Microscopy Sciences, cat. no. 70312-24) and sequential immersion in the following 5 solutions for 2 minutes each: 70% ethanol, 95% ethanol, 95% ethanol/HCl (50 mL 12 N HCl per 450 mL 95% ethanol), 95% ethanol, 95% ethanol. The slides are then placed in an oven at 53°C overnight to dry and transferred into Coplin jars (eight slides per jar). The slides are immersed in Coplin jars with the gelatin solution for one hour. The slides are then removed from the Coplin jars and placed into staining trays and allowed to air dry overnight.

3.6   Staining

As with fixation techniques, we utilize a variety of staining techniques. Our favorite, Von Kossa with tetrachrome counterstain, stains bone black, gives beautiful detail to cellular structures (Fig. 3) and is described here in detail. The sections need to have the plastic dissolved before staining. This is best accomplished by placing them in methoxyethyl acetate in Coplin jars for 3 d.

Solution 1–4 are prepared as described on page 328–329. Because solution 4 stirs the longest, it is prepared first followed by preparation of the other solutions. One hour prior to finishing stirring of solution 4, the specimens are placed in staining trays and immersed in the following solutions for 2 min each: 100% alcohol, 95% alcohol, 70% alcohol, 40% alcohol, and dH$_2$O. The specimens are then stained in solution 1 for 15 min (in the aluminum covered beaker), rinsed in 3 changes (1 min each) of dH$_2$O, stained in solution 2 for 2 min (the second set of slides is placed in solution 1 at this time), again rinsed in 2 changes (1 min each) of dH$_2$O, stained in solution 3 for 20 s, rinsed in running tap H$_2$O for 30 mins, and run through 2 changes (1 min each) of dH$_2$O. The specimens are then stained in solution 4 (in the aluminum foil covered beaker) for 2 min and immersed in 3–5 changes (1 min each) of dH$_2$O to remove excess stain. Afterwards, the specimens are placed in 70% ethanol for at least 1 min and the color of the stain evaluated using a microscope. If the color is too dark, the specimens are returned to the 70% ethanol until the desired color reduction is achieved. Once the color is optimized, the specimens are placed in 100% ethanol for at least 1 minute, immersed in 2 changes (3 min each) of xylenes, and coverslipped. For coverslipping, one to two drops of Permount (Fisher, cat. no. SP15-100) are deposited on the slide over the section using a transfer pipet and a glass coverslip (Fisher, cat. no. 12-548-5A) is placed over it. Enough Permount is deposited to completely fill the space between the coverslip and slide. To avoid bubbles, one edge of the coverslip is placed onto the slide while supporting the coverslip vertically with the transfer pipet and slowly laying the coverslip onto the Permount. After the coverslip is allowed to settle over the section, it is clamped (black binder clips from Office Max work well) and allowed to dry overnight. The clamps are then removed and the specimens allowed to air dry at room temperature for several days before excess Permount is removed from the slide with

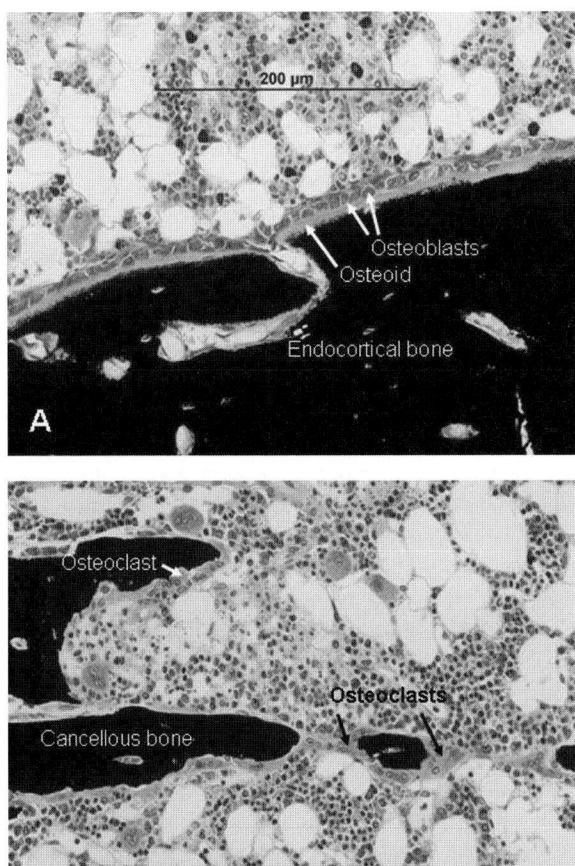

Fig. 3 Bone sections fixed in alcohol and stained according to the von Kossa method with a tetrachrome counterstain. Osteoblasts lining the endocortical bone surface are shown in A and numerous osteoclasts can be seen in B. Note the unmineralized matrix (osteoid) underlying the osteoblasts (**A**) and the absence of osteoid on inactive and resorbing surfaces (**B**)

a razor blade. Sections for assessment of fluorochromes are coverslipped with Permount unstained. Prior to coverslipping, the slides are cleaned around the periphery of the section with a razor and 40–70% ethanol.

3.7 Ground Cortical Sections

Obtaining quality cross sections of undemineralized cortical bone for dynamic histomorphometry using a conventional microtome is challenging. We embed cortical bone (usually distal part of tibia with tibio-fibular junction intact) in plastic and use a rotary saw to cut cross sectional slices through the bone that are approximately 150 μm thick. The specimens are then ground to a thickness of approx 20 μm.

Before embedding, the bones are dehydrated in 70% alcohol (1 day), 95% alcohol (1 day), 100% alcohol (10 changes over 2–3 d, 1–2 h/change), and acetone (10 changes over 2–3 d, 1–2 h/change). To facilitate cutting of embedded samples, a basal layer of bioplastic is prepared while the bones are dehydrating. Six to seven drops of catalyst are added to 100 mL of liquid plastic and stirred for 1 hour. Approximately 2.5 mL of the solution is then poured into 20-mL glass scintillation vials. The vials are capped (loosely) and placed in a water bath at 43–45°C overnight for polymerization to occur.

On the day of embedding, fresh plastic solution is prepared. The dehydrated bones are placed horizontally, with the flat medial side of the bone facing downwards, in a vial (one bone/vial) on top of the hardened basal layer of plastic. A second layer of the bioplastic solution is added to the vial to cover the bone. The vials (uncapped) are then placed in a vacuum dessicator connected to a vacuum line for 6–8 hours with intermittent release of the vacuum at thirty-minute intervals. At the end of the day, the vials (capped loosely) are placed in a water bath at 43–45°C overnight for the plastic to polymerize.

The plastic blocks with bone are mounted on a rotary saw and cut into 150 μm-thick sections. One section, proximal to the tibio-fibular junction, is chosen for grinding. The grinding can be performed using a ground glass plate that is roughened with aluminum oxide powder (#400 grit) or sand paper. Bone cross sections are hand ground between the glass and a cork (from a wine bottle) using a circular motion with water as a lubricant. The proper thickness is achieved when the bone sections become transparent. The ground sections are mounted onto slides, coverslipped, and used to measure the dimensions of the cortical bone tissue (e.g., medullary area and bone cross sectional area) and fluorochrome labels. Alternatively, the bone can be cut with a rotary saw and ground unembedded. This alternative procedure produced the cortical bone section shown in Fig. 2.

3.8   Histomorphometry

All variables listed below are measured at a magnification of 100–200x. Standard bone histomorphometry nomenclature is used (*11*). Semiautomated digitizing software systems are available and calculate the histomorphometric endpoint described below (e.g., Osteomeasure, OsteoMetrics, Inc., Atlanta, GA). However, the measurements can be performed manually using a Merz grid and the endpoints calculated manually. We typically evaluate cancellous bone in the lumbar vertebra and the proximal metaphysis of the tibia. The region of interest for data collection in lumbar vertebrae consists of secondary spongiosa at distances greater than 0.5 mm from the cranial and caudal growth plates. In the proximal tibial metaphysis, the region of interest begins 0.5–1.0 mm distal to the growth plate/metaphyseal junction and includes secondary spongiosa only.

1. Cancellous bone volume/total tissue volume (%): The percentage of mineralized cancellous bone within a specified volume of total tissue (bone + bone marrow). Cancellous bone volume/total volume can be measured in either stained or unstained sections.

2. Trabecular thickness (µm): The thickness of individual trabeculae; calculated as 2/(bone surface/bone volume) *(11)*.

3. Trabecular number (/mm): The number of trabeculae intersected by a randomLy drawn line; calculated as (bone volume/total volume)/ trabecular thickness *(11)*.

4. Trabecular separation (µm): The mean distance between trabeculae; calculated as (1/trabecular number) – trabecular thickness *(11,12)*.

5. Osteoid surface/bone surface (%): The percentage of cancellous bone surface covered with osteoid. Increased osteoid surface may be due to increased bone formation or impaired bone mineralization.

6. Osteoblast surface/bone surface (%): The percentage of cancellous bone surface covered with osteoblasts. Osteoblast surface is an index of bone formation. Osteoblasts are identified as an array of columnar shaped cells located adjacent to bone surfaces. Usually, a thin layer of osteoid can be observed beneath the osteoblasts. The appearance of osteoblasts changes with age. The osteoblasts are much larger in young animals than in aged animals.

7. Osteoclast surface/bone surface (%): The percentage of cancellous bone surface covered with osteoclasts. Osteoclasts are identified as multinucleated cells having a foamy appearing cytoplasm located within a resorption cavity on a bone surface. Osteoclast surface is an index of bone resorption.

8. Single labeled surface/bone surface (%): The percentage of cancellous bone surface with a single fluorochrome label.

9. Double-labeled surface/bone surface (%): The percentage of cancellous bone surface with a double fluorochrome label.

10. Mineralizing surface/bone surface (%): The percentage of cancellous bone surface with a double fluorochrome label plus half the percentage of cancellous bone surface with a single label. Mineralizing surface is a dynamic index of bone formation. Alternatively, mineralizing surface is simply defined as double label surface/bone surface.

11. Mineral apposition rate (µm/d): The distance between the two fluorochrome markers that comprise a double label divided by the number of days between label administrations. Mineral apposition rate is an index of osteoblast activity.

12. Bone formation rate ($\mu m^3/\mu m^2/d$): Bone formation rate is equivalent to the volume of new bone formed per unit of total bone surface per unit time and is calculated by multiplying mineralizing surface by mineral apposition rate. Bone formation rate is perhaps the most meaningful index of bone turnover in general and bone formation in particular.

13. Longitudinal bone growth (µm/d) in rats <7 mo old: The average distance (measured at 5 equidistant points) between the growth plate-metaphyseal junction and the second fluorochrome label administered before necropsy divided by the time interval between the second label administration and necropsy.

Fig. 4 An embedded bone block that has been trimmed for sectioning

Cortical bone areas and perimeters are generally measured at a magnification of 20x while fluorochrome-based indices of bone formation are measured at magnifications of 100–200x.

1. Total tissue area (mm^2): Combined area of cortical bone and bone marrow in a cross section of the tibial diaphysis.
2. Marrow area (mm^2): The area within the cross section occupied by marrow.
3. Cortical bone area (mm^2): The area within the cross-section occupied by cortical bone. Cortical area is calculated by subtracting marrow area from total tissue area.
4. Cortical width (mm): The distance between the periosteal and endocortical surfaces is measured in each cross section of the tibial diaphysis at four equally spaced sites (cranial, caudal, medial, and lateral). The sums of the measurements are averaged to calculate a mean cortical width.

The following indices of bone formation are measured at the periosteal and endocortical surfaces as described above for cancellous bone: single labeled surface/bone surface (%), double labeled surface/bone surface (%), mineralizing surface/bone surface (%), mineral apposition rate (µm/d), and bone formation rate (µm^3/µm^2/d).

3.9   Conclusion

The methods detailed in this paper provide tools for characterization of the effects of alcohol on cancellous and cortical bone growth, architecture, and turnover. They also provide reliable methods for measuring osteoblast and osteoclast number. The activity of osteoblasts is determined from measurements of mineral apposition rate and bone formation rate. The bone resorption rate is difficult to measure directly, but

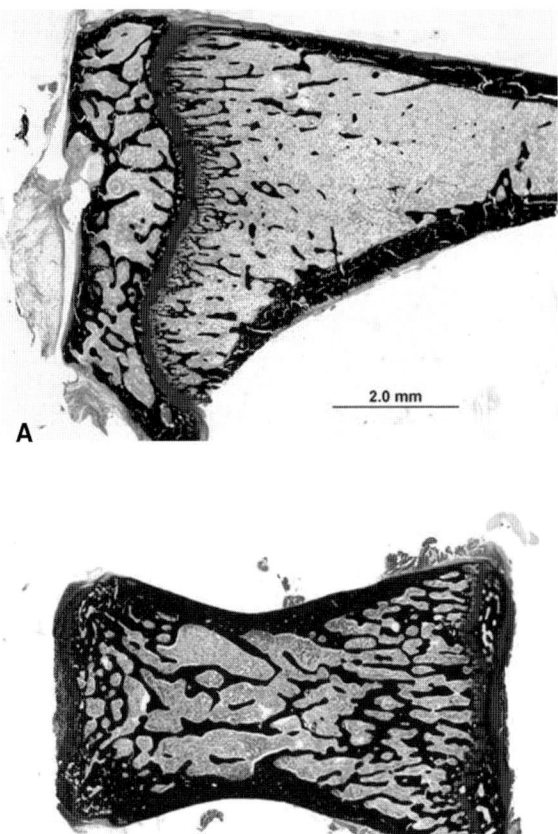

Fig. 5 Representative sections from the proximal tibia (**A**) and lumbar vertebra (**B**) of a rat stained as in **Fig. 3**. The anisotropic nature of the distribution of cancellous bone in both the tibia and vertebra demonstrates that extreme care must be executed to evaluate well-defined and reproducible sampling sites

changes in levels of bone resorption can often be inferred from osteoclast number, bone formation rate, and change in bone volume. Although described predominantly for the rat, the methods in this chapter are easily adapted for smaller and larger animals. Nondestructive analysis of bone using, DXA and microCT can provide additional data related to bone mass and three-dimensional microarchitecture, respectively. Biochemical blood and urine markers of bone turnover can be used to evaluate global changes in bone resorption and formation. These other methods compliment, but are rarely an adequate substitute for bone histomorphometry. In terms of sensitivity and specificity, bone histomorphometry remains the gold standard for evaluating bone.

4   Notes

1. We normally prepare the calcein solution on the day of injection in a sterile beaker covered with aluminum foil (calcein is light sensitive).
2. Because solution 4 stirs the longest, it is prepared first followed by preparation of the other solutions.
3. Calcein should result in a very bright green label. The temporal order of the two labels in a histological section is known at sites where both labels are present. However, at sites where only one label is present, it may not be possible to determine whether the label was the first or the second administered. If this is crucial for the experimental design, an alternative procedure must be employed. The simplest alternative is to use two bone-seeking fluorochromes that fluoresce different colors. We recommend the combination of declomycin (first label), which fluoresces yellow, and calcein (second label), which fluoresces green. There are other fluorochromes that can be used (e.g., alizarin red). However, it is important to consider the excitation and emission wavelengths of the labels. Calcein and declomycin can generally be visualized simultaneously (using a single filter), which facilitates measurement of double labels. In contrast, it is difficult to simultaneously visualize declomycin and alizarin red. Declomycin fluoresces less brightly than calcein. We, therefore, recommend using a higher dose (20 mg/kg) and subcutaneous administration over two days. Alternatively, declomycin and calcein can be administered juxta tail vein. This route may improve bioavailability. However, if the tail vein route is used, it is recommended that the fluorochrome concentrations be increased such that no more than 0.1 mL of total volume is administered. Although the label should fluoresce well, UV filter efficiencies and detector sensitives (if an image analysis system is used) vary greatly. Therefore, it is also advised that a small pilot study be performed to assure the proper dose of declomycin to be used for labeling the bone.
4. We, at times, inject labels on two successive days (e.g., on day 11 and 10 and then on day 3 and 2) or administer the labels twice in the same day (e.g., AM and PM). Although generally unnecessary, these alternatives reduce the risk that an animal will need to be excluded from a study because of poor fluorochrome labeling. The aforementioned schedule recommendation does not apply to mice. Mice have a very high turnover rate for cancellous bone and as a consequence, the label intervals need to be decreased. The optimal labeling schedule in mice is age- and strain-dependent.
5. If the vials have tightly attached lids without holes, the plastic embedding solution will overheat and bubbling may occur. The polymerization is performed in a hood because the fumes are toxic and have an unpleasant odor.
6. Weakly adhering sections are more likely to fall off a slide or develop cracks in bone during staining. Wavy sections are more difficult to measure because only a portion of the region on the slide being measured will be in focus. In order to prevent waves/wrinkles when placing sections on slides, the surface of the slide for section placement is saturated with 40% alcohol.

References

1. Turner, R. T. (2000) Skeletal response to alcohol. *Alcohol Clin. Exp. Res.* **24,** 1693–1701.
2. Turner, R. T., Greene, V. S., and Bell, N. H. (1987) Demonstration that ethanol inhibits bone matrix synthesis and mineralization in the rat. *J. Bone Miner. Res.* **2,** 61–66.
3. Turner, R. T., Aloia, R. C., Segel, L. D., Hannon, K. S., and Bell, N. H. (1988) Chronic alcohol treatment results in disturbed vitamin D metabolism and skeletal abnormalities in rats. *Alcohol Clin. Exp. Res.* **12,** 159–162.
4. Hogan, H. A., Sampson, H. W., Cashier, E., and Ledoux, N. (1997) Alcohol consumption by young actively growing rats: a study of cortical bone histomorphometry and mechanical properties. *Alcohol Clin. Exp. Res.* **21,** 809–816.
5. Rapuri, P. B., Gallagher, J. C., Balhorn, K. E., and Ryschon, K. L. (2000) Alcohol intake and bone metabolism in elderly women. *Am. J. Clin. Nutr.* **72,** 1206–1213.
6. An, Y. H., and Martin, K. L. (2003) *Handbook of Histology Methods for Bone and Cartilage.* Humana Press, Totowa, NJ.
7. Erben, R. G. (2003) Bone-labeling techniques, in An, Y. H., Martin, K. L. (eds.) *Handbook of Histology Methods for Bone and Cartilage.* Humana Press, Totowa, NJ, p 99–117.
8. Kremer, M., Quintanilla-Martinez, L., Nahrig, J., von Schilling, C., and Fend, F. (2005) Immunohistochemistry in bone marrow pathology: a useful adjunct for morphologic diagnosis. *Virchows Arch.* **447,** 920–937.
9. Boivin, G., Anthoine-Terrier, C., and Obrant, K. J. (1990) Transmission electron microscopy of bone tissue. A review. *Acta Orthop. Scand.* **61,** 170–180.
10. Turner, R. T., Evans, G. L., and Wakley, G. K. (1994) Reduced chondroclast differentiation results in increased cancellous bone volume in estrogen-treated growing rats. *Endocrinology* **134,** 461–466.
11. Parfitt, A. M., Drezner, M. K., Glorieux, F. H., Kanis, J. A., Malluche, H., Meunier, P. J., Ott, S. M., and Recker, R. R. (1987) Bone histomorphometry: Standardization of nomenclature, symbols, and units. *J. Bone Miner. Res.* **2,** 595–610.
12. Parfitt, A. M., Mathews, C. H.E, Villanueva, A. R., Kleerekoper, M., Frame, B., and Rao, D. S. (1983) Relationship between surface, volume, and thickness of iliac trabecular bone in aging and osteoporosis. *J. Clin. Invest.* **72,** 1396–1409.

22

Assessing Effects of Alcohol Consumption on Protein Synthesis in Striated Muscles

Thomas C. Vary and Charles H. Lang

Summary The development of alcoholic muscle disease, which affects both cardiac and skeletal muscle, leads to increased morbidity and mortality in patients who abuse alcohol. The disease pathology includes myocyte degeneration, loss of striations, and myofilament dissolution, which is consistent with alterations in structural and myofibrillar proteins. One explanation for the changes in myofibrillar architecture is that the expression of cellular proteins may be compromised by ethanol consumption. The dynamic balance of proteins in striated muscle is dependent upon rates of protein synthesis and protein degradation. We have shown that protein synthesis is depressed in striated muscle after either acute alcohol intoxication or chronic alcohol ingestion. The loss of myofibrillar proteins occurs prior to any detection of abnormal muscle function in vivo. It is therefore of major importance to evaluate the regulation of protein turnover after ethanol consumption. This review describes protocols to study protein synthesis either in vivo or under in vitro conditions. The methods can be modified for studies involving transgenic mice allowing mechanisms responsible for the defects in protein synthesis to be dissected.

Keywords Phenylalanine; flooding-dose; gastrocnemius; heart; ethanol.

L. E. Nagy (ed.), *Alcohol: Methods and Protocols*
© Humana Press 2008

1   Introduction

The loss of muscle mass and protein that characterizes alcoholic muscle disease is accompanied by radical alterations in the fluxes of metabolites between and within organs. Although whole-body studies can describe the alterations in the overall metabolic flux, they do not provide any information as to which particular organ system is affected. Knowledge of changes in individual tissue beds, which involves both protein and energy metabolism, was gained initially by studying substrate concentrations and net fluxes across organs. However, these techniques have limited applicability in rodent models of alcoholism because cannulation of blood vessels in individual muscles is difficult and the amount of blood that can be withdrawn for measurement of substrates without affecting metabolism is limited. Furthermore, there are numerous assumptions relating to pool sizes and compartmentation that seriously impact on the accuracy of the methods. To circumvent these problems, the incorporation of radioactive amino acids into tissue proteins can be used *(1)*.

The main difficulty in using radioisotopes lies in the heterogeneity of the amino acid precursor pool. This problem was recognized early on *(2)*. The specific activity of amino acid in tissues was lower than that in the plasma because of dilution of the label with unlabelled amino acid originating from tissue protein degradation. Since that time, there have been many studies suggesting the transfer ribonucleic acid (tRNA), the direct precursor of protein synthesis, is charged from a pool of different isotopic enrichment from that in either the plasma or the tissue free amino acid pool *(see [3,4])*. Direct measurement of tRNA is not generally feasible for routine studies because the pool is extremely small and labile. One method that largely avoids this problem is to inject the label not in tracer amounts but together with a large bolus dose of unlabelled amino acid *(5,6)*. This bolus is much larger than the endogenous pool of free amino acid, which becomes inundated by the exogenous amino acid of uniform specific radioactivity activity. This procedure, referred to as the "flooding-dose" method, therefore aims to equalize the radioisotopic enrichment in all free amino acid pools, thus simplifying the measurement of precursor specific radioactivity.

The "flooding-dose" procedure has been widely adopted for studies in animals *(see [4])* because it involves less restraint of the animal and does not require the maintenance of a metabolic steady state during extended periods of infusion, in addition to the advantages outlined above. We and others have used this protocol to demonstrate both acute and chronic ethanol intoxication limits protein synthesis in cardiac and skeletal muscle *(7–16)*.

2   Materials

1. Radioactive phenylalanine dilution: slowly add 1.242 g of L-phenylalanine to 50 mL of 0.9% (wt/vol) NaCl (physiological saline) with stirring. Warm the solution to ensure L-phenylalanine completely goes into solution. Add 3 mCi of L-[^{3}H]-phenylalanine to mixture.

2. Heparinized saline: Add 1250 units of heparin (1000 USP units/mL) to a 1-L bottle of sterilized 0.9% (wt/vol) NaCl.

3. Biruet reaction mixture: Add 30 g of NaOH to 300 mL and stir until dissolved. Once the NaOH is dissolved, bring up to 900 mL. Add 6 g of NaK Tartrate. Add 1 g of KI. Add 1.5 g of $CuSO_4 \cdot 5H_2O$, which had previously been dissolved in 5 mL of H_2O. Add 15 g of deoxycholic acid. Bring up volume to 1 L and store at room temperature.

4. Dabsyl chloride reagent: rinse two borosilicate 20-mL scintillation vials with high-performance liquid chromatography (HPLC)-grade acetonitrile (made up fresh on day the dabyslation is performed). Dissolve 13 mg of dabsyl chloride (Pierce Chemical Co.) (used without additional recrystallization) in 10 mL of acetonitrile in one of the vials. Put cap on vial and vortex vigorously to get dabsyl chloride into solution (gently heat if necessary). Filter using a syringe with a Nalgene™ (0.2-µm cellulose acetate membrane) attached into the other vial. Make sure the solution is clear. If not, then repeat vortxing, heating, and filtering steps.

5. HPLC buffer:

   a. Buffer A: Add 400 mL of a 100 mM sodium citrate (pH 6.5) (made up in HPLC water) with 600 mL of HPLC water in a brown glass bottle. Add 1 L of HPLC water while stirring. Add 160 mL of dimethylformamide and bring up to 3 L with HPLC water while stirring. Bring up to 4 L with HPLC water while stirring. Filter using Nalgene cellulose membrane (0.22 µm, 47 mm) and degas. Store in a brown bottle.

   b. Buffer B: Add 1200 mL of Buffer A (*see* Subheading 2, step 5a) with 800 mL of HPLC-grade acetonitrile. Add 112 mL of dimethylformamide and bring up to 3 L with HPLC-grade acetonitrile. Add another liter of HPLC grade acetonitrile. Filter using Nalgene nylon membrane filter (0.2 µm, 47 mm) and degas. Store in a brown bottle.

3   Methods

Rates of protein synthesis can be assessed in striated muscle in vivo using the protocols described to follow. These measurements provide an accurate picture of effects of both hormones and metabolites on rates of protein synthesis at any given time. Rates of protein synthesis in vivo are sensitive to nutritional and hormonal status of the animal (17). This procedure has been used extensively in dogs (18), pigs (19), and rats (6,7,13,14,20–23) (*see* Note 1). This approach has also been adapted to use stable isotope-labeled phenylalanine for measurement of protein synthesis in humans (1).

However, when performing in vivo experiments, it may be necessary to either 1) limit the in vivo hormonal or metabolite influences or 2) accentuate a response independent of other changes. Both can be accomplished by using in vitro preparations such as the incubated epitrochlearis muscle wherein the particular metabolite or

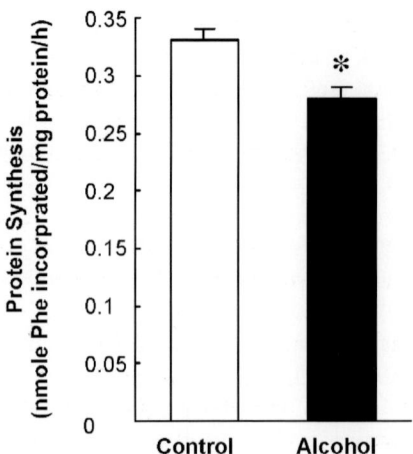

Fig. 1 Effect of ethanol on protein synthesis in incubated epitrochlearis muscle. Epitrochlearis from control rats were excised and incubated in vitro in the presence (50 mM) (Alcohol) or absence (Control) of ethanol as described in the Methods section. Rates of protein synthesis were measured by the incorporation of ^{3}H-phenylalanine into muscle proteins over a 1-h time interval. Values shown are means ± SE for n = 17 in each incubation condition. *p < 0.005 vs Control (no ethanol)

hormone is added without corresponding changes in the whole-body fluxes of other substances. For example, insulin or insulin-like growth factor can be injected, but this causes corresponding decreases in plasma amino acid and glucose concentrations in vivo. The incubated epitrochlearis muscle has been used to analyze the effects of various hormones on protein synthesis without changes in metabolic substrates or compensatory hormones *(24–27)*. In the present studies, we used this technique to analyze the effect of ethanol per se on protein synthesis in skeletal muscle (Fig. 1). The addition of 50 mM ethanol to the incubation medium results in a 15% decrease in protein synthesis in epitrochlearis, indicating that ethanol per se has a direct effect on the protein synthetic machinery in skeletal muscle. One advantage to this method is that it can also be used to quantitate rates of myofibrillar protein breakdown by measuring the release of tyrosine and/or 3-methylhistidine *(24,26,28)*.

3.1   *Measurement of Protein Synthesis In Vivo in Rats*

1. Prefill 1-mL syringe with saline.
2. Attach PE 50 tubing to a 22-gauge needle, cut free end of tubing on a bevel, and connect needle to a 1-mL syringe filled with heparinized saline. Remove all air from needle and tubing.

3. Anesthetize rat with Nembutal or other appropriate agent.
4. Make a mid-neck incision through the skin.
5. Using blunt dissection, isolate the left carotid artery and place a straight iris forceps under vessel.
6. Loop a piece of 2-O nylon suture thread around proximal and distal ends of carotid artery.
7. Tie off the distal end of the carotid artery and put tension on the vessel by gently pulling the nylon suture and clamping onto the skin with a mosquito forceps.
8. Place a microvascular clamp on the carotid artery proximal to the looped nylon suture material.
9. Using microvascular scissors, nick the carotid artery approximately a third of the way through the vessel.
10. Gently cannulate the vessel and advance the cannulate caudally through the suture loop to microvascular clamp and tie suture around cannula (2–3 cm).
11. Release the microvascular clamp. There should not be any bleeding and the arterial pulse should be visible in the saline filled cannula.
12. Pull back on the syringe until only blood enters the syringe.
13. Rapidly switch syringes on the cannula to another 1-mL syringe and remove 1 mL of blood. Retain the blood-filled syringe.
14. Rapidly switch syringes on the cannula to another 1-mL syringe that had been rinsed with heparin. This blood sample represents the baseline for blood constituents. The blood should be put into an Eppendorf tube, capped, and stored on ice until centrifuged.
15. Replace the 1 mL of blood withdrawn in step 13 by infusing another 1 mL of physiologic saline.
16. Isolate the jugular vein.
17. Using a 26-gauge needle, infuse the L-[^{3}H]-phenylalanine solution (1 mL/100 g body weight) into the jugular vein and start a timer (e.g., time "0").
18. At 1 min, 45 s, repeat steps 12 through 15. This is the 2-min blood sample.
19. At 5 min, 45 s, repeat steps 12 through 15. This is the 6-min blood sample.
20. At 8 min, snip the skin around the ankles and gently tease the skin away from underlining fascia.
20. At 9 min, 45 s, repeat steps 12 through 15. This is the 10-min blood sample.
21. Dissect the soleus, cut, and freeze between clamps precooled to the temperature of liquid nitrogen. Record the exact time the soleus was frozen. Quickly, weigh the soleus and wrap the frozen muscle in aluminum foil and store at −80°C.
22. Repeat step 21 except dissect the gastrocnemius.
23. Make an incision below the zyphoid plexus, and cut the diaphragm. Open the chest cavity.
24. Remove the heart and freeze between clamps precooled to the temperature of liquid nitrogen. Record the exact time the heart was frozen. Quickly, weigh the heart and wrap the frozen heart in aluminum foil and store at −80°C.
25. Centrifuge blood samples from times T_2, T_6, and T_{10} and save the plasma.

3.2   Measurement of Protein Synthesis In Vivo in Mice

1. Mice are injected ip with L-[³H]-phenylalanine solution (1 mL/100 g body weight).
2. Wait 15 min.
3. Decapitate mice using a guillotine and collect truncal blood in pediatric ethylene diamine tetraacetic acid-coated tubes.
4. Dissect muscles as described above for rats and freeze between clamps pre-cooled to the temperature of liquid nitrogen.

3.3   Preparation of [³H]Phenylalanine Plasma Samples for HPLC

1. Mark and weigh Eppendorf tubes A and B for each plasma sample.
2. Add 100 μL of 10% TCA solution to tube A and weigh.
3. Add 100 μL of each plasma sample and vortex.
4. Reweigh.
5. Centrifuge for 3–4 min at the maximum speed.
6. Decant supernatant into corresponding tube B.
7. Reweigh tubes A and B.
8. pH supernatant to 9.0 (range 8.9–9.2).
9. Reweigh Tube B.

3.4   Dabsylation of Plasma Amino Acids

1. Add 40 μL of 100 mM sodium bicarbonate to each glass tube (50 mm × 6 mm). Add 5 μL of 2 mM norleucine as internal standard.
2. Add 40 μL of the pH 9.0 sample or 5 mM phenylalanine solution (standard) to corresponding tube.
3. Add 80 μL of dabsyl chloride solution and place a silicon stopper in the tube.
4. Place stoppered tubes in heating block at 70°C.
5. Vortex after 1 min.
6. Vortex after 4 min.
7. After 12 min remove from heating block and let cool for 5 min.
8. Add 220 μL of dilution buffer (equal volumes of 50 mM Na$_2$HPO$_4$ [pH 7.0] mixed in HPLC-grade water and 100% ethanol; mixed and filtered).
9. Invert tubes to mix.
10. Begin HPLC analysis as soon as possible (samples remain stable for 5 d).

3.5   HPLC Method

1. The HPLC system (Beckman Instruments) consists of dual pumps with an in-line visible light absorbance detector connected to the outflow tubing of the analytical column. The separation of dabsylated amino acids is accomplished using a guard column composed of a 5-µm Adsorbosphere C18 universal connect and an analytical column (250 mm × 4.6 mm) composed of 5 µm Econosphere (Grace Davison Discovery Sciences, Deerfield, IL). The in-line scintillation radioactivity counter (Beckman Model 171) is connected to the outflow track of the absorbance detector. The scintillant (Ready Flow II, Beckman Instruments) is pumped at a ratio of 7 (scintillant):1 and radioactivity is reported as DPM using proper corrections to account for quenching.
2. The dabsylated amino acids are separated at a flow rate of 1.4 mL/min. The flow rate is maintained constant throughout the program.
3. Initially, the mobile phase is 60% Buffer A and 40% Buffer B.
4. Approximately 200–300 µL of sample is injected and the volume is dependent upon radioactivity in the sample.
5. The % Buffer B is increased over 12 minutes to 100% using a linear gradient.
6. The % Buffer B is maintained at 100% for an additional 3 min.
7. After 15 min, the % Buffer B is reduced to 40%.
8. The next sample is ready to be injected after 20 min.
9. Phenylalanine elutes at approx 6.8 min. The data are quantified as peak area and standard curve is generated after the injection of known standards of dabsylated phenylalanine.
10. The specific radioactivity is determined by dividing the DPM of phenylalanine divided by the amount of Phenylalanine injected in the plasma sample. The specific radioactivities from the three time points (T_2, T_6, and T_{10}) are averaged (*see* **Note 2**).

3.6   Measurement of Incorporation of Radioactivity into Muscle Protein

1. Weigh out 0.2 g of tissue and homogenize in 1.5 mL of physiologic saline.
2. Remove 0.05 mL of homogenate in duplicate using clipped pipet tips to measure total tissue protein using Biuret.
3. Immediately add 3 mL of 10% TCA to homogenates and vortex.
a. Leave on ice or put at 4°C until Biuret is completed.
4. Centrifuge homogenate samples (from step 3) – 1250 g for 10 min at 4°C.
5. Decant superntant into radioactive waste container (slowly, do not disrupt pellet).
6. Add 2 mL of 10% TCA and vortex (Use glass stir rod if pellet does not break up).

7. Centrifuge at 1250 g for 10 min at 4°C (Repeat if much of precipitate has not pelleted).
8. Decant supernatant into radioactive waste container.
9. Repeat steps 6–8 two more times (Decant last wash carefully as pellet may not be pelleted strongly).
10. Add 2 mL of 1 N NaOH.
11. Incubate at 37°C with shaking for a minimum of 30 min (normally 45 min or greater).
12. Vortex samples approximately every 15 min during incubation (if pellet remains homogenize again for approximately 5 s).
13. Remove 0.1 mL sample in duplicate for Biuret assay of protein.
14. Perform Biuret assay.
15. To scintillation vial, add 500 µL of dH$_2$O, 500 µL of sample, and 10 mL of scintillation fluid and then add 1 drop of 1 N HCl (blank contains 500 µL of 1 N NaOH instead of sample; rest is the same).

3.7 Rates of Protein Synthesis in Myofibrillar and Sacroplasmic Proteins

3.7.1 Separation of Myofibrillar and Sacroplasmic Proteins

1. Homogenize 0.5 g of frozen powdered muscle tissue in 2 mL of 10 mM KH$_2$PO$_4$ (pH 7.4) using a motor-driven glass-on-glass tissue homogenizer.
2. Pour homogenate into a preweighed (M1) 15 mL Corex tube and reweigh (M2).
3. Centrifuge homogenate at 2500 g at 4°C for 20 min.
4. Pipet supernatant (do not pour) (sarcoplasmic protein fraction) into preweighed (S1) 15-mL Corex tube and place on ice.
5. Retain pellet for myofibrillar proteins.

3.7.2 Enrichment of Myofibrillar Protein Fraction

1. To myofibrillar pellet (Subheading 3.7.1, step 5) add 2 mL of 100 mM KH$_2$PO$_4$ (pH 7.4) and vortex (use glass stirring rod while vortexing to break up pellet).
2. Centrifuge homogenate at 2500 g at 4°C for 20 min.
3. Pipette supernatant (do not pour) and combine with supernatant obtained in Subheading 3.7.1., step 4.
4. Repeat Subheading 3.7.2., steps 1 through 3.
5. To pellet, add 2 mL 0.6 M NaCl and vortex (use glass stirring rod while vortexing to break up pellet).
6. Centrifuge homogenate at 550 g at 4°C for 1 min.
7. Pour supernatant into a preweighed 15-mL Corex tube (M3) and discard pellet.

8. Remove 50 µL in duplicate for Biruet assay for total protein.
9. Reweigh Corex tube (M4).
10. Add volume of 20% TCA equal to the weight of supernatant in Corex tube (i.e., M4 minus M3).
11. Continue with Subheading 3.6., step 4.

3.7.3 Enrichment of Sarcoplasmic Protein Fraction

1. Take 200-µL sample of the sarcoplasmic protein supernatant fraction from Subheading 3.7.2., step 4 for Biruet protein assay.
2. Reweigh Corex tube containing sarcoplasmic protein supernatant fraction from Subheading 3.7.2., step 4 (S2).
3. Add volume of 20% trichloroacetic acid equal to the weight of supernatant in Corex tube (i.e., S2-S1).
4. Continue with Subheading 3.6., step 4.

3.8 Incubated Epitrochlearis Preparation

1. Rats are anesthetized with pentobarbital or other appropriate agent.
2. The skin on each of the forelimbs is removed.
3. The epitrochlearis muscles are excised intact, pinned at resting length to an opened piece of Tygon tubing, and immediately placed in Krebs-Henseleit bicarbonate buffer (pH 7.4) (120 mM NaCl, 4.8 mM KCl, 25 mM NaHCO$_3$, 2.5 mM CaCl, 1.2 mM KH$_2$PO$_4$, and 1.2 mM MgSO$_4$) supplemented with 5 mM HEPES, 0.1% bovine serum albumin (99% fatty acid free), 5 mM glucose, 0.17 mM leucine, 0.20 mM valine, and 0.10 mM isoleucine.
4. The muscles are quickly rinsed and transferred to plastic tubes containing 2 mL of Krebs-Henseleit bicarbonate buffer that has been prewarmed and preoxygenated to 37°C.
5. The tubes are capped and 95% O$_2$/5% CO$_2$ is superfused into the buffer through the rubber cap.
6. Epitrochlearis muscles are first pre-incubated for 30 min.
7. After the preincubation period, muscles are transferred to fresh Krebs-Henseleit bicarbonate buffer (2 mL) and incubated for a further 180 min, with a change of buffer every 60 min.
8. During the final 60 min of the incubation period, the Krebs-Henseleit bicarbonate buffer is supplemented with 1 mM L-[^{14}C]phenylalanine (0.20 µCi/mL).
9. At the end of the incubation, muscles are removed from the incubation buffer, trimmed of connective tissue, immersed into 2 mL of ice-cold 10% (wt/vol) TCA, and weighed.
10. The incubation medium is frozen and stored at −20°C for analysis of tyrosine and the specific radioactivity of phenylalanine.

11. The muscles are treated as in Subheading 3.6. to measure incorporation of radioactive phenylalanine into proteins.

3.9 Discussion

We *(7,9,10,13–15,23,29)* and others *(16,30–32)* have used these methods to define
the inhibitory effects of both acute intoxication (binge drinking) and chronic ethanol
consumption (alcoholism) on protein synthesis in striated muscles. The diminished
rates of protein synthesis occur despite equal caloric and nitrogen intake as in pair-
fed controls *(13,29)*. Furthermore the rate of synthesis of proteins in the myofibrillar
and sarcoplasmic fractions was diminished proportionately after chronic alcohol
feeding in heart *(23)* and gastrocnemius (*see* Fig. 2). However, the soleus appears to
be spared the detrimental effects of chronic alcohol ingestion on protein synthesis,
indicating the effects of ethanol are tissue specific primarily affecting muscles

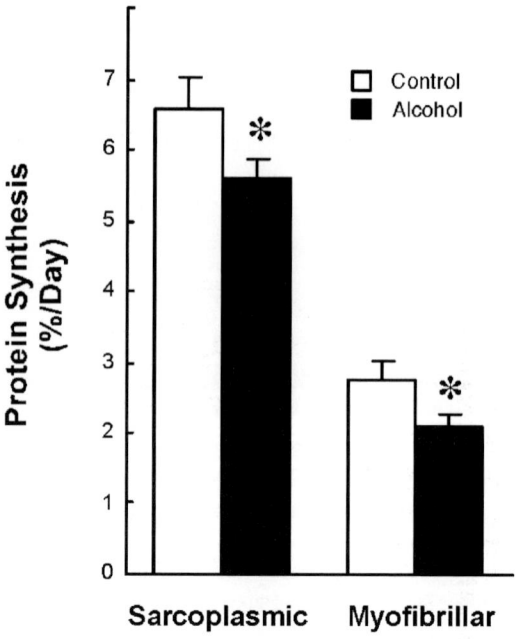

Fig. 2 Chronic (14 wk) ethanol consumption inhibits synthesis of both sarcoplasmic and myofibrillar proteins in gastrocnemius. The rate of synthesis of sarcoplasmic and myofibrillar proteins
in hearts of rats fed a diet containing ethanol for 14 wk (Alcohol) or rats pair fed a diet that was
isonitogenous and isocaloric with respect to the alcohol group (Control). Rates of protein synthesis were measured in vivo after intravenous injection with saline containing L-[³H]phenylalanine
as described in the Methods section. Sarcoplasmic and myofibrillar proteins were separated as
described in the Methods section. Values shown are means ± SE for 8–10 animals in each group.
*p < 0.01 vs. Control *p < 0.05 vs Control

composed of fast-twitch fibers and heart. The inhibition of protein synthesis correlates with a 25% decrease in the protein content of muscles after a lengthy (14 wk) dietary input of ethanol *(13)* and diminished ventricular wall thickness *(9)*.

Prognosis improves in patients with alcoholic cardiomyopathy if complete abstinence from alcohol is accomplished. These techniques have also allowed us to determine that rates of protein synthesis can be restored after the withdrawal of alcohol from the diet *(14)*. Thus, changes in protein metabolism in cardiac muscle observed during chronic alcohol intake remain reversible during this stage, indicating irreversible alterations have not occurred. However, these hearts have not undergone dilation but show signs of ventricular dysfunction (i.e., diminished cardiac output secondary to reduced stroke volume) *(9,15)*. The inhibition of protein synthesis represents a relatively early derangement in cellular homeostasis and provides a link to functional and structural abnormalities induced by alcoholism in muscle.

Acknowledgments This work was supported in part by National Institute on Alcohol Abuse and Alcoholism Grants AA-12814(TCV) and AA-11290(CHL) and a National Institute of General Medical Services Grant GM-39277(TCV).

4 Notes

1. There is growing appreciation for the power of manipulating the genome to examine the ablation or overexpression of specific proteins. Part of the importance of these mouse transgenic models involves characterization of phenotypes

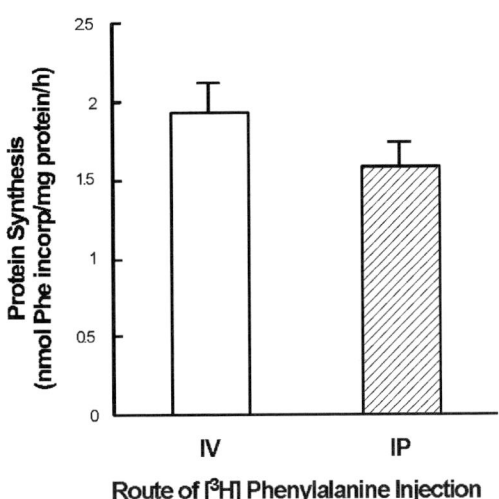

Fig. 3 Comparison of protein synthesis measured in vivo with the [³H]phenylalanine administered either via intravenous injection in the vena cava or via an intraperitoneal injection in gastrocnemius from C57B/6 mice. Similar results were obtained in cardiac muscle (data not shown)

generated in response to the genetic manipulation. Techniques to measure rates of protein synthesis in vivo are being used. However, because of the small size of the mouse, cannulation of blood vessels can become problematic. We have developed a method that allows for the measurement of protein synthesis in the mouse where we do not need to cannulate blood vessels. We have compared the new method with the previous method of injecting the radioactive phenylalanine intravenously to demonstrate essentially the same rates of protein synthesis are obtained (Fig. 3).

2. For mouse studies, only the 15-min blood sample is used as the specific radio-activity to account for the difference in time to equilibrate the plasma pool following ip injection.

References

1. Garlick, P. J., and Cerosimo, E. (1997) Techniques for assessing protein and glucose kinetics. *Bailliere Clin. Endocrinol Metab.* **11**, 629–644.
2. Loftfield, R. B., and Harris, A. (1956) Participation of free amino acids in protein synthesis. *J. Biol. Chem.* **219**, 151–159.
3. McKee, E., Cheung, J.-Y., Rannels, D. E., and Morgan, H. E. (1978) Measurement of the rate of protein synthesis and compartmentation of heart phenylalanine. *J. Biol. Chem.* **253**, 1030–1040.
4. Garlick, P. J., McNurlan, M. A., Essen, P., and Wernerman, J. (1994) Measurement of tissue protein synthesis rates in vivo: a critical analysis of contrasting methods. *Am. J. Physiol. Endocrinol. Metab.* **266**, E287-E297.
5. Loftfield, R. B., and Eigner, E. A. (1958) The time required for the synthesis of a ferritin molecule in rat liver. *J. Biol. Chem.* **231**, 925–943.
6. Garlick, P. J., McNurlan, M. A., and Preedy, V. R. (1980) A rapid and convenient technique for measuring the rate of protein synthesis in tissue by injection of [³H]phenylalanine. *Biochem. J.* **192**, 719–723.
7. Lang, C. H., Frost, R. A., Kumar, V., and Vary, T. C. (2000) Impaired myocardial protein synthesis induced by acute alcohol intoxication is associated with changes in eIF4F. *Am J. Physiol. Endocrinol. Metab.* **279**, E1029–E1038.
8. Lang, C. H., Frost, R. A., Kumar, V., Wu, D., and Vary, T. C. (2000) Impaired protein synthesis induced by acute alcohol intoxication is associated with changes in eIF4E in muscle and eIF2B in liver. *Alcohol. Clin. Exp. Res.* **24**, 322–331.
9. Lang, C. H., Frost, R. A., Sumner, A. D., and Vary, T. C. (2005) Molecular mechanisms responsible for alcohol-induced myopathy in skeletal muscle and heart. *Int. J. Biochem. Cell Biol.* **37**, 2180–2195.
10. Lang, C. H., Kimball, S. R., Frost, R. A., and Vary, T. C. (2001) Alcohol myopathy: impairment of protein synthesis and translation initiation. *Int. J. Biochem. Cell Biol.* **33**, 457–473.
11. Lang, C. H., Wu, D., Frost, R. A., Jefferson, L. S., Vary, T. C., and Kimball, S. R. (1999) Chronic alcohol feeding impairs hepatic translational initiation by modulating eIF2 and eIF4E. *Am. J. Physiol. Endocrinol. Metabol.* **277**, E805-E814.
12. Vary, T. C., and Deiter, G. (2005) Long-term alcohol administration inhibits synthesis of both myofibrillar and sarcoplasmic proteins in heart. *Metabolism Clin. Exp.* **54**, 212–219.
13. Vary, T. C., Lynch, C. J., and Lang, C. H. (2001) Effects of chronic alcohol consumption on regulation of myocardial protein synthesis. *Am. J. Physiol. Heart. Circ.* **281**, H1242-H1251.
14. Vary, T. C., Nairn, A. C., and Lang, C. H. (2004) Restoration of protein synthesis in heart and skeletal muscle after withdrawal of alcohol. *Alcohol. Clin. Exp. Res.* **28**, 517–525.

15. Vary, T. C., and Sumner, A. D. (2004) Deleterious effects of alcohol intoxication on heart: arrhythmias to cardiomyopathies, in *Nutrition and Alcohol: Linking Nutrient Interactions and Dietary Intake*. (eds. R. R. Watson and V. Preedy), CRC Press, Boca Raton, FL, pp. 117–141.

16. Preedy, V. R., and Peters, T. (1992) Protein metabolism in alcoholism. In: *Nutrition and Alcohol*. (eds. R. Watson and B. Watzl) CRC Press, Boca Raton, FL, pp. 143–189.

17. Balage, M., Sinaud, S., Prod'Homme, M., Dardevet, D., Vary, T. C., Kimball, S. R., et al. (2001) Amino acids and insulin are both required to regulate assembly of eIF4EseIF4G complex in rat skeletal muscle. *Am. J. Physiol. Endocrinol. Metab.* **281,** E565-E574.

18. Caso, G., Ford, C., Nair, K. S., Garlick, P. J., and McNurlan, M. A. (2002) Aminoacyl-tRNA enrichment after a flood of labeled phenylalanine: insulin effects of protein synthesis. *Am. J. Physiol. Endocrinol. Metab.* **282,** E1029–1038.

19. Davis, T. A., Fiorotto, M. L., Burrin, D. G., Reeds, P. J., Hanh V. Nguyen, H. V., Beckett, P. R., Vann, R. C., and O'Connor, P. M. J. (2002) Stimulation of protein synthesis by both insulin and amino acids is unique to skeletal muscle in neonatal pigs. *Am. J. Physiol. Endocrinol. Metab.* **282,** E880–E890.

20. Anthony, J. C., Gautsch, T., Kimball, S. R., Vary, T. C., and Jefferson, L. S. (2000) Orally administered leucine stimulates protein synthesis in skeletal muscle of postabsorptive rats in association with increased eIF4F formation. *J. Nutr.* **130,** 139–145.

21. Anthony, J. C., Yoshizawa, F., Anthony, T. G., Vary, T. C., Jefferson, L. S., and Kimball, S. R. (2000) Leucine stimulates translation initiation in skeletal muscle of postabsorptive rats via a rapamycin-sensitive pathway. *J. Nutr.* **130,** 2413–2419.

22. Garlick, P. J., McNurlan, M. A., Bark, T. H., Lang, C. H., and Gelato, M. C. (1998) Hormonal regulation of protein metabolism in relation to nutrition and disease. *J. Nutr.* **128,** 356S-359S.

23. Vary, T. C., and Deiter, G. (2005) Chronic alcohol administration inhibits synthesis of both myofibrillar and sarcoplasmic proteins in heart. *Metab. Clin. Exp.* **54,** 212–219.

24. Dardevet, D., Sornet, C., Vary, T. C., and Grizard, J. (1996) Phosphotidylinositol 3-kinase and p70 S6 kinase participate in regulation of protein turnover in skeletal muscle by insulin and IGF-I. *Endocrinology* **137,** 4087–4094.

25. Fedele, M. J., Vary, T. C., and Farrell, P. A. (2001) IGF-I antibody prevents increases in protein synthesis in epitrochlearis muscles from refed, diabetic rats. *J. Appl. Physiol.* **90,** 1166–1173.

26. Vary, T., Dardevet, D., Grizard, J., Voisin, L., Buffiere, C., Denis, P., et al. (1998) Differential regulation of skeletal muscle protein turnover by insulin and IGF-I following bacteremia. *Am. J. Physiol. Endocrinol. Metab.* **275,** E584-E593.

27. Brozinick, J. T., and Birnbaum, M. J. (1998) Insulin, but not contraction, activates Akt/PKB in isolated rat skeletal muscle. *J. Biol. Chem.* **273,** 14679–14682.

28. Vary, T. C., Dardevet, D., Voisin, L., Grizard, J., Buffiere, C., Denis, P., et al. (1999) Pentoxifylline improves insulin action limiting skeletal muscle catabolism following infection. *J. Endocrinol.* **163,** 15–24.

29. Lang, C., Wu, D., Frost, R., Jefferson, L., Kimball, S., and Vary, T. (1999) Inhibition of muscle protein synthesis by alcohol is associated with modulation of eIF2B and eIF4E. *Am. J. Physiol. Endocrinol. Metab.* **277,** E268–E276.

30. Preedy, V. R., Patel, V. B., Why, H. J. F., Corbett, J. M., Dunn, M. J., and Richardson, P. J. (1996) Alcohol and the heart: biochemical alterations. *Cardiovasc. Res.* **31,** 139–147.

31. Siddiq, T., Richardson, P. J., Mitchell, W., Teare, J., and Preedy, V. R. (1993) Ethanol-induced inhibition of ventricular protein synthesis in vivo and possible role of acetylaldehyde. *Cell Biochem. Function* **11,** 45–54.

32. Siddiq, T., Richardson, P. J., Morton, J., Smith, B., Sherwood, R., Marway, J. S., et al. (1993) Rates of protein synthesis in different regions of the normotensive and hypertensive heart in response to acute alcohol toxicity. *Alcohol Alcohol.* **28,** 297–310.

23

Methods to Investigate the Effects of Chronic Ethanol on Adipocytes

Becky M. Sebastian, Li Kang, Xiaocong Chen, and Laura E. Nagy

Summary Chronic ethanol consumption dysregulates glucose and lipid homeostasis, is associated with insulin resistance, and alters serum levels of adipokines including adiponectin and tumor necrosis factor-α. However, the mechanisms involved in these chronic ethanol-induced pathologies are not fully understood. Adipose tissue has been implicated as an important contributor to chronic ethanol-induced disease states and, therefore, the effects of chronic ethanol feeding in rats on adipocytes has been investigated. Three major functions of the adipocyte include glucose transport, adipokine secretion, and triglyceride breakdown via lipolysis. Included in this chapter are protocols for studying the effect of chronic ethanol feeding on these adipocyte functions.

Keywords Adipose; fat depots; insulin resistance; glucose transport; β-adrenergic agonist; lipolysis; adipokine secretion; adiponectin.

L. E. Nagy (ed.), *Alcohol: Methods and Protocols*
© Humana Press 2008

1   Introduction

Adipose tissue has long been known as the major depot for fat storage. However, it is now well recognized as an important endocrine organ in which the secretion of peptides, called adipokines, occurs, regulating various pathways in other organs, such as the liver (reviewed in *[1,2]*). Adipose tissue is also important in the regulation of whole body glucose and lipid homeostasis. Recently, an increasing amount of literature suggests a critical role for adipose in the pathophysiology of diseases, including insulin resistance and alcoholic liver disease *(3,4)*. Chronic ethanol consumption leads to hepatic steatosis and is an independent risk factor for type 2 diabetes. In rats, adipocytes contribute to chronic ethanol-induced insulin resistance, which is a major characteristic of type 2 diabetes. Specifically, chronic ethanol feeding induces glucose intolerance and decreases insulin-stimulated glucose uptake in adipocytes *(5,6)*, which is associated with disrupted insulin signaling through the Cbl-TC10 pathway *(7)*. Moreover, chronic ethanol consumption in rats reduces circulating adiponectin by impairing its secretion from adipocytes *(8)* and dysregulates lipid homeostasis by decreasing adipocyte lipolysis in response to β-adrenergic receptor activation *(9)*.

   Adipocytes are morphologically very different than most other cell types; lipid occupies the majority of the cytosol and, therefore, adipocytes require special considerations in most protocols. In this chapter, we discuss the isolation of adipocytes and the intricacies of working with them in three different assays: insulin-stimulated glucose transport, secretion of adipokines, and β-adrenergic receptor-activated lipolysis.

2   Materials

2.1   Adipocyte Isolation

1. Hanks Wash Buffer: 25 mM HEPES in Hanks Balanced Salt Solution (without sodium bicarbonate). Stable at room temperature 1–2 mo or longer at 4°C.
2. Sharp scissors.
3. Adenosine is dissolved at 1.2 mM in Milli-Q water and stored at 4°C.
4. Digestion Buffer: 25 mM HEPES, 10 mg/mL bovine serum albumin (BSA; RIA grade), and 200 nM adenosine in Hanks Balanced Salt Solution (without sodium bicarbonate), adjusted to pH 7.4 (*see* **Note 1**). Make fresh daily.
5. Collagenase, Type II for adipocyte isolation.
6. Autoclaved polypropylene flasks, 250 mL.
7. 37°C shaking water bath.
8. Mesh filter, 224 μm (Sefar America Inc., cat. no. 03-225-42).
9. Hemostats.
10. 20-mL syringe with a 21-gauge needle.

11. Adipocyte Wash Buffer: Phosphate-buffered saline (PBS) with 1 mM MgCl$_2$, 0.68 mM CaCl$_2$, 25 mM HEPES, 1 mg/mL BSA (RIA grade), 1 mM pyruvic acid (sodium salt) as an energy source (*see* **Note 2**), and 200 nM adenosine, adjusted to pH 7.4 (*see* **Note 1**). Stable (without adenosine) for 1 wk at 4°C.
12. Hemocytometer.
13. Adenosine deaminase.

2.2 Glucose Transport

1. 110 nM Insulin (11X): Insulin from bovine pancreas (Sigma cat. no. I6634) is dissolved in a small amount of 20% acetic acid in Milli-Q water (e.g., 500 μL, or more if necessary, for a 50-mg vial), then brought to a stock concentration of 830 μM with Milli-Q water, aliquoted and stored at −20°C. An 11X working stock (110 nM) is prepared and kept on ice just before cell stimulation.
2. 3.3 mM phloretin (11X): Phloretin is dissolved with ethanol to a stock concentration of 100 mM and stored at −20°C. An 11X working stock (3.3 mM) is prepared just before use by combining 230 μL of ethanol and 33 μL of 100 mM phloretin stock then adding 747 μL of Milli-Q water. Phloretin is light sensitive; therefore, avoid exposing it to light.
3. 10 mM 2-deoxyglucose in PBS = "cold" glucose. Store at room temperature.
4. 2-[1,2-^{3}H(N)]-deoxy-D-glucose = "hot" glucose.
5. Hot/cold stock mix: add 5 μL of "hot" glucose to 750 μL of "cold" glucose, making the final concentration to cells 2.5 mM, 0.33 μCi/tube. (Make enough for 50 μL per tube, which is three times each sample, plus 50 μL for measuring total cpm; *see* **Note 3**). Store at room temperature.
6. Oil mixture: Light mineral oil (Fisher cat. no. O121) and silicone oil (Sigma cat. no. 175633) in a 57:43 (vol:vol) ratio. Store at room temperature.
7. 17 × 100-mm polystyrene round bottom tubes.
8. 0.25-mL polyethylene microcentrifuge tubes (*see* **Note 4**).
9. Eppendorf Repeater pipet.
10. 7-mL scintillation vials.
11. ScintiSafe Econo 1.
12. Beckman Coulter LS 6500 multipurpose scintillation counter.

2.3 Adiponectin Secretion

1. DMEM/F12.
2. Adenosine is dissolved at 1.2 mM in Milli-Q water, stored at 4°C.
3. BSA (RIA grade).
4. 12 × 75-mm polypropylene round-bottom tubes.
5. Adenosine deaminase.

6. RIPA buffer: 50 mM Tris-HCl, pH 7.4, 1% Nonidet P40, 0.25% Na-deoxycholate, 150 mM NaCl, 1 mM ethylene diamine tetraacetic acid (EDTA), and protease inhibitor cocktail (1 tablet/50 mL; Roche cat. no. 1-873-580), adjusted to pH 7.4. Store at 4°C.

2.4 Adipocyte Lipolysis

1. Dulbecco's phosphate-buffered saline plus (PBS$^+$), pH 7.4: 137 mM NaCl, 2.68 mM KCl, 15 mM Na$_2$HPO$_4$, 1.47 mM KH$_2$PO$_4$, 1.05 mM MgCl$_2$, 0.68 mM CaCl$_2$. This solution is stable at room temperature for up to 3 mo.
2. PBS$^+$/HEPES, pH 7.4: 25 mM HEPES, in PBS$^+$. This solution is stable at room temperature for up to 1 mo.
3. PBS$^+$/HEPES/BSA, pH 7.4: PBS$^+$ supplemented with 25 mM HEPES and 10 mg/mL BSA (RIA grade). Make fresh daily.
4. PBS$^+$/glucose/BSA, pH 7.4: PBS$^+$ supplemented with 6 mM glucose and 1 mg/mL BSA (RIA grade). This solution is stable at 4°C for 1 wk.
5. Adenosine is dissolved at 1.2 mM in Milli-Q water; store at 4°C.
6. 12 × 75-mm polypropylene round-bottom tubes.
7. Adenosine deaminase.
8. 25 mM Isoproterenol: Isoproterenol is first dissolved at 3.8 mM in PBS containing 20 mM citrate, and a 25 mM working solution is prepared in 2 mM ascorbic acid in Milli-Q water. Isoproterenol solution is prepared fresh daily, and protected from light.
9. 250 nM Insulin: Insulin from bovine pancreas (Sigma cat. no. I6634) is dissolved in a small amount of diluted acetic acid (20% in Milli-Q water) then brought to a stock concentration of 830 µM with Milli-Q water, aliquoted and stored at −20°C. A 250 nM working solution is prepared in PBS.
10. (-)-N^6-(2-phenylisopropyl) adenosine (R-PIA; Sigma cat. no. P4532-25) is dissolved at 2.4 mM in 0.1 N HCl, and made fresh daily. A 250 nM working solution is prepared in Milli-Q water.
11. Free glycerol reagent (Sigma cat. no. F6428).
12. Flat bottom 96-well plates.
13. Microplate reader.

3 Methods

3.1 Adipocyte Isolation

Although adipocytes make up the bulk of the mass of adipose, most cells are actually nonadipocytes. These stromal vascular cells include endothelial cells, preadipocytes, and resident macrophages. The isolation of adipocytes from these other cell types relies on the ability of the adipocytes to float because of their high lipid content *(10,11)*. The following protocol has been used for the isolation of adipocytes from epididymal, retroperitoneal, subcutaneous and omental fat from rats yielding a near pure adipocyte suspension.

1. Warm the Hanks Wash Buffer, Digestion Buffer, and Adipocyte Wash Buffer to 37°C.
2. After the 4-wk feeding period, rats are anesthetized and fat depots are excised and weighed (*see* **Note 5**).
3. Fat pads are washed two to three times with the Hanks Wash Buffer and minced with sharp scissors, approximately 50 times. Scissors are wiped clean with a Kimwipe between samples.
4. The minced adipose tissue is added to Digestion Buffer (2 mL/g fat) in autoclaved polypropylene flasks covered with foil and incubated 37°C in a shaking water bath (80 rpm) for 50 min.
5. The digested adipose tissue is filtered through a coarse mesh held in place onto 50-mL polypropylene conical tubes with hemostats.
6. The adipocytes are allowed to float for 2 min, and the infranatant is removed using a 20-mL syringe with an 18-gauge needle attached (to pierce through the layer of floating adipocytes) and discarded.
7. Adipocytes are washed three times using 20 mL of Adipocyte Wash Buffer, waiting 2 min each time to allow cells to float. Infranatants are discarded.
8. The isolated adipocytes are suspended in Adipocyte Wash Buffer to a total volume of 10–20 mL and counted using a hemocytometer.
9. After cells are diluted or concentrated to the desired concentration, adenosine deaminase is added at 1 U/mL to degrade the adenosine added earlier (*see* **Note 6**).

3.2 Glucose Transport

3.2.1 Assay

Muscle and adipose are the major organs that take up circulating glucose to maintain homeostasis. To study the effects of chronic ethanol feeding on glucose transport in isolated adipocytes, the following protocol has been used in epididymal, retroperitoneal, subcutaneous, and omental adipocytes.

1. Adipocytes are isolated as described above in Subheading 3.1, diluted or concentrated to 1×10^6 cells/mL and aliquoted into 17×100-mm polystyrene round bottom tubes, 1 mL each tube (*see* **Note 7**). For each adipocyte sample, there will be three tubes: phloretin-treated (*see* **Note 8**), untreated (basal), and insulin-treated.
2. Insulin-treated cells are stimulated by adding 100 μL of 110 nM stock insulin per milliliter of cells and all samples are incubated 30 min at 37°C in a shaking water bath, 100 rpm. Untreated cells receive 100 μL of adipocyte wash buffer to equalize the volume.

3. After 20 min of insulin stimulation, 100 μL of 3.3 mM stock phloretin is added to the phloretin-treated cells and incubated another 10 min.
4. During the last ~5 min of insulin stimulation, three 150-μL aliquots from each tube is layered atop 75 μL of oil mixture in each of three 0.25-mL polyethylene tubes, that is, the assay is performed in triplicate for each sample (*see* **Note 9**).
5. Transport is initiated by the addition of 50 μL of the hot/cold stock mix to all tubes carefully using an Eppendorf Repeater pipette, allowed to continue for exactly 3 min and stopped by centrifuging the tubes for 6 s at ~16,000 g (*see* **Note 10**).
6. The tubes are cut in the middle of the oil layer and the top half of each tube containing the adipocytes are placed in 7-mL scintillation vials (*see* **Note 11**). The bottom half of the tubes are discarded as radioactive waste.
7. Four milliliters of ScintiSafe Econo 1 is added to each vial and the ^{3}H counted using a Beckman LS 6500 Multipurpose scintillation counter (*see* **Note 3**).

3.2.2 Analysis of Data

1. Calculate the cpm/nmol 2-deoxyglucose used in the assay:

$$\frac{\text{Total cpm in 50 μL hot/cold stock mix}}{\text{(Final concentration of 2DG)(μL volume in each assay tube excluding the oil)}}$$

For example:

$$\frac{\text{total cpm}}{(2.5 \text{ mM})(0.2 \text{ mL})} = \text{total cpm/0.5 mmol}$$

$$= \text{total cpm/500 nmol}$$

2. Calculate the picomols 2DG per 10^5 cells in each sample:

$$\frac{\text{Average sample cpm} - \text{average phloretin cpm}}{\text{(cpm/nmol calculated above) (no. cells per tube)}} = \text{nmol/cell}$$

Then convert to pmol/10^5 cells:

$$\frac{[(\text{nmol/cell})*100000]}{1000} = \text{pmol/}10^5 \text{ cells}$$

3.3 Adiponectin Secretion

Chronic ethanol feeding has been reported to modulate the serum concentration of different adipokines, including adiponectin and tumor necrosis factor-α *(8,13)*.

Adiponectin is an abundant adipokine that is reported to be important in the pathogenesis of alcoholic and non-alcoholic liver diseases *(4)*. Studying the secretion of adiponectin *ex vivo* can further our understanding of the effects of chronic ethanol on adipocyte function.

1. Warm all DMEM/F12 buffers to 37°C.
2. After isolation through step 5 in Subheading 3.1, wash the adipocytes with DMEM/F12 containing 0.1% RIA-grade BSA and 200 nM adenosine (*see* **Note 1**) two to three times, allowing cells to float 2 min between washes. Infranatant is removed with a 20-mL syringe with an 18-gauge needle attached (to pierce through the layer of floating adipocytes) and discarded (*see* **Note 12**).
3. Prepare the treatment solution, (e.g. insulin, β-adrenergic receptor agonist), in DMEM/F12 containing 0.1% RIA-grade BSA and add 50 μL into 12 × 75-mm polypropylene round-bottom tubes. Add 50 μL of DMEM/F12 containing 0.1% RIA-grade BSA to untreated tubes.
4. Dilute the cell suspension to 0.5×10^6 cells/mL in DMEM/F12 containing 0.1% RIA-grade BSA and 200 nM adenosine (*see* **Note 1**). Adenosine deaminase is added at 1 U/mL to degrade the adenosine added earlier (*see* **Note 6**).
5. Transfer 200 μL of diluted cell suspension to each tube for a total volume of 250 μL (*see* **Note 7**).
6. Incubate the cells at 37°C in a shaking water bath (100 rpm). Media are collected at various time points (i.e., 0, 30, 60 and 120 min) by allowing cells to float and the media below are sampled by piercing through the cell layer using a 200-μL tip. Adipokine concentration in the media can be detected with enzyme-linked immunosorbent assay (ELISA; *see* **Note 13**).
7. To measure intracellular adipokines, the remaining adipocytes are transferred to microcentrifuge tubes and washed two to three times with ice-cold PBS (250 μL each and centrifuging 4.5 g, 10 s to force cells to float). Lyse the concentrated cells by adding 150 μL of ice-cold RIPA buffer, vortex, and rotate at 4°C for 15 min. Lysates are centrifuged at 4.5 g, 5 min, and the upper layer of fat is aspirated off. Lysates can be stored at −20°C. Adipokine concentration can be determined with ELISA.

3.4 Adipocyte Lipolysis

Lipolysis is the breakdown of stored triglycerides, leading to the release of glycerol and free fatty acids into the circulation. Chronic ethanol feeding in rats increases plasma free fatty acid concentration *(14)*, and increased circulating free fatty acids is associated with the development of obesity, insulin resistance, as well as type 2 diabetes *(15–17)*. Measuring lipolysis *ex vivo* is a useful way of studying the effects of chronic ethanol consumption on the regulation of lipolysis.

1. Warm PBS+/HEPES, PBS+/HEPES/BSA with or without 200 nM adenosine and PBS+/glucose/BSA with or without 200 nM adenosine to 37°C.

2. Adipocytes are isolated from epididymal adipose tissue in pair- and ethanol-fed rats as described above in Subheading 3.1 except using different buffers. Fat pads are washed with PBS+/HEPES and digested in PBS+/HEPES/BSA with or without 200 nM adenosine (*see* **Note 14**). Isolated cells are washed in PBS+/glucose/BSA with or without 200 nM adenosine (*see* **Note 14**).

3. Adipocytes are counted in a hemocytometer and cell concentration is adjusted to 1×10^6 cells/mL (*see* **Note 15**).

4. 200-μL aliquots of cell suspension are placed into 12×75-mm polypropylene tubes and incubated with or without 1 μM isoproterenol and/or 10 nM insulin in the presence or absence of 0.4 U/mL adenosine deaminase (ADA) and 10 nM R-PIA (*see* **Note 14**) for 1 h (*see* **Note 15**) at 37°C in a shaking water bath (100 rpm).

5. After incubation, cells are centrifuged briefly, and the cell medium is collected for analysis of glycerol concentration using free glycerol reagent in a flat-bottom 96-well plate.

6. Optical density at 540 nm is measured with a microplate reader.

Acknowledgments The work in this chapter was supported by National Institute on Alcohol Abuse and Alcoholism Grant no. AA-11876.

4 Notes

1. Adenosine is added to inhibit lipolysis during isolation. This maintains the adipocyte's ability to float, as well as avoiding potential metabolic effects of exogenous free fatty acids. Adenosine is added right before usage and is only stable for approx 1 h.

2. Sodium pyruvate is used as the energy source in glucose transport assays rather than glucose because the exogenous glucose would compete with the "hot" glucose for transport into the cells.

3. Include a scintillation vial containing 50 μL of the hot/cold mix plus 4 mL ScintiSafe Econo 1 for calculation of cpm/nmol 2-deoxyglucose used in the data analysis.

4. The small diameter of these tubes allow for clear separation of phases and help create a strong oil barrier that is not disrupted during cutting due to the oil mixture's cohesive properties. Additionally, the polyethylene nature of these tubes facilitates cutting.

5. Adipose is located throughout the body; however, certain fat depots exist and are easily accessible. These various depots not only differ in there location, but may also differ in function (*8,12*). The depots used in these assays include epididymal fat, the gonadal fat of the male (a large source of adipocytes containing the least amount of contaminating tissues, located in the groin area and attached to the epididymis), retroperitoneal fat (also a large source of adipocytes, located and attached to the inner dorsal surface of the peritoneum), omental fat (the smallest depot used in these assays, attached to the bottom of

the greater omentum, has a web-like appearance), and subcutaneous fat (the most vascularized of the four depots, often requiring extra washes to remove contaminating red blood cells, located under the skin in the lower abdomen and femoral area).

6. Adenosine deaminase is added to degrade adenosine and counteract the inhibition of lipolysis.

7. Adipocytes must be aliquoted carefully to ensure that the cell suspension is homogenous. This requires orbital shaking or inverting tube before removing each aliquot.

8. Phloretin is added to inhibit nonspecific glucose transport.

9. After adding 75 μL of oil mixture to 0.25-mL tubes, a quick spin may be necessary to remove any air at the bottom of the tube. This step is imperative to fit the total volume of the assay mixture into the tube.

10. Centrifugation separates the extracellular medium containing exogenous glucose (bottom phase, containing majority of "hot" glucose) from the adipocytes (top layer, containing intracellular "hot" glucose), with a layer of the oil mixture in between (not radioactive).

11. Because the oil mixture does not contain "hot" glucose, wiping the razor between samples with a kimwipe is sufficient to avoid contamination sample to sample.

12. DMEM/F12 was used for a nutrient-rich medium more suitable for longer incubations required for adipokine secretion studies.

13. Adipocytes secrete a relatively high level of adiponectin compared to other adipokines, such as TNF-α. To optimize the conditions for ELISAs of other adipokines, more media or RIPA lysate may be needed. As a result, more cells incubated in larger volume tubes may be required.

14. If adenosine is used during the adipocyte isolation, ADA (0.4 U/mL) is added at the end of the isolation procedure to counteract the inhibition of lipolysis. However, ADA also degrades the endogenous adenosine and therefore increases "basal" rates of lipolysis; therefore, 10 nM R-PIA, an adenosine receptor agonist, is added to the incubation mixture to normalize basal rates of lipolysis. Control experiments show that the addition adenosine, followed by adding ADA and R-PIA, had no effect on glycerol release.

15. Control experiments show glycerol release from pair- and ethanol-fed adipocytes is dependent on cell concentration and incubation time.

References

1. Ahima, R. S. (2006) Adipose tissue as an endocrine organ. *Obesity (Silver Spring)* 14(Suppl 5), 242S–249S.
2. Tilg, H., and Moschen, A. R. (2006) Adipocytokines: mediators linking adipose tissue, inflammation and immunity. *Nat. Rev. Immunol.* **6.** 772–783.
3. Carvalho, E., Kotani, K., Peroni, O. D., and Kahn, B. B. (2005) Adipose-specific overexpression of GLUT4 reverses insulin resistance and diabetes in mice lacking GLUT4 selectively in muscle. *Am. J. Physiol. Endocrinol. Metab.* **289,** E551–E561.

4. Xu, A., Wang, Y., Keshaw, H., Xu, L. Y., Lam, K. S., and Cooper, G. J. (2003) The fat-derived hormone adiponectin alleviates alcoholic and nonalcoholic fatty liver diseases in mice. *J. Clin. Invest.* **112,** 91–100.

5. Wilkes, J. J., DeForrest, L. L., and Nagy, L. E. (1996) Chronic ethanol feeding in a high-fat diet decreases insulin-stimulated glucose transport in rat adipocytes. *Am. J. Physiol.* **271,** E477–E484.

6. Wilkes, J. J., and Nagy, L. E. (1996) Chronic ethanol feeding impairs glucose tolerance but does not produce skeletal muscle insulin resistance in rat epitrochlearis muscle. *Alcohol. Clin. Exp. Res.* **20,** 1016–1022.

7. Sebastian, B. M., and Nagy, L. E. (2005) Decreased insulin-dependent glucose transport by chronic ethanol feeding is associated with dysregulation of the Cbl/TC10 pathway in rat adipocytes. *Am. J. Physiol. Endocrinol. Metab.* **289,** E1077–E1084.

8. Chen, X., Sebastian, B. M., and Nagy, L. E. (2007) Chronic ethanol feeding to rats decreases adiponectin secretion by subcutaneous adipocytes. *Am. J. Physiol. Endocrinol. Metab.* **292,** E621–E628.

9. Kang, L., and Nagy, L. E. (2006) Chronic ethanol feeding suppresses beta-adrenergic receptor-stimulated lipolysis in adipocytes isolated from epididymal fat. *Endocrinology* **147,** 4330–4338.

10. Honnor, R. C., Dhillon, G. S., and Londos, C. (1985) cAMP-dependent protein kinase and lipolysis in rat adipocytes. I. Cell preparation, manipulation, and predictability in behavior. *J. Biol. Chem.* **260,** 15122–15129.

11. Rodbell, M. (1964) Metabolism of isolated fat cells. I. Effects of hormones on glucose metabolism and lipolysis. *J. Biol. Chem.* **239,** 375–380.

12. Thakur, V., Pritchard, M. T., McMullen, M. R., and Nagy, L. E. (2006) Adiponectin normalizes LPS-stimulated TNF-alpha production by rat Kupffer cells after chronic ethanol feeding. *Am. J. Physiol. Gastrointest. Liver Physiol.* **290,** G998–1007.

13. Balasubramaniyan, V., and Nalini, N. (2003) The potential beneficial effect of leptin on an experimental model of hyperlipidemia, induced by chronic ethanol treatment. *Clin. Chim. Acta* **337,** 85–91.

14. Blaak, E. E. (2003) Fatty acid metabolism in obesity and type 2 diabetes mellitus. *Proc. Nutr. Soc.* **62,** 753–760.

15. Boden, G. (1997) Role of fatty acids in the pathogenesis of insulin resistance and NIDDM. *Diabetes* **46,** 3–10.

16. Unger, R. H. (1995) Lipotoxicity in the pathogenesis of obesity-dependent NIDDM. Genetic and clinical implications. *Diabetes* **44,** 863–870.

17. Einstein, F. H., Atzmon, G., Yang, X. M., Ma, X. H., Rincon, M., Rudin, E., Muzumdar, R., and Barzilai, N. (2005) Differential responses of visceral and subcutaneous fat depots to nutrients. *Diabetes* **54,** 672–678.

Part V
Proteomic and Genomic Approaches to Study Response to Ethanol

24

Proteomic Approaches to Identify and Characterize Alterations to the Mitochondrial Proteome in Alcoholic Liver Disease

Shannon M. Bailey, Kelly K. Andringa, Aimee Landar, and Victor M. Darley-Usmar

Summary Mitochondrial dysfunction is recognized as a contributing factor to a number of diseases, including chronic alcohol-induced hepatotoxicity. Although there is a detailed understanding of the metabolic pathways and proteins of the liver mitochondrion, little is known of how changes in the mitochondrial proteome contribute to the development of hepatic pathologies. In this short overview the insights gained from study of changes in the mitochondrial proteome in alcoholic liver disease will be described. Profiling the liver mitochondrial proteome has the potential to shed light on the alcohol-mediated molecular defects responsible for mitochondrial and cellular dysfunction. The methods presented herein demonstrate the power of using complementary proteomics approaches, that is, 2-D IEF/SDS-PAGE and BN-PAGE, to identify changes in the abundance of mitochondrial proteins after chronic alcohol consumption. These proteomic data can then be integrated into a logical and mechanistic framework to further our understanding of the role of mitochondrial dysfunction in the pathogenesis of alcohol-induced liver disease.

Keywords Alcohol; liver; mitochondria; proteomics; isoelectric-focusing; blue native gel electrophoresis (BN-PAGE).

L. E. Nagy (ed.), *Alcohol: Methods and Protocols*
© Humana Press 2008

1   Introduction

It is widely recognized that a key component in the development of alcohol-induced liver disease is the diminished capacity of liver to generate and maintain sufficient levels of ATP. This resulting decrease in bioenergetic capacity is thought to contribute to decreased hepatocyte viability and ultimately the pathological changes that occur in alcohol-dependent hepatotoxicity. The importance of mitochondrial dysfunction in this disease process has long been appreciated due to chronic alcohol-mediated defects on the oxidative phosphorylation system, which leads to decreased ATP synthesis *(1,2)*. Moreover, evidence indicates that these chronic alcohol-induced alterations in mitochondria structure and function also contribute to increased production of reactive oxygen and nitrogen species in mitochondria *(3)*. This increase in mitochondrial oxidant production could result in post-translational oxidative modifications and inactivation of mitochondrial macromolecules, particularly proteins, thereby further compromising mitochondria function in the chronic alcohol consumer.

Although the mechanisms responsible for alcohol-dependent mitochondrial dysfunction have been studied, the impact of chronic alcohol consumption on the overall content of mitochondrial proteins, that is, the global mitochondrial proteome, has only recently been investigated *(4,5)*. Early studies by Cunningham and colleagues demonstrated that chronic alcohol consumption decreases the synthesis of the 13 mitochondrial-encoded proteins that are components of complexes I, III, IV, and V *(6,7)*, as the result of defects in mtDNA *(4,8,9)* and a decrease in functional mitochondrial ribosomes *(10,11)*. It is important to note, however, that there are well over 600 proteins that comprise the mitochondrial proteome *(12,13)*, and that close to 100 of these are components of the oxidative phosphorylation system, most of which are encoded by the nuclear genome. Because very little information is available on the effects of chronic alcohol consumption on the total mitochondrial proteome, especially effects on nuclear encoded proteins, we have used two complementary proteomics approaches, two-dimensional isoelectric focusing (IEF)/SDS-PAGE and blue native-PAGE (BN-PAGE) to improve our understanding of the impact these changes have on mitochondrial functionality and their contribution to the development of alcohol induced liver pathology.

2   Materials

2.1   Isoelectric Focusing

1. Buffer for rehydration of IEF gel strips: $7M$ urea, $2M$ thiourea, 2% CHAPS, 0.5% *N*-dodecyl-β-D maltoside (i.e.. lauryl maltoside), and 0.002% bromophenol blue. Store at −20°C in 1.0-mL aliquots.
2. Equilibration buffer for IEF gel strips: $6M$ urea, 2.0% sodium dodecyl sulfate (SDS), $0.375M$ Tris-HCl, 20% glycerol, and 0.002% bromophenol blue, pH 8.8

(with HCl). Store at −20°C in 2.0-mL aliquots. Before using, warm to 37°C to redissolve urea into solution.

3. DTT stock solution: 1 M DTT in water. Store at −20°C in 50-μL aliquots. Do not refreeze and reuse DTT aliquots for future experiments.

4. Ultrapure, low melting-temperature agarose (Aqua*Pōr*™ LM, National Diagnostics, cat. no. EC-204) for sealing IEF gel strips onto SDS-polyacrylamide gel electrophoresis (PAGE) gels: 1.0% agarose (w/v) in 1X SDS-PAGE running buffer. Make fresh on day of experiment.

5. The Invitrogen ZOOM IPG runner (cat. no. ZM0001) is used to perform IEF in combination with Invitrogen ZOOM strips (cat. no. ZM0011, pH 3–10), ZOOM IPG Runner Cassettes (cat. no. ZM0003), and the ZOOM Dual Power Supply (cat. no. ZP10002). Please refer to manufacturer's manual for additional details on set-up and use.

6. The ampholyte carriers used for IEF are Ampholines Electrophoresis Reagent (Sigma, cat. no. A5174, pH 3–10). Other carrier ampholines or ampholytes can be substituted.

2.2 Blue Native-Polyacrylamide Gel Electrophoresis

2.2.1 One-Dimensional Blue Native-Polyacrylamide Gel Electrophoresis

1. Cathode buffers: a) High-blue cathode buffer: 50 mM tricine, 15 mM BisTris, and 0.02% Coomassie brilliant blue G-250, pH 7.0. b) Low-blue cathode buffer: 50 mM tricine, 15 mM BisTris, and 0.002% coomassie brilliant blue G-250, pH 7.0. Store at 4°C (*see* **Note 1**).

2. Anode buffer: 50 mM BisTris, pH 7.0. Store at 4°C (*see* **Note 1**).

3. 3X gel buffer: 1.5 M aminocaproic acid, 150 mM BisTris, pH 7.0. Store at room temperature (*see* **Note 2**).

4. Extraction buffer: 0.75 M aminocaproic acid, 50 mM BisTris, pH 7.0. Store at 4°C (*see* **Note 1**).

5. Coomassie brilliant blue G-250 suspension: 0.5 M aminocaproic acid and 5% Coomassie brilliant blue G-250, pH 7.0. Store at 4°C (*see* **Note 1**).

6. Lauryl maltoside solution: 10% (w/v) *N*-dodecyl-β-D-maltoside in water. Store at −20°C.

7. Molecular weight standards for 1-D BN-PAGE: High-molecular weight native marker kit (Amersham Biosciences, cat. no. 17–0445–01). Dissolve contents of 1 vial into 200 μL of extraction buffer (#4), 25 μL lauryl maltoside (#6), and 12 μL of Coomassie brilliant blue (#5). Store at −20°C.

2.2.2 Two-Dimensional Blue Native-Polyacrylamide Gel Electrophoresis

1. Cathode buffer: 100 mM Tris, 100 mM tricine, and 0.1% SDS, pH 8.25. Store at room temperature.

2. Anode buffer: 200 mM Tris, pH 8.9 with HCl. Store at room temperature.
3. 2-D BN-PAGE Gel buffer: 3M Tris-HCl and 0.3% SDS, pH 8.45 with HCl. Store at room temperature.
4. SDS/β-Mercaptoethanol solution: 20 μL of β-mercaptoethanol, 200 μL of 10% SDS into 1780 μL of ultrapure H$_2$O. Make fresh on day of experiment.
5. Agarose solution for sealing 1-D BN-PAGE gel strips onto SDS-PAGE gels: 100 mg of low melting temperature agarose (Aqua $Pōr$™ LM, National Diagnostics, cat. no. EC-204), 1 mL 10% SDS, 100 μL of β-mercaptoethanol into 9.0 mL of ultrapure H$_2$O. Make fresh on day of experiment.

2.3   Total Protein Staining for Gels

1. Coomassie blue stain: 0.3 g of Coomassie blue R-250, 100 mL of glacial acetic acid, 250 mL of isopropanol, and 650 mL of ultrapure H$_2$O. Use 100 mL per one mini-gel. Stain gel overnight and do not reuse stain. Gels are destained the next day using 10% glacial acetic acid solution. A small piece of a paper towel is added to the gel container to absorb the Coomassie blue as it "leaches" from gel. Change blue-stained paper towels every 2–4 h until clear background on gels is achieved.
2. SYPRO® Ruby Protein Gel Stain (Invitrogen, cat. no. S-12000): After electrophoresis, fix gel in 100 mL of a 40% methanol and 10% glacial acetic acid solution for 1 h. After fixation step, discard fix solution and incubate gel with 60 mL SYPRO Ruby solution overnight. After staining gel, discard SYPRO Ruby solution and destain gel in a 10% methanol/7% glacial acetic acid solution for 1–2 h before visualizing gels. It is recommended that gels are destained for at least 16–24 h using multiple changes in destain solution to minimize background (i.e., decrease speckling) of gels. A number of imaging platforms have been validated for visualizing SYPRO Ruby-stained gels. Please refer to the SYPRO Ruby product information sheet provided by Invitrogen to obtain the excitation sources and emission filters for optimal visualization of protein in gels.

3   Methods

It is hypothesized that chronic alcohol-induced alterations to the mitochondrial proteome negatively affects mitochondrial bioenergetics and signaling, which contribute, in part, to the development of liver pathology. To address this question, our laboratories have begun to profile the mitochondrial proteome using both conventional 2-D IEF/SDS-PAGE and BN-PAGE. Recent advancements in the field have now made it possible to identify post-translational modifications to mitochondrial proteins by ROS/RNS, identify defects in the assembly of multi-protein complexes in mitochondria, and separate the more hydrophobic respiratory complex proteins of the inner membrane using newly refined gel electrophoresis methods *(14)*.

A scheme depicting our strategy for detecting alterations to the mitochondrial proteome in models of chronic alcohol-induced liver injury is shown in Fig. 1. Briefly, mitochondria isolated from the liver of control and alcohol-fed animals can be subjected to conventional 2-D IEF/SDS-PAGE and BN-PAGE. The high-resolution protein "maps" generated via these electrophoresis methods can then

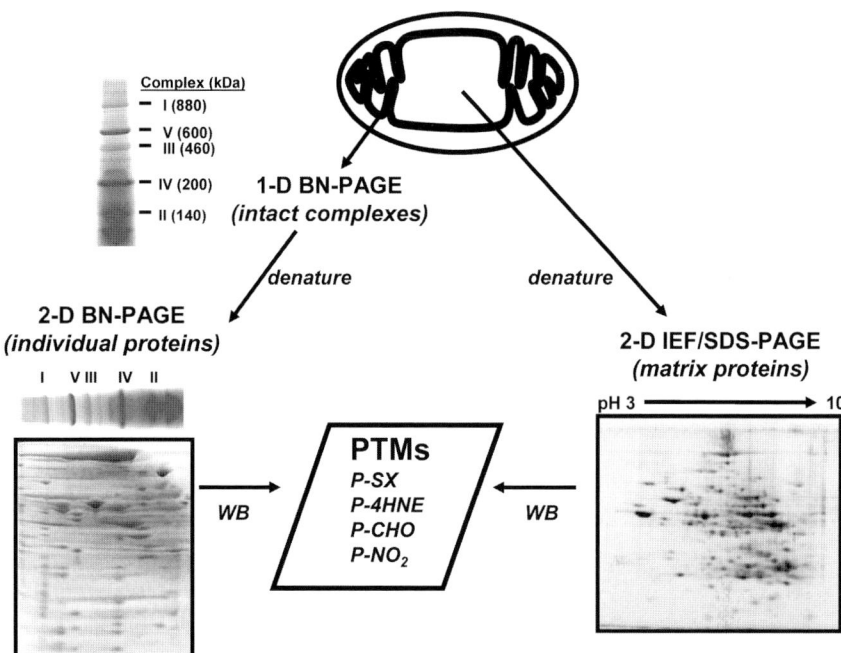

Fig. 1 Analysis of the mitochondrial proteome using 2-D IEF/SDS-PAGE and BN-PAGE proteomics. 2-D IEF/SDS-PAGE is used to generate high resolution maps of the soluble matrix proteins (**right panel**), whereas 1-D and 2-D BN-PAGE is used to define changes to respiratory chain proteins (**left panels**). The right-side panel illustrates a 2-D IEF/SDS-PAGE separation of proteins from whole liver mitochondria using the Invitrogen IPGRunner® System in combination with pH 3–10 IEF gel strips, whereas the left side panels demonstrate the BN-PAGE separation of inner membrane proteins. Roman numerals are used to identify the five oxidative phosphorylation complexes in the BN-PAGE gels. During nondenaturing 1-D BN-PAGE, the five complexes remain intact (first step) whereas under the denaturing conditions used for 2-D BN-PAGE the individual proteins of each complex are resolved by size (second step). Thus, those proteins that comprise each complex are aligned vertically within the gel. Note that both gel platforms can be subjected to image analysis to look for changes in protein abundances or subjected to western blotting (WB) to reveal post-translational modification (PTMs) by reactive oxygen and nitrogen species, as well as electrophilic lipids, as a number of antibodies have been developed against these modifications. P-SX, protein thiol modifications; P-4HNE, 4-hydroxynonenal protein adduct; P-CHO, protein carbonyls; P-NO$_2$ 3-nitrotyrosine protein modification

be analyzed for changes in protein abundance or a variety of well-defined ROS/ RNS-mediated post-translational modifications using immunoblotting techniques. Finally, proteins of interest can then be identified using several mass spectrometry techniques. Proteomic studies from our laboratories are considered significant in that novel changes in several key energy metabolism pathways, including fatty acid oxidation, TCA cycle, and oxidative phosphorylation system were found to be altered in mitochondria isolated from animals exposed to ethanol chronically *(4,5,15)*. These alterations were in response to alcoholic-mediated metabolic stress and correlated to those pathways with a direct impact on the etiology of disease. A detailed discussion of these techniques is provided in the sections to follow.

3.1   Preparation of Mitochondrial Samples

1. Mitochondria are prepared from fresh liver tissue by standard differential centrifugation techniques *(16)* and should exhibit tightly coupled respiration with respiratory control ratios in the range of 4–8. Moreover, mitochondrial protein yield should average 25–30 mg of mitochondrial protein/g wet weight liver and cytochrome *c* oxidase and citrate synthase activities should be determined as additional markers of mitochondria purity and yield *(4,15)*. It is imperative that mitochondria used for proteomics studies are isolated by methodologies that result in functional and pure mitochondrial preparations. After isolation and determination of the mitochondrial protein concentration, the appropriate volume to achieve 1.0 mg of protein is pipetted into a 1.5-mL microcentrifuge tube and mitochondria are spun at $15,000 \times g$ for 10 min to form a 1.0 mg protein pellet. The supernatant is removed and pellets are stored and dried at $-80°C$ before used in experiments. Mitochondria samples can then be resuspended in the appropriate buffer for BN-PAGE proteomic experiments. For IEF focusing studies mitochondria are typically stored in suspension at a concentration of at least 10 mg of protein per milliliter.

3.2   Running One-Dimensional Isoelectric Focusing Gels

1. Mitochondria samples are thawed and kept on ice immediately before use. The optimal protein concentration for a 2-D IEF/SDS-PAGE gel is 50–200 μg of protein per gel to achieve maximal resolution of proteins with minimal streaking. Repeat protein determination using standard methods (i.e., Lowry or Bradford protein assay) as protein concentration can change over time. A protein concentration of no less than 5 mg/mL is preferred as this will allow minimal dilution of reagents in IEF rehydration buffer.

2. Remove rehydration buffer and DTT aliquots from freezer and thaw to room temperature. Do not keep these reagents on ice as urea and DTT will precipitate from solution.
3. Into 1.0 mL of rehydration buffer add 10 μL of pH 3–10 ampholines, 10 μL of 200 mM tributylphosphine (BioRad, cat. no. 163–2101), and 40 μL of 1 M DTT. Mix well and keep at room temperature.
4. IEF strips can be rehydrated overnight with a total volume of 150–160 μL containing the mitochondrial sample. The following example conditions are given below for the rehydration of strips with buffer and a protein sample that has a concentration of 10 mg/mL. Samples are allowed to extract in rehydration buffer for at least 30 min at room temperature before loading samples onto IEF gel strips. Gently vortex samples every 5–10 min during extraction.

Protein amount	Quantity of sample	Quantity of rehydration buffer
50 μg	5 μL	155 μL
100 μg	10 μL	150 μL
200 μg	20 μL	140 μL

5. Following the procedure described by the specific manufacturer for the instrumentation being used, slowly load 160 μL of sample onto the strip. Samples will load along the entire lane by capillary action. Pipet slowly to minimize the introduction of bubbles into the gel lane as this could result in uneven rehydration of the IEF strips and loss of sample. Load all samples into cassette lanes before removing IEF gel strips from freezer.
6. After loading samples, remove the IEF strip pack from the freezer and keep strips on ice to minimize condensation. For each sample, carefully remove an IEF strip from the plastic backing, holding it with clean forceps by the marked negative (−) end. Holding the IEF strip from the (−) end and with the gel side up, gently slide the IEF gel strip into the cassette well until the strip reaches the (+) end of the cassette.
7. After rehydration of the IEF gel strips is complete, assemble the electrophoresis apparatus according to the manufacturer's instructions.
8. Perform IEF gel electrophoresis using the following conditions: 175 V for 20 min, ramp to 2000 V for 45 min, 2000 V for 30 min, ramp down to 500 V for 30 min, and 500 V for 2 h. As IEF proceeds, the bromophenol blue dye front will migrate from the top to the bottom of the IEF strips and a green/blue/yellow band will appear at the anode (+) after IEF.
9. After IEF, seal cassettes in plastic wrap and freeze at −80°C until ready to perform second dimension SDS-PAGE.

3.3 Running 2-D IEF/SDS-PAGE Gels

1. For second dimension SDS-PAGE gels a wide range of equipment is commercially available that is compatible with the first dimension IEF. Homogenous

(10 or 12%) or gradient (8–15%) acrylamide gels can be used for 2-D SDS-PAGE. It is important that spacer plates are selected that provide sufficient width for inserting the IEF gel strips.

2. Remove IEF gel strips from freezer and allow IEF gel strips to thaw for a few minutes before beginning the equilibration step. For equilibration, prepare 1.0-mL equilibration buffer per IEF gel strip. Remove equilibration buffer and DTT aliquots from freezer and thaw at room temperature. To 1.0 mL of equilibration buffer, add 50 μL of 1 M DTT.

3. To remove the IEF gel strips from cassette, remove plastic covering gently and using forceps place each strip into a 15-mL conical tube, gel side up, and cover with 1.0 mL of equilibration buffer. Gently rock strips with equilibration buffer for 15 min. During the equilibration step, prepare the agarose solution and keep warm. The agarose solution is used to "seal" the IEF gel strips to the top of SDS-PAGE gel.

4. After equilibration, remove strips from conical tubes, cut small plastic end from the anode (−) side, rinse strips in 1X SDS-PAGE running buffer, and slide the IEF gel strip between the plates of the SDS-PAGE gel. Gently pipet the warm agarose solution over strip to "seal" the IEF gel strip on top of the SDS-PAGE gel. Do not allow bubbles to form between the IEF gel strip and the top of the SDS-PAGE gel. Slide a "tooth" from a 1.5-mm gel comb between glass plates at the (−) end of the gel strip. This will serve as the "well" for molecular weight markers. Gels are placed upright for 15–30 min to allow the agarose to set before beginning electrophoresis.

5. Once the agarose has set, gently remove the gel comb "tooth" and load broad range molecular weight markers to the well. Assemble the gel apparatus per the manufacturer's instructions, fill inner and outer chambers with standard 1 X SDS-PAGE gel electrophoresis running buffer, and run gels for 1–2 h at 100 V or until the dye front reaches the bottom of gel.

6. After electrophoresis, gels can be stained for total protein using conditions described above (Coomassie Blue or SYPRO Ruby) or subjected to western blotting. Typically, we stain the 2-D IEF/SDS-PAGE gels with SYPRO Ruby stain to increase sensitivity of low abundance proteins. An example of a 2-D IEF/SDS-PAGE gel stained with SYPRO Ruby stain is presented in Fig. 1.

3.4   Running 1-D BN-PAGE Gels

1. BN-PAGE is a specialized method of electrophoresis that facilitates the high-resolution analysis of membrane protein complexes in their native state from biological membranes. A detailed discussion of the development and theory of BN-PAGE for analysis of the oxidative phosphorylation system proteins is presented in the seminal papers of Schägger and von Jagow (17,18).

2. For BN-PAGE, the procedure more recently described by Brookes et al. (19) is used with the 1-D BN-PAGE, typically performed with 5–12% or 5–16.5% gradient gels with 1.5-mm spacer plates. Recipes for gels are as follows (see Note 3).

		Resolving gel solutions		Stacking gel solution
	Light-5%	Heavy-12%	or Heavy-16.5%	4%
Protogel	0.66 mL	1.60 mL	1.91 mL	0.60 mL
Water	1.97 mL	0.56 mL	–	2.40 mL
3X Gel Buffer	1.34 mL	1.34 mL	1.16 mL	1.50 mL
Glycerol	–	0.47 mL	0.40 mL	–
10% AMPS	26.0 μL	26.0 μL	23.0 μL	70.0 μL
TEMED	4.0 μL	4.0 μL	3.50 μL	9.0 μL

3. On the day of the experiment, remove the 1.0 mg mitochondrial pellets from the −80°C freezer and keep on ice. Add 100 μL of extraction buffer and 12.5 μL of 10% N-dodecyl-β-D-maltoside to each 1.0 mg of mitochondrial pellet. Gently resuspend mitochondrial protein by pipetting to extract and dissociate the protein. Place samples on ice for 30–60 min with gentle vortexing every 5–10 min. After this extraction step, centrifuge samples at 15,000 × g for 5 min at 4°C to pellet any non-dissolved material, remove supernatant to a fresh microcentrifuge, and then determine the protein concentration of the extract. A typical protein concentration of sample prepared by the procedure described here is 5–8 mg/mL.

4. After determining the protein concentration in each sample, add 6.3 μL of the cold coomassie brilliant blue G-250 suspension (Solution #5, Subheading 2.2.1.) to each tube of mitochondrial extract and gently vortex. Load samples immediately onto gel for 1-D BN-PAGE (*see* **Note 4**).

5. After assembling the gel apparatus per manufacturer's directions, fill wells of the stacking gel with cold Hi-Blue cathode buffer (Solution #1a, Subheading 2.2.1.), and load 75–250 μg of protein per lane. Fill the inner (i.e. upper) buffer chamber with cold Hi-Blue cathode buffer (Solution #1a, Subheading 2.2.1.) and the outer chamber with cold anode buffer (Solution #2, Subheading 2.2.1.). Electrophoresis is performed in the cold room at 40 V for 1 h or until samples have migrated into the resolving gels. At this time the cathode buffer is changed to cold Low-Blue cathode buffer (Solution #1b, Subheading 2.2.1.) and electrophoresis is continued for an additional 3–4 h or until dye front reaches the bottom of the gel.

6. At this point, gels can be stained with Coomassie blue to visualize the content of the intact oxidative phosphorylation complexes in samples or gels can be processed for 2-D BN-PAGE by the procedures described below. A representative 1-D BN-PAGE gel is shown in the upper left panel of Fig. 1.

3.5 Running 2-D BN-PAGE Gels

1. To resolve the individual polypeptides that comprise each oxidative phosphorylation complex, the intact gel lane containing all five complexes is cut from the 1-D gel, rotated 90°, and laid on top of a Tris-Tricine/SDS-PAGE gel. The rec-

ipe for preparing one, 1.5-mm thick 2-D BN-PAGE gel is as follows leaving a 1.0-cm gap at the top of the gel.

2-D BN-PAGE Gel Buffer	2.98 mL
Protogel	2.98 mL
H_2O	2.31 mL
Glycerol	0.72 mL
10% AMPS	60.0 μL
TEMED	6.00 μL

2. To apply the 1-D BN-PAGE gel strip on to the top of the SDS-PAGE resolving gel, raise-up the gel plates in the clamps and place the top, that is, a 1- to 2-cm gap on an angled hot plate or heat block. Pour approximately 4–5 mL of hot agarose (Solution #5, Subheading 2.2.2.) into the gap at the top of the gel sandwich, and using the back plate as a "staging area" and the hot agarose as a "lubricant", gently slide the gel lane down between the plates until the strip lays on top of the SDS-PAGE resolving gel. Once the gel is in place, insert a single "tooth" of a gel comb on the end to serve as a well for molecular weight markers.
3. Remove the gel plate assembly from the hot plate and allow the agarose to set for 20–30 min. Excess agarose is then removed from the top of the gel with a scalpel blade and then the gel is overlaid with a thin layer of the SDS/β-mercaptoethanol solution every 5–10 min for 30–45 min to denature proteins.
4. Fill the inner chamber with Cathode buffer (Solution #1, Subheading 2.2.2.) and the outer chamber with the anode buffer (Solution #2, Subheading 2.2.2.), load molecular weight markers into well, and run gels at 30 V for 45 min followed by 110 V for 1.5 to 2 h.
5. After electrophoresis, gels can be stained for total protein using the conditions described in Section 2.2 (Coomassie Blue or SYPRO Ruby) or subjected to western blotting. Typically we stain 2D BN Page gels with Coomassie Blue stain. An example of a 2D BN Page gel is given in Figure 1.

3.6   Image Analysis of Gels

Instructions for image analysis of gels are detailed in the following publications from our laboratories (4,5,15). In brief, scanned TIFF images for 1-D and 2-D BN-PAGE gels can be analyzed using either Scion Image Beta 4.02 (Scion Corp) or Quantity One (BioRad Laboratories) using the directions provided for each software program. For 2-D IEF/SDS-PAGE image analysis, gels are scanned, saved as TIFF files, and analyzed for differences in protein abundance between control and ethanol samples using PDQuest Image Analysis software (BioRad Laboratories).

3.7   Mass Spectrometry for Identification of Proteins

Proteins found by image analysis to change in relative abundance in response to a treatment are cut from gels and subjected to matrix-assisted laser desorption ioniza- tion time-of-flight (MALDI-TOF) mass spectrometry by standard methodologies as described in *(4,19)* to identify proteins. The peptide masses identified using MALDI-TOF mass spectrometry are entered into the MASCOT search engine (see www.matrixscience.com) and the NCBI data base is searched to match the tryptic peptide fingerprint with a parent protein.

Acknowledgments   The authors would like to thank Dr. Paul S. Brookes, University of Rochester, for advice and assistance in establishing the BN-PAGE technique in our laboratories. This work was supported by AA15172 (SMB), AA13395 (VDU), and DK 073116 (AL). The MALDI-TOF mass spectrometer in the UAB mass spectrometry shared facility was purchased with funds provided by NCRR Grant S10 RR-11329. Operation of the mass spectrometry shared facility came, in part, from the UAB Comprehensive Cancer Center Core Support Grant P30 CA-13148.

4   Notes

1. Adjust the pH of these solutions at 4°C because gels are run in the cold room (4°C).
2. The 3X gel buffer can be stored at room temperature. However, the pH must be adjusted at 4°C because gels are run in the cold room (4°C).
3. These gel volumes can be proportionally increased to prepare a resolving gel that has an extended length. This is typically done if the 1-D BN-PAGE gels are to be run overnight to improve the resolution (i.e., separation) of the com- plexes. Similarly, when gels are run overnight the cathode buffer can be switched to a "blue-free" buffer, that is, 50 mM Tricine, 15 mM BisTris, pH 7.0, with no added coomassie blue G-250. This helps to minimize the high blue background when 1-D BN-PAGE gels are destained for imaging and densitometry purposes.
4. For the separation of large membrane protein complexes it is recommended that the concentration of the coomassie blue G-250 in the sample be one quarter of the lauryl maltoside concentration for electrophoresis *(17,18)*. Therefore, based on the given stock solutions and protocol for the extraction of inner membrane proteins this works out to be approximately 6.3 μL of a 5% (w/v) coomassie blue G-250 solution. Addition of a molar excess of dye to protein also helps to rem- edy one problem of native electrophoresis, that is, the tendency of membrane proteins to form aggregates in the presence of detergents. However, the problem of aggregation is minimized by the inclusion of the coomassie blue G-250. This dye binds to the hydrophobic regions on the proteins surface, which induces a negative surface charge thereby reducing protein aggregation. In addition, the anionic nature of coomassie blue G-250 induces the needed "charge shift" on proteins such that they will migrate to the anode at pH 7.5 *(17,18)*.

References

1. Cunningham, C. C., and Bailey, S. M. (2001) Ethanol consumption and liver mitochondria function. *Biol. Sig. Recept.* **10**, 271–282.
2. Hoek, J. B., Cahill, A., and Pastorino, J. G. (2002) Alcohol and mitochondria: A dysfunctional relationship. *Gastroenterology* **122**, 2049–2063.
3. Bailey, S. M. (2003) A review of the role of reactive oxygen and nitrogen species in alcohol-induced mitochondrial dysfunction. *Free Radic. Res.* **37**, 585–596.
4. Venkatraman, A., Landar, A., Davis, A. J., Chamlee, L., Sanderson, T., Kim, H., Page, G., Pompilius, M., Ballinger, S., Darley-Usmar, V., et al. (2004) Modification of the mitochondrial proteome in response to the stress of ethanol-dependent hepatotoxicity. *J. Biol. Chem.* **279**, 22092–22101.
5. Venkatraman, A., Landar, A., Davis, A. J., Ulasova, E., Page, G., Murphy, M. P., Darley-Usmar, V., and Bailey, S. M. (2004) Oxidative modification of hepatic mitochondria protein thiols: effect of chronic alcohol consumption. *Am. J. Physiol. Gastrointest. Liver Physiol.* **286**, G521–G527.
6. Coleman, W. B., and Cunningham, C. C. (1990) Effects of chronic ethanol consumption on the synthesis of polypeptides encoded by the hepatic mitochondrial genome. *Biochim. Biophys. Acta* **1019**, 142–150.
7. Coleman, W. B., and Cunningham, C. C. (1991) Effect of chronic ethanol consumption on hepatic mitochondrial transcription and translation. *Biochim. Biophys. Acta* **1058**, 178–186.
8. Cahill, A., Stabley, G. J., Wang, X., and Hoek, J. B. (1999) Chronic ethanol consumption causes alterations in the structural integrity of mitochondrial DNA in aged rats. *Hepatology* **30**, 881–888.
9. Cahill, A., Wang, X., and Hoek, J. B. (1997) Increased oxidative damage to mitochondrial DNA following chronic ethanol consumption. *Biochem. Biophys. Res. Commun.* **235**, 286–290.
10. Cahill, A., Baio, D. L., Ivester, P. I., and Cunningham, C. C. (1996) Differential effects of chronic ethanol consumption on hepatic mitochondrial and cytoplasmic ribosomes. *Alcohol. Clin. Exp. Res.* **20**, 1362–1367.
11. Patel, V. B., and Cunningham, C. C. (2002) Altered hepatic mitochondrial ribosome structure following chronic ethanol consumption. *Arch. Biochem. Biophys.* **398**, 41–50.
12. Mootha, V. K., Bunkenborg, J., Olsen, J. V., Hjerrild, M., Wisniewski, J. R., Stahl, E., Bolouri, M. S., Ray, H. N., Sihag, S., Kamal, M., et al. (2003) Integrated analysis of protein composition, tissue diversity, and gene regulation in mouse mitochondria. *Cell* **115**, 629–640.
13. Taylor, S. W., Fahy, E., Zhang, B., Glenn, G. M., Warnock, D. E., Wiley, S., Murphy, A. N., Gaucher, S. P., Capaldi, R. A., Gibson, B. W., and et al. (2003) Characterization of the human heart mitochondrial proteome. *Nat. Biotechol.* **21**, 281–286.
14. Bailey, S. M., Landar, A., and Darley-Usmar, V. (2005) Mitochondrial proteomics in free radical research. *Free Radic. Biol. Med.* **38**, 175–188.
15. Bailey, S. M., Robinson, G., Pinner, A., Chamlee, L., Ulasova, E., Pompilius, M., Page, G. P., Chhieng, D., Jhala, N., Landar, A., et al. (2006) S-adenosylmethionine prevents chronic alcohol-induced mitochondrial dysfunction in the rat liver. *Am. J. Physiol. Gastrointest. Liver Physiol.* **291**, G857–G867.
16. Bailey, S. M., Patel, V. B., Young, T. A., Asayama, K., and Cunningham, C. C. (2001) Chronic ethanol consumption alters the glutathione/glutathione peroxidase-1 system and protein oxidation status in rat liver. *Alcohol. Clin. Exp. Res.* **25**, 726–733.
17. Schägger, H., Cramer, W. A., and von Jagow, G. (1994) Analysis of molecular masses and oligomeric states of protein complexes by blue native electrophoresis and isolation of membrane protein complexes by two-dimensional native electrophoresis. *Anal. Biochem.* **217**, 220–230.
18. Schägger, H., and von Jagow, G. (1991) Blue native electrophoresis for isolation of membrane protein complexes in enzymatically active form. *Anal. Biochem.* **199**, 223–231.
19. Brookes, P. S., Pinner, A., Ramachandran, A., Coward, L., Barnes, S., Kim, H., and Darley-Usmar, V. M. (2002) High throughput two-dimensional blue-native electrophoresis: a tool for functional proteomics of mitochondria and signaling complexes. *Proteomics* **2**, 969–977.

25

Alcoholic Liver Disease and the Mitochondrial Ribosome

Methods of Analysis

Alan Cahill and Peter Sykora

Summary Chronic alcohol consumption has been shown to severely compromise mitochondrial protein synthesis. Hepatic mitochondria isolated from alcoholic animals contain decreased levels of respiratory complexes and display depressed respiration rates when compared to pair-fed controls. One underlying mechanism for this involves ethanol-elicited alterations in the structural and functional integrity of the mitochondrial ribosome. Ethanol feeding results in ribosomal changes that include decreased sedimentation rates, larger hydrodynamic volumes, increased levels of unassociated subunits and changes in the levels of specific ribosomal proteins. The methods presented in this chapter detail how to isolate mitochondrial ribosomes, determine ribosomal activity, separate ribosomes into nucleic acid and protein, and perform two-dimensional nonequilibrium pH gradient electrophoretic polyacrylamide gel electrophoresis to separate and subsequently identify mitochondrial ribosomal proteins.

Keywords Ethanol, liver; mitochondrial protein synthesis; mitochondrial ribosomes; ribosomal proteins; NEPHGE.

L. E. Nagy (ed.), *Alcohol: Methods and Protocols*
© Humana Press 2008

1   Introduction

Mitochondrial ribosomes (mitoribosomes) are responsible for the translation of 13 proteins coded for by the mitochondrial genome. These proteins (ND1, ND2, ND3, ND4, ND4L, ND5, ND6, Cyt b, COI, COII, COIII, ATPase6, and ATPase8) are required for the successful assembly of the respiratory complexes NADH dehydrogenase, ubiquinol: cytochrome c oxidoreductase, cytochrome c oxidase, and ATP synthase, respectively. Accurate assembly and subsequent activity of mitoribosomes is therefore essential for the maintenance and correct functioning of the electron transport chain. The rat liver mitochondrial ribosome has a molecular mass of approximately 3.57 MDa, dimensions of 26.2 nm × 23.6 nm, and a sedimentation coefficient of 55S *(1)*. It has a chemical composition of 75% protein and 25% RNA and comprises a large 39S subunit (LSU) and a small 28S subunit (SSU). Subunits are assembled inside the mitochondrial matrix from 12S (SSU) and 16S (LSU) rRNA, encoded for by the mitochondrial genome, and ribosomal proteins, encoded for by the nucleus and imported into the mitochondrion.

One of the earliest observations seen in the liver during chronic ethanol feeding is decreased mitochondrial protein synthesis *(2,3)*. Although a number of potential molecular mechanisms may explain this decrease, one specific mechanism involves the impaired assembly of mitoribosomes. Investigations conducted by the Cunningham lab have revealed that mitoribosomes isolated from young male rats (200 g) fed ethanol (Lieber-DiCarli diet) *(4)* for 31 d exhibit decreased sedimentation rates through sucrose density gradients when compared with their paired controls *(5,6)*. Further, sedimentation velocity experiments reveal a significant decrease in the average sedimentation coefficient for the intact ethanol mitoribosome (52.2, ethanol-fed; 53.7, control) *(6)* as well as for the small ribosomal subunit (27.0, ethanol-fed; 28.3, control) *(6)*. In addition, mitoribosomes isolated from ethanol-fed animals exhibit significant decreases in their translational diffusion coefficients ($1.02 \times 10^{-7} cm^2 s^{-1}$, ethanol-fed; 1.10, control) *(6)*, and larger hydrodynamic diameters (42.1 nm, ethanol-fed; 39.1, control) *(6)*. These ethanol-elicited alterations in the physicochemical properties of the hepatic mitoribosome are accompanied by significant amounts of disassociation of the intact 55S monosome (8–14%) into its constituent subunits *(5,6)*. Further, in vitro studies on the translation capacity of mitoribosomes have revealed that while total translation activity is depressed by 29% in the ethanol-fed animals relative to the controls, no difference is detected between the two treatment groups in the activity of the intact monosomes *(6)*, which suggests that the decrease in translation activity observed in ethanol-fed animals is caused by either increased disassociation or decreased association of functional ribosomes rather than impaired activity of intact monosomes. Whether this means that ethanol feeding induces increased disassociation or decreased association of mitoribosomes in vivo or whether it causes a more relaxed and loosely assembled monosome which becomes disassociated upon isolation has yet to be determined.

The sedimentation properties of mitochondrial ribosomes are affected not only by the size of the ribosomes but also by their shape. Any ethanol-elicited alterations in the ratios of constitutive mitochondrial ribosomal proteins (MRPs) may lead to a change in the shape of the ribosome and consequently a decrease in sedimentation rates. Additionally, altered MRP levels may lead to formation of inaccurately assembled ribosomes that are easily dissociated into subunits or unable to bind to functionally important docking sites on the inner mitochondrial membrane (IMM). Further, it is unclear as to whether any of the MRPs are susceptible to the kinds of protein damage commonly incurred during chronic ethanol feeding, that is oxidative modification, nitrosative damage and formation of adducts such as acetaldehyde (AA), malondialdehye-acetaldehyde (MAA), and the hydroxyethyl radical (HER). Ethanol-elicited modifications of MRPs may result in an alteration in their binding to other MRPs, rRNA or IMM components, as well as alter their rates of degradation. In addition to altering the overall structure of the mitochondrial ribosome, ethanol feeding has been shown to significantly decrease the levels of a specific population of MRPs *(5,7)*. The reasons for the ethanol-elicited decreases in a specific population of MRPs as opposed to a global down-regulation are unclear; however, a recent proteomics study by Chevallet et al. *(8)* has demonstrated the selective depression of nine MRPs upon depletion of mitochondrial DNA (mtDNA) in 143B osteosarcoma cells. Our laboratory has shown that mtDNA depletion is a common occurrence during chronic ethanol feeding *(9–11)*.

In this chapter, we detail methodologies for the isolation of mitochondrial ribosomes from hepatic mitochondria, determination of their activities, extraction of MRPs from the ribosomes and two-dimensional nonequilibrium pH gradient electrophoretic (NEPHGE) analysis of MRPs. Separated proteins can then be excised from the gels and subjected to mass spectrometry or transferred to nitrocellulose membranes and probed for oxidative and nitrosative modifications by western blotting.

2 Materials

2.1 Isolation of Mitochondrial Ribosomes From Mitochondria

1. Isolation buffer: $0.25\,M$ sucrose and $2\,mM$ HEPES, pH 7.4 (pH with $6\,M$ KOH). Store at 4°C.
2. RC DC Protein Assay Reagents (BioRad, cat. no. 500–0120).
3. Digitonin (solid, free of water-insoluble constituents, Sigma cat. no. D-141).
4. Buffer A: $0.26\,M$ sucrose, $40\,mM$ KCl, $15\,mM$ $MgCl_2$, $14\,mM$ Tris-HCl, pH 7.5, and $0.8\,mM$ ethylene diamine tetraacetic acid (EDTA; *see* **Note 1**). Store at 4°C. Just before use add β-mercaptoethanol to a final concentration of $5\,mM$.
5. Buffer B: $100\,mM$ KCl, $20\,mM$ $MgCl_2$, and $20\,mM$ triethanolamine pH 7.5. Store at 4°C. Just before use, add β-mercaptoethanol to a final concentration of $5\,mM$.
6. Puromycin dihydrochloride (solid, Sigma, cat. no. P 7255).
7. Magnesium-deficient buffer B: buffer B without the $MgCl_2$ but including $1\,mM$ EDTA.

2.2   Ribosome Activity Assay

1. Ribosome activity buffer: 50 mM KCl, 50 mM Tris-HCl, pH 7.8, and 20 mM MgCl$_2$. Store at 4°C.
2. Transfer RNA (from baker's yeast, Type, X., Sigma cat. no. R9001).
3. Pyruvate kinase (from rabbit muscle, Type VII, cat. no. P7768).
4. Phospho*enol*pyruvate (Sigma, cat. no. P 7127).
5. ATP (Sigma, cat. no. A 7699) and GTP (Sigma, cat. no. G 5884).
6. L-[4-³H]Phenylalanine (Amersham, cat. no. TRK204–250UC).
7. Econo-Safe (RPI, cat. no. 111175).

2.3   Preparation of Soluble Mitochondrial Translation Factors

1. Hypotonic buffer: 20 mM Tris-HCl, pH 7.5, 20 mM MgCl$_2$, and 40 mM KCl. Store at 4°C. Just before, use add β-mercaptoethanol to a final concentration of 5 mM.
2. 2 M KCl.
3. Sephadex G-25 (Amersham, cat. no. 17–0031–02).

2.4   Extraction of Ribosomal Proteins From Pelleted Mitochondrial Ribosomes

Both MRP extraction procedures produce similar 2DE profiles.

2.4.1   Lithium Chloride-Urea Procedure

This procedures is used in the lab when purified ribosomes or subunits are first pelleted; starting material usually two livers.

1. Ribosomal protein solubilization solution: 6 M urea, 3 M LiCl, 50 mM KCl, and 5 mM β-mercaptoethanol, pH 3.5–5.5, with 3 N HCl. Can be stored at −20°C in 500 μL aliquots.
2. 3 N HCl.
3. 20% (w/v) trichloroacetic acid (TCA).
4. Ethanol/ether (1:1 vol/vol).

2.4.2   Acetic Acid-Acetone Procedure

This procedure is used when MRPs are extracted directly from sucrose density gradient fractions without prior sedimentation of purified ribosomes or subunits; starting material can be one liver.

1. Glacial acetic acid.
2. 1 *M* magnesium chloride.
3. Acetone.

2.5   2D-Electrophoretic Analyses Of Mitochondrial Ribosomal Proteins (NEPHGE)

1. 40% Biolyte 3/10 ampholyte (BioRad, cat. no. 163–1113) and 40% Biolyte 5/7 ampholyte (BioRad, cat. no. 163–1153).
2. Ribosomal protein lysis buffer: 9.5 M urea (ultrapure™ urea, Invitrogen, cat #. 15505–035), 2% (w/v) Triton X-100, 2% ampholytes (1% Biolyte 3/10 ampholyte, 1% Biolyte 5/7 ampholyte). Store at −20°C as 500 µL aliquots.
3. 30% acrylamide / bis solution, 29, 1 ratio (BioRad, cat. no. 161–0156). Store at 4°C.
4. 10% (w/v) Triton X-100.
5. 10% (w/v) ammonium persulfate (Sigma, cat #. A3678). Make up fresh every 2 wk.
6. TEMED (BioRad, cat. no. 161–0801).
7. Equilibration buffer: 10% (w/v) glycerol, 2.3% (w/v) sodium dodecyl sulfate (SDS), 0.002% bromophenol blue, and 62.5 m*M* Tris-HCl, pH 6.8.
8. 0.5% (w/v) agarose (ultrapure™ agarose, Invitrogen, cat. no. 15110–019) in equilibration buffer.
9. Dithiothreitol (DTT), solid (BioRad, cat. no. 161–0610).
10. Possible staining procedures: SYPRO® Ruby protein gel stain (Invitrogen, cat. no. S-12000), Brilliant Blue R staining solution (Sigma, cat. no. B6529) and SilverQuest™ Silver Staining kit (Invitrogen, cat. no. LC6070).
11. Destaining solution (for Brilliant Blue R staining only): 30% (v/v) ethanol and 10% (v/v) acetic acid.

2.6   Equipment

1. Nitrocellulose filter disks (Millipore); used in Subheading 3.2.
2. Gel extrusion needles (BioRad, cat. no. 165–1944); used in Subheading 3.5.
3. Appropriate ultracentrifuge, rotors and centrifuge tubes.
4. Microcentrifuge.
5. UV plate reader or spectrophotometer; used in Subheading 3.1.
6. Scintillation counter; used in Subheading 3.2.
7. Gradient fractionator (e.g. Labconco, Auto-Densiflow), 3.1.
8. Liquid nitrogen; used in Subheading 3.3.
9. Probe sonicator (e.g., Fisher Sonic Dismembrator 300, with Microtip); used in Subheading 3.3.

10. Lyophilizer; used in Subheading 3.3.
11. Glass tubes for NEPHGE (e.g., 7 mm OD, 5 mm ID, 15 cm length); used in Subheading 3.5.
12. Electrophoresis unit (e.g., Hoefer Scientific Instruments, GT1); used in Subheading 3.5.
13. Image analysis system and suitable software; used in Subheading 3.6.
14. Microwave (optional); used in Subheading 3.1.

3 Methods

3.1 Isolation of Mitochondrial Ribosomes From Mitochondria

This procedure provides sufficient yields of mitochondrial ribosomes for sedimentation profile analyses and, when used in conjunction with silver staining, 2D/SDS-PAGE mini-gel analyses of ribosomal proteins. If 55S ribosomal activity is to be determined or 2D electrophoretic analysis of ribosomal proteins using larger gels is intended we suggest using two rat livers as the starting tissue.

1. Isolate hepatic mitochondria using standard procedures (12).
2. Perform a protein assay on the isolated mitochondria. We use the Lowry protein assay but any detergent compatible assay will work just as well. Digitonin needs to be added to the mitochondria at a ratio of 0.11 mg of digitonin per milligram of mitochondrial protein. Dissolve a suitable amount of digitonin (see Note 2) in mitochondrial isolation buffer such that when the digitonin solution is added to the mitochondria it not only gives the required detergent to protein ratio but that it also dilutes the mitochondrial protein concentration down to 20 mg/mL.
3. Let the digitonin solution cool down before adding it to the mitochondria on ice. Stir the mitochondrial suspension gently for exactly 15 min (see Note 3).
4. Dilute the suspension down 10-fold with ice-cold isolation buffer, allow to continue stirring for a further 10 s, then centrifuge immediately at 12,000 g for 10 min.
5. Collect the pelleted mitoplasts and resuspend them in isolation buffer to wash them. Re-pellet the mitoplasts and resuspend them in buffer A. Mitoplasts can be stored overnight at −70°C.
6. Perform a protein assay on the mitoplasts and dilute them down with buffer A to a concentration of 20 mg/mL. Lyse them (see Note 4) with Triton X-100 (2% w/v final concentration).
7. Clarify the suspension at 12,000 g for 45 min. Remove the supernatant and layer it onto 15 ml cushions of 1 M sucrose and 1% (w/v) Triton X-100 dissolved in buffer B. Centrifuge at 200,000 g) for 4 h (a 50.2Ti is a good choice of rotor). The pellets consist of mitochondrial ribosomes and some mitoplast membrane fragments.

8. Decant the supernatant, resuspend the crude mitochondrial ribosomal pellets in 1 ml of buffer B and layer them on top of a 10–30% linear sucrose gradient (30 mL) in buffer B and centrifuge at 53,000 g for 14 h (SW27 rotor). Fractionate the gradients and monitor the A$_{260nm}$ of the fractions for ribosomal RNA. We detect the A$_{260nm}$ of 50-μL aliquots of our gradient fractions using a 96-well fluorescent plate reader that is also capable of reading U. V. Alternatively, fractions can be monitored spectrophotometrically. Absorbance peaks of free ribosomal particles corresponding to the small (28S) and large (39S) ribosomal subunits and the intact 55S monosome should be detected (*see* Fig. 1). A small peak is also detected at the bottom of the gradient. This corresponds to a population of ribosomes that may be attached to inner mitochondrial membrane fragments via unique attachments, for example, nascent polypeptide chains being inserted into the inner mitochondrial membrane.

9. If the aim is to isolate mitochondrial ribosomal proteins without any contaminating nascent polypeptides or, alternatively, if it is required to strip from the IMM those ribosomes attached via newly synthesized polypeptides, then incubate the crude ribosomal pellet in 1 ml buffer B containing 1 m*M* puromycin (*see* **Note 5**) for 15 min at 37°C before layering on top of the sucrose gradients.

10. Fractions corresponding to the intact ribosomes or the ribosomal subunits can be centrifuged at 230,000 g for 5 h (60Ti rotor) and stored as a pellet at −70°C. Alternatively, the proteins associated with the intact 55S ribosome or the respective subunits can be subjected directly to electrophoresis and detected by silver staining.

11. If it is desired to separate 55S mitochondrial ribosomes into their 28S and 39S, small and large ribosomal subunits, respectively, resuspend the pelleted 55S monosomes or crude ribosomal pellets in magnesium-deficient buffer B and sediment them through 10–30% sucrose gradients made up in the same buffer (*see* **Note 6**).

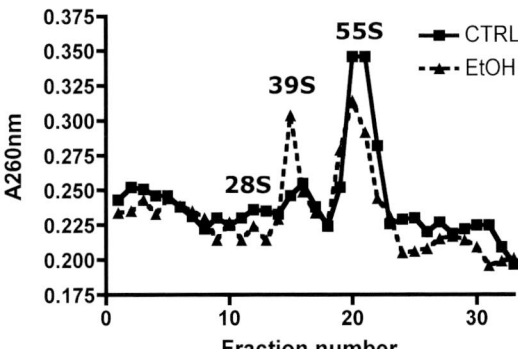

Fig. 1 Effects of chronic ethanol feeding (12 mo) upon the sucrose density gradient profiles of mitochondrial ribosomes and their subunits isolated from a single control and alcoholic liver. Sucrose cushion purified hepatic mitochondrial ribosomes were separated through 10–30% sucrose gradients into monosomes and subunits. The A$_{260nm}$ of the 50-μL fraction aliquots was monitored using a 96-well fluorescent plate-reader (in UV mode)

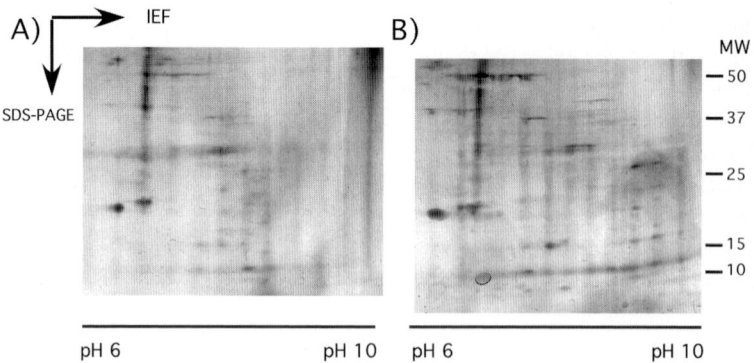

Fig. 2 Two-dimensional (IEF/SDS-PAGE) proteomic analysis of the hepatic mitochondrial LSU. Proteins were extracted from the LSU 39S peaks shown in Fig. 1 using acetic acid and acetone (*see* Subheading 3.4.2.), separated in the first dimension on pH 6–10 ZOOM IPG strips (see chapter by Bailey et al. in this volume) and in the second dimension by 12% SDS-PAGE. Proteins were stained with silver. (**A**) Proteins were extracted from control 39S peak and (**B**) from ethanol 39S peak

3.2   Ribosome Activity Assay

1. Incubate 2–5 pmol of 55S mitochondrial ribosomes (*see* **Note 7**) in ribosome activity buffer supplemented with (final concentrations) 500 mg/mL tRNA, 100 mg/m*l* pyruvate kinase, 2 m*M* phospho*enol*pyruvate, 5 m*M* ATP, 0.3 m*M* GTP, 0.2 mg/mL translation factors (*see* Subheading 3.3) and 10 μCi [^{3}H]phenylalanine in a final volume of 50 μL.
2. Remove 5 μL of the reaction mixture and spot onto a nitrocellulose filter disk. This represents time zero. Start the reaction by adding poly(U) to a final concentration of 1 mg/ml, remove 5 μL aliquots at suitable time points up to 1 h (we find that radiolabel incorporation is linear for this time period) and spot them onto nitrocellulose filter disks (1 cm^{2}).
3. Incubate the filter disks in ice-cold 10% TCA (w/v, 2 mL per 1 cm^{2} filter) for 30 min with gentle swirling to stop the reaction and to precipitate the protein onto the filter.
4. Wash the disks in fresh ice-cold 10% TCA for 10 min with gentle swirling.
5. Place the disks in 5% TCA and heat to 90°C for 15 min.
6 Quantify the radioactivity present on the filters by adding 5 mL of a scintillation cocktail of choice (Econo-Safe works well for us). Express the phenylalanine polymerization as pmol [^{3}H]phenylalanine/mg RNA/pmol mitochondrial ribosomes.

3.3   Preparation of Soluble Mitochondrial Translation Factors

1. Resuspend mitochondria in hypotonic buffer at a concentration of 5 mg/mL and incubate on ice for 15 min to allow the mitochondria to swell.

2. Subject the mitochondria to three cycles of freeze-thawing using liquid nitrogen then sonicate (we use 4 × 30 s pulses at 50%; *see* Subheading 2.5.) to disrupt the mitochondria.

3. Clarify the mitochondrial suspension at 16,000 g for 20 min at 4°C. Remove the supernatant and adjust it to 500 mM KCl.

4. Centrifuge at 200,000 g for 4 h at 4°C to pellet mitochondrial ribosomes.

5. Pass the supernatant over a Sephadex G-25 column (to remove endogenous amino acids) and collect the void volume. This contains the mitochondrial translation factors. Lyophilize the translation factors and store at −70°C.

3.4   Extraction of MRPs From Pelleted Mitochondrial Ribosomes

3.4.1   Lithium Chloride-Urea Procedure

1. Resuspend 55S mitochondrial ribosomes or mitoribosomal subunits in 150 μL of ribosomal protein solubilization solution in a 1.5 mL microcentrifuge tube.

2. Place a micro stir bar in the centrifuge tube (the stir bar will stand on its end) and stir the ribosomes overnight at 4°C (make sure the sample is stirring and it is not the tube that is spinning).

3. Pellet the RNA at 200,000 g (we use a TLA-100 rotor with 200 μl capacity tubes) for 1 h. Remove the supernatant and reserve. Re-extract the RNA pellet by adding another 150 μL of ribosomal protein solubilizing solution. Stir for 6 h and pellet the RNA at 200,000 g for 1 h (*see* **Note 7**). Remove the supernatant and combine it with the supernatant from the first extraction.

4. The RNA pellet should contain rRNA and any mRNA that remains associated with the ribosome.

5. Precipitate the ribosomal proteins overnight at 4°C by adding 2 volumes of 20% TCA. Centrifuge at 10,000 g in a microcentrifuge for 15 min. Remove the supernatant and wash the pellet in 1:1 (v/v) ethanol:ether.

6. Allow to air-dry for 5–10 min before resuspending the ribosomal proteins in an electrophoresis buffer of choice. We suggest using ribosomal protein lysis buffer supplemented with DTT to a final concentration of 1% (w/v) just before use.

3.4.2   Acetic Acid-Acetone Procedure

We have used this procedure on as little as 0.1 $A_{260\,nm}$ units of mitochondrial ribosomes.

1. Collect fractions from sucrose density gradients (*see* Fig. 1) corresponding to intact 55S monosomes or ribosomal subunits.

2. Add magnesium chloride to a final concentration of 100 mM.

3. Add 2 volumes of glacial acetic acid. Invert a couple of times. Leave for 45 min on ice.

4. Pellet precipitated RNA at 5000 g for 10 min.
5. Add 5 volumes of ice-cold acetone. Invert a couple of times. Leave at −20°C overnight.
6. Pellet precipitated protein at 5000 g for 10 min.
7. Wash pellets with ice-cold acetone. Re-pellet protein at 5000 g for 10 min.
8. Remove supernatant and allow pellet to air-dry (approximately 10 min).
9. Resuspend pellet in ribosomal protein lysis buffer supplemented with DTT to a final concentration of 1% (w/v) just before use (*see* Subheading 2.4.).

3.5   *2D-Electrophoretic Analyses Of Mitochondrial Ribosomal Proteins (NEPHGE)*

Early isoelectric focusing methodologies in which ampholytes were used to form pH gradients suffered from gradient drift. The development in recent years of immobilines (buffering acrylamide derivatives that contain either a free carboxylic acid group or a tertiary amino group) has allowed the formation of stable pH gradients in the range of pH 3–12. These immobilized pH gradients (IPGs) are reproducible and insensitive to disturbances from sample components. We have used the ZOOM® IPGRunner™ system from Invitrogen in conjunction with silver staining to analyze the protein content of the large ribosomal subunit from control and eth-anol mitochondrial ribosomes. Figure 1 shows a sucrose density gradient profile of mitochondrial ribosomes isolated from control and ethanol-fed animals. Ethanol-fed animals show decreased levels of intact 55S ribosomes but an increase in the level of 39S ribosomal subunits. This is suggestive of an ethanol-elicited increase in ribosome dissociation or a decrease in ribosome association. The 39S peaks from Fig. 1 were extracted using the acetic acid-acetone procedure (*see* Subheading 2.3) and separated in the first dimension by isoelectric focusing using IPGs (ZOOM strips, Invitrogen) of range pH 6–10 followed by second dimension SDS-polyacrylamide gel electrophoresis (PAGE; Fig. 2). A significantly greater amount of ribosomal protein was extracted from the 39S peak in the ethanol-fed animals when compared with the paired control. This greater yield of ribosomal protein reflects the increase in 39S subunit seen in Fig. 1 and suggests that the increase in 39S fraction as measured by $A_{260\,nm}$ is the result of increased levels of intact 39S subunits and not due to increased levels of mRNA and rRNA associated with the 39S fraction.

The methodologies involved in using Invitrogen's ZOOM® IPGRunner™ system are described in detail in the chapter "Proteomic Approaches to Identify and Characterize Alterations to the Mitochondrial Proteome in Alcoholic Liver Disease" by Bailey et al. in this volume. We will therefore focus here on an alter-native and excellent method of separating ribosomal proteins, that is, NEPHGE. Ribosomal proteins are, for the most part, extremely basic in charge. The 39S ribosomal subunit for example is believed to comprise 48 proteins with pIs ranging from 6.6 to 12.3 with an average of 9.6 *(13)*. Even if these proteins were to be focused using IPGs, it would be difficult to include them all on the same 2D/IEF/

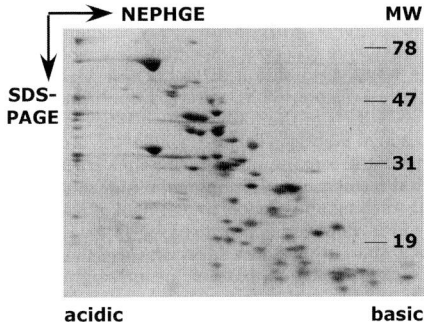

Fig. 3 Two-dimensional (NEPHGE/SDS-PAGE) proteomic analysis of mitochondrial ribosomal proteins extracted from intact 55S monosomes (modified from [7]). Proteins were extracted from the 55S monosome using lithium chloride and urea (*see* **Subheading 3.4.1.**), separated in the first dimension by NEPHGE (pH 3–10) and in the second dimension by 12% SDS-PAGE. Proteins were stained with Coomassie Brilliant Blue

SDS-PAGE gel. During NEPHGE proteins are not necessarily focused to their pIs but instead migrate at different rates across the gel due to their charge. At the end of the run, some proteins will have focused to their pIs while the others will still be migrating. The protein pattern produced on the gel is dependent on the accumulated volt hours so to ensure reproducibility this must remain constant for a specific population of proteins. The method described below is intended for use with the larger 2D system, that is, 15-cm tube gels and 16 cm × 16 cm SDS-PAGE slab gels which we find to be excellent for resolving the complete population of ribosomal proteins. Mini-systems such as the Mini-PROTEAN® Tube Cell Module from Bio-Rad can be used but we find the larger set-up easier to manipulate (Fig. 3).

1. Rinse out the glass tubes with water followed by acetone. Allow to air dry.
2. Stopper one end of the tubes with parafilm.
3. To make 10 mL of gel mix, place 5.5 g of urea, 1.33 mL of 30% acrylamide mix, 2 mL of 10% Triton X-100, 1.97 mL of dH₂O, 0.25 mL of Biolyte 3/10 ampholyte (40%), and 0.25 mL of Biolyte 5/7 ampholyte (40%) into a 125-mL side-arm flask.
4. Heat slightly to dissolve urea.
5. Add 30 μL of AMPS.
6. De-gas for 1 min (*see* **Note 9**).
7. Add 21 μL of TEMED.
8. Pour tube gels using a suitable syringe to approx 1 cm from the top. Overlay with dH₂O. Leave to polymerize for 1–2 h. To test whether the gel mix has polymerized successfully we suggest sucking up any remaining gel mix, after the tubes have been poured, into a plastic pipet. Once it has polymerized inside the pipet it is usually safe to assume that it has also polymerized inside the tubes.
9. Fill the lower reservoir of the electrophoresis apparatus with 0.02 N NaOH. Remove the parafilm from the bottom of the tubes. Place the tubes inside the tube holder according to apparatus instructions (*see* **Note 10**).

10. Load ribosomal protein sample solubilized in ribosomal protein lysis buffer (50 μL is a good volume) onto polymerized gel. For Coomassie blue staining of gels, we suggest 200 μg per tube gel, and for silver staining of gels we suggest 50 μg. We suggest using the RC DC Protein Assay (BioRad, cat. no. 500–0120) to quantify sample protein levels. This assay is compatible with a number of components of the ribosomal protein lysis buffer, for example, ampholytes and thiol reducing agents that would normally interfere with protein determination.

11. Overlay with ribosomal protein lysis buffer diluted 1 to 4 leaving a small space at the top.

12. Fill the tube to the top with upper reservoir solution, that is, $0.01\,M$ phosphoric acid.

13. Run electrophoresis with the polarity reversed i.e. red lead from apparatus into black port on power pack, black into red. Run at 400 V for 1 h and 30 min depending upon the desired separation profile. This corresponds to 600 Vh.

14. Extrude gel from glass tube using a syringe with a long thin gel extrusion needle. Fill the syringe with dH_2O and insert it gentle inside the glass tube down the side of the gel. The needles recommended can easily be slid inside the tubes to just over half way. Pipet water gently while sliding the needle around the outside of the gel. Then withdraw the needle, turn the glass tube around and reinsert the needle into the other side of the tube. Gently pipet water as before. On withdrawing the needle from the tube this time, the gel inside should slide easily out. If not, determine which end of the gel is sticking and reinsert the needle into that side and pipette. We suggest placing a sheet of parafilm or aluminum foil on the lab bench to catch the tube gel should it slide out unexpectedly during extrusion.

15. Place tube gel into 15 mL conical tube and add 5 mL of equilibration buffer supplemented with DTT (0.5% w/v) just before use. Incubate with gentle rocking for 30 min.

16. Pour an SDS slab gel (8–20% resolving gel works well, 5% stacking gel) according to standard procedures. Assemble the electrophoresis apparatus according to manufacturer's instructions.

17. Add a small amount of 1% agarose dissolved in equilibration buffer containing 0.5% (w/v) DTT to the top of the gel to serve as a platform for the tube gel to sit on. Once the agarose has set, lay the tube gel on top. Allow some room at one end to add molecular weight markers. We normally embed a small volume of markers in about 20 μL of 1% agarose in equilibration buffer containing 0.5% DTT in the bottom of a 1.5 mL centrifuge tube. After the agarose has set pop out the markers plug and sit it next to the tube gel on the agarose platform.

18. Pipet more of the agarose/equilibration buffer solution over the top of the tube gel and markers plug to hold them in place (*see* **Note 11**).

19. Run the gel at 25 mA for 7 h or until the Bromophenol blue has migrated to the bottom of the gel.

20. Detect ribosomal proteins using a suitable staining procedure. We have successfully used Coomassie Brilliant Blue R250, SYPRO® Ruby protein gel stain and silver staining procedures. All these staining systems are compatible with the

excision of protein spots from the gels, tryptic digestion of the proteins and analysis by mass spectrometry.
21. Alternatively, the proteins can be transferred from the gel to nitrocellulose membranes and analyzed for specific protein modifications by western blotting.

3.6 Analysis of Proteins Separated by 2DG

A number of current systems and software products are available for protein analyses in 2D gels. Our laboratory possesses a Kodak ImageStation 440CF with Kodak Digital Science 1D & 2D Software. This is convenient for protein staining via Coomassie Brilliant Blue, SYPRO® Ruby and silver.

Acknowledgments This work was supported by grant AA14151 from the National Institute of Alcohol and Alcohol Abuse.

4   Notes

1. A high level of magnesium ions (15–20 mM) is necessary to ensure that mitochondrial ribosomal subunits remain associated during the isolation procedure.
2. To dissolve digitonin, place it in a suitable amount of buffer in a glass beaker and heat it up in a microwave. Watch for the buffer to just begin to boil then stop the microwave and remove the beaker.
3. Digitonin is a detergent and is being added in a precise amount to the mitochondria to remove the outer mitochondrial membrane and in doing so eliminate the possibility of contamination by cytoplasmic ribosomes. If it is left in contact with the mitochondria for too long it will begin to permeabilize the inner mitochondrial membrane.
4. Solubilization of the mitoplasts occurs immediately and is accompanied by a darkening of the mitoplast suspension as matrix contents are released.
5. Puromycin is an antibiotic compound that binds to the ribosome as an analogue of aminoacyl-tRNA and arrests the formation of nascent polypeptide chains. The polypeptides chains are then liberated from the ribosome. If using puromycin be warned that the compound also has an absorbance at 260 nm and this will mask the presence of the 28S subunit (*see* Fig. 1). If the 28S subunit is required we suggest removing the puromycin by pelleting the crude ribosomes through a second sucrose cushion before layering them onto the sucrose gradients.
6. We have found that complete dissociation of rat liver hepatic mitochondrial ribosomes is achieved by incubating them in a buffer containing no Mg^{2+} rather than low levels of Mg^{2+} as is suggested in some publications. The inclusion of 1 mM EDTA into the Mg^{2+}-deficient buffer B ensures a complete absence of Mg^{2+}.
7. On the basis of the molecular weight of the 55S ribosome 1 $A_{260\,nm}$ unit approximates to 40 pmol *(14)*.

8. It is necessary to remove nucleic acid from protein samples that are to be used in isoelectric focusing or NEPHGE analyses because it can bind to the proteins and cause artifactual migration and streaking.
9. Air bubbles can inhibit gel polymerization.
10. Make sure that no air bubbles are present at the bottom of the tubes as they sit in the lower reservoir solution. If there are, they can easily be removed by blowing reservoir solution across the bottom of the tubes using a glass pipet that has been heated gently and curved at its narrow end.
11. It is very important to pipette the agarose solution gently to avoid the formation of bubbles. These can impede electrophoresis and cause distortions of the protein patterns. If air bubbles form, suck them out with a glass pipette. Also make sure that the marker plugs have no air bubbles present.

References

1. Patel, V. B., Cunningham, C. C., and Hantgan, R. R. (2001) Physiochemical properties of rat liver mitochondrial ribosomes. *J. Biol. Chem.* **276,** 6739–6746.
2. Coleman, W. B., and Cunningham, C. C. (1990) Effects of chronic ethanol consumption on the synthesis of polypeptides encoded by the hepatic mitochondrial genome. *Biochim. Biophys. Acta* **1019,** 142–150.
3. Coleman, W. B., and Cunningham, C. C. (1991) Effect of chronic ethanol consumption on hepatic mitochondrial transcription and translation. *Biochim. Biophys. Acta* **1058,** 178–186.
4. Lieber, C. S., and DeCarli, L. M. (1982) The feeding of alcohol in liquid diets: Two decades of applications and 1982 update. *Alcohol. Clin. Exp. Res.* **6,** 523–531.
5. Cahill, A., Baio, D. L., Ivester, P., and Cunningham, C. C. (1996) Differential effects of chronic ethanol consumption on hepatic mitochondrial and cytoplasmic ribosomes. *Alcohol. Clin. Exp. Res.* **20,** 1362.
6. Patel, V. B., and Cunningham, C. C. (2002) Altered hepatic mitochondrial ribosome structure following chronic ethanol consumption. *Arch. Biochem. Biophys.* **398,** 41–50.
7. Cahill, A., and Cunningham, C. C. (2000) Effects of chronic ethanol feeding on the protein composition of mitochondrial ribosomes. *Electrophoresis* **21,** 3420–3426.
8. Chevallet, M., Lescuyer, P., Diemer, H., van Dorsselaer, A., Leize-Wagner, E., and Rabilloud, T. (2006) Alterations of the mitochondrial proteome caused by the absence of mitochondrial DNA: A proteomic view. *Electrophoresis* **27,** 1574–1583.
9. Cahill, A., Wang, X., and Hoek, J. B. (1997) Increased oxidative damage to mitochondrial DNA following chronic ethanol consumption. *Biochem. Biophys. Res. Commun.* **235,** 286–290.
10. Cahill, A., Stabley, G. J., Wang, X., and Hoek, J. B. (1999) Chronic ethanol consumption causes alterations in the structural integrity of mitochondrial DNA in aged rats. *Hepatology* **30,** 881–888.
11. Cahill, A., Hershman, S., Davies, A., and Sykora, P. (2005) Ethanol feeding enhances age-related deterioration of the rat hepatic mitochondrion. *Am. J. Physiol. Gastrointest. Liver Physiol.* **289,** G1115–G1123.
12. Cohen, N. S., Kyan, F. S., Kyan, S. S., Cheung, C. W., and Raijman, L. (1985) The apparent Km of ammonia for carbamoyl phosphate synthetase (ammonia) in situ. *Biochem. J.* **229,** 205–211.
13. Koc, E. C., Burkhart, W., Blackburn, K., Moyer, M. B., Schlatzer, D. M., Moseley, A., and Spremulli, L. L. (2001) The large subunit of the mammalian mitochondrial ribosome. Analysis of the complement of ribosomal proteins present. *J. Biol. Chem.* **276,** 43958–4369.
14. Ulbrich, B., Czempiel, W., and Bass, R. Mammalian mitochondrial ribosomes. (1980) Studies on the exchangeability of polypeptide chain elongation factors from bacterial and mitochondrial systems. *Eur. J. Biochem.* **108,** 337–343.

26
Microarray Analysis of Ethanol-Induced Changes in Gene Expression

Robnet T. Kerns and Michael F. Miles

Summary DNA microarray studies offer a robust method for nonbiased analysis of whole genome messenger ribonucleic acid expression patterns. A growing number of studies have applied this experimental approach to studies on ethanol either in cell culture of animal models of ethanol exposure or self-administration. Expression profiling has identified novel gene networks responding to ethanol or differing across animal strains with differing responses to ethanol. Recent studies have shown benefit for meta-analysis of microarray data across different laboratories. Gene network analysis offers unique opportunities for understanding the molecular mechanisms of ethanol responses, toxicity and addiction. Eventually, such work may generate novel targets for future pharmacotherapy. To fully capitalize on the prom ise alluded to above, particularly in regard to meta-analysis of microarray data, it is critical that high quality standards are followed in the generation and analysis of microarray studies. This chapter will discuss experience of our laboratory in performing and analyzing microarray studies on ethanol, focusing discussion mainly on short oligonucleotide microarrays (Affymetrix). However, the general principals of technique and analysis that are discussed have broad applicability to other types of microarray platforms and experimental designs.

L. E. Nagy (ed.), *Alcohol: Methods and Protocols*
© Humana Press 2008

Keywords Microarray; ethanol; gene expression; analysis; bioinformatics; neurobiology.

1 Introduction

Alcoholism is generated from long term exposure and the development of this disease may be due to corresponding changes in molecular signaling events. Critical molecular events underlying this neuronal plasticity are thought to include changes in gene expression *(1)*. High-density DNA microarrays allow for simultaneous, nonbiased measurement of the expression of thousands of genes *(2,3)* and the identification of patterns of genes with similar function or regulation *(4,5)*. Instead of quantifying single genes, microarray analysis can focus on entire gene networks and their associated biological pathways and functions *(6)*. Such an approach allow a nonbiased analysis of complex biological events involved in brain responses to ethanol or other drugs of abuse and can potentially generate novel hypotheses about mechanisms and treatment of these diseases.

Expression profiling with DNA microarrays has been used to identify molecular network responses to ethanol in cell culture *(7,8)*, animal models *(9–15)*, and humans *(16,17)*. Most of this work has been in neural cells or brain tissue. This enormous amount of complex data has grown to the point of allowing sophisticated meta-analyses of expression patterns across multiple laboratories *(18)* and the compilation of microarray databases across large genetic models of inbred mouse strains *(19)*. Despite this progress, the precision and sensitivity of microarray data analysis are critical determinants on whether downstream analysis and interpretation will yield significant results. Because of the inherent complexity of the central nervous system, expression profiling in neurobiology is even more challenging. Studies of whole brain neglect the elegant construction of distinct brain regions, yet microdissection can introduce errors from casual cross-contamination and dissection variability as well as the inclusion of multiple cell types in any dissected tissue. Studies of the central nervous system are further complicated by the need for accurate detection of low-abundance genes that are postulated to have significant impact on neuronal phenotype.

Regardless of the aforementioned complications that are involved in performing microarray studies on brain tissue, more than 1000 citations are currently listed in a PubMed search for "brain" and "microarray". A growing number of these involve studies on ethanol and other drugs of abuse. Although the inherent complexity of brain regional and cellular makeup cannot be avoided, careful attention to experimental design and technical issues can greatly improve the quality and impact of microarray studies on brain tissue. This methods review will focus largely on techniques for oligonucleotide microarray analysis of brain responses to ethanol in the hope of encouraging future work in this challenging research area.

2 Materials

2.1 DNA Microarrays

The two principle types of deoxyribonucleic acid (DNA) microarrays are spotted complementary (c)DNA and oligonucleotide arrays *(3,20)*. Oligonucleotide arrays can be spotted by robotic liquid handlers, as with spotted cDNA arrays, synthesized *in situ* or immobilized on specialized bead supports. The spotted cDNA or oligonucleotide arrays can be synthesized in-house or acquired commercially (*see* **Note 1**), and many research universities now have core facilities that offer these services. Although in-house spotted cDNA arrays are relatively inexpensive and allow the researcher to dictate the sequences to be probed on the array, commercial oligonucleotide arrays offer higher standards of quality assurance and consistency, and are rapidly declining in price. Major commercial sources of oligonucleotide arrays are Affymetrix, Agilent, and Illumina, and each of these has its own unique analysis features. Additionally, 15 National Institute of Health divisions have provided microarray core facilities that allow grantees to have analyses done at greatly reduced costs with several different microarray platforms including Affymetrix and Illumina (http://arrayconsortium.tgen.org/np2/home.do).

A description of features and use of multiple array types is beyond the scope of this chapter. Therefore, we will focus on the use of Affymetrix GeneChip™ oligonucleotide arrays in studies on ethanol. This type of commercial microarray is the most commonly used and offers considerable resources in terms of microarray availability, annotation of sequences on the arrays, standard reagents and reaction protocols, and availability of multiple statistical tools for array analysis. In particular, because the reagents and protocols for preparation of probes and hybridization of Affymetrix arrays are highly standardized, we will largely limit discussion to experimental design features and analysis of Affymetrix arrays. It should be stressed that alternative approaches to Affymetrix microarrays can produce comparable data, often at a considerably reduced price *(21)*.

Some caveats about Affymetix microarrays should be considered. All Affymetrix-produced microarrays are manufactured through a novel photo-lithography technique for simultaneous synthesis of very high density (currently up to >600,000 probes/array) short oligonucleotide (approx 25 nucleotides in length) probes. In general, there are now two major types of "expression arrays" from Affymetrix, 3'-end biased arrays and exon-arrays. The former type uses multiple oligonucleotides derived mainly from the 3'-untranslated end of a target messenger ribonucleic acid (mRNA). This produces a composite picture of the abundance for most transcripts from a given gene. However, because many genes produce multiple mRNA species either through alternative splicing or use of multiple promoters, a more complete (albeit, exceedingly complex) picture of gene expression can be obtained through measurement of transcript abundance for individual exons. The extremely high density of the photolithography arrays allows production of chips sampling most exons for the entire expressed genome.

Another crucial feature of using Affymetrix microarrays for gene expression studies concerns the need to produce a complementary (c)RNA probe for hybridization to the arrays. This process is a three-step protocol where total RNA is first converted to a double-stranded cDNA containing a T7-polymerase recognition site at the 3′-end of the molecule. This allows production of large amounts of biotin-labeled cRNA from the cDNA (general yields are 5–10X the initial amount of input total RNA). This amplification process is linear and produces a highly faithful representation of the original mRNA abundance *(22)*. Moreover, the product cRNA can actually be re-converted to double-stranded cDNA and the process repeated to generate adequate amounts of cRNA for microarray hybridization even with sub-microgram amounts of starting total RNA *(23)*. This two-step amplification process, however, biases the product cRNA toward the 3′-end of the mRNA molecules due to incomplete extension during the cycles of cDNA and cRNA synthesis. Thus, such two-cycle approaches do not routinely allow use of exon chips since more distal exons would be severely under-represented in the hybridization mixture.

2.2   Reagents

Because of the complex hybridization kinetics and multiple steps entailed in preparation of probes and array hybridization, reagents recommended by the microarray manufacturer should be used. These reagents can either be assembled by buying individual reagents from various suppliers or by using various kits developed and quality control checked by Affymetrix. Because of the sensitivity of the cRNA synthesis and labeling steps, reagents lots should not be mixed during an experiment. Doing so will produce large "batch effects" such that the cRNA reagent kit defines more variance of the array results than resulting from the desired experimental variables themselves. Thus, arrays from different samples but run with one batch of cRNA reagents could appear more similar to one another than arrays run on identical biological samples but with differing reagent lots. Statistical approaches for minimizing such batch effects can be used, but it is preferable to avoid as many such confounds as possible during the experimental design stage. If more cRNA samples are to be prepared than can be made from a single lot of reagents, multiple reagent kits should be combined at the beginning of the experiment so that enough reagents are available as a single "batch."

3   Methods

3.1   Experimental Design

The experimental design step of a microarray experiment is crucial to the generation and analysis of data and overall success of the experiment. Even if an array experiment produces high quality data, improper design can make the experiment extremely difficult to analyze and interpret, so much so that the results may be inconclusive and the hypothesis untested. Because microarray experiments are so expensive in terms of labor hours, reagents, biological materials, and the chips themselves, care should be exercised from the start to ensure success. Extensive discussions of statistical considerations of microarray experimental design have been published (24). The following factors should be considered during the design of microarray experiments:

1. Number of replicates: As with any experiment, an number of independent biological replicates are required to ensure statistical power for microarray data analysis. Traditional power analysis of microarray experiments suggests that upwards of ten replicates is required for each group (25), in part because of statistical issues caused by multiple testing when studying the entire genome simultaneously. Although the more replicates the better, in practice far fewer replicates are necessary to achieve statistical significance and quality data. In addition, the number of replicates required is partly dictated by the type of experiment because the needs of a time course or dose response experiment might differ from an experiment on a panel of recombinant inbred strains. Anything less than three replicates makes downstream statistical filtering methods nearly useless, except for particular experimental designs such as QTL mapping or the application of advanced approaches such as the Bioconductor "Timecourse" package (http://www.bioconductor.org). In our experience four biological replicates produces much higher quality data for statistical filtering and clustering than three replicates, and also allows for the occasional loss of a sample caused by error or chance.

   It is also important to distinguish biological samples derived from separate animals or cultures, from technical replicates, which are repeated measurements (microarrays) of the same sample (mRNA prep). Due to the very high reproducibility of Affymetrix microarrays, technical replicates run on the same biological sample are usually avoided in favor of running single arrays on more biological replicates. Extensive statistical discussions on the use of technical versus biological replicates in microarray experiments have been published (24,26).

2. Amount of biological material: experimental design should account for the amount of material necessary for arrays and subsequent validation experiments. Affymetrix array protocols call for a minimum of 10 μg of total RNA for standard methods of producing cRNA for chip hybridization. However, as mentioned previously, 2-cycle amplification protocols exist and are useful for instances where much lower amounts of starting RNA are available, such as when using laser capture microdissection (27). Although commercial RNA isolation kits and reagents specify a minimum amount for starting material, high-quality RNA is more easily obtainable when processing larger amounts of biological sample. Thus, pooling of animals or cultures may be necessary to obtain sufficient material for RNA isolation and subsequent microarray analysis. This is particularly important for isolated brain region RNA samples from small experimental animals such as mice, where dissections of nucleus accumbens or prefrontal cortex

which may be 30–50 mg in mass from a single animal. Pooling of animals/ samples also has the advantage of reducing the influence of random environmental influences that can complicate microarray studies on ethanol with whole animals or cell cultures. Obviously, multiple biological replicates of "pooled" samples are necessary for statistical analysis as mentioned above. This pooling approach can decrease the statistical power of some experimental designs where it is desired to correlate a phenotypic variable (e.g., behavior) with gene expression. In such cases, analysis of individual animals and a relatively larger number of microarrays are needed. In general, 1 µg of total RNA is obtained from 1 mg (wet weight) of brain tissue. Different tissues or cell cultures will produce different yields. Human autopsy brain tissue will generally produce lower yields. For single-cycle cRNA synthesis protocols, 5–10 µg of starting total RNA is needed. Thus, as a general rule for best results with brain tissue, this requires starting with >50 mg of tissue (wet weight). As mentioned previously, two-cycle cRNA amplification protocols can use much less starting material. In addition, our laboratory has obtained high quality microarray data starting with as little as 1 µg of total RNA and using a single-cycle cRNA protocol (Wolstenholme and Miles, unpublished results), but such designs are not used routinely. If a given sample contains a very low amount of total RNA (but has passed RNA integrity checks) we routinely use that amount of input RNA for all cRNA reactions. Keeping the amount of input total RNA constant across the cRNA reactions is an important factor in avoiding variability due to cRNA synthesis. Similarly, if less than the recommended amount of cRNA is available for array hybridization, then it is critical that all arrays are at least hybridized with the same amount of cRNA, In extreme cases, samples with greatly reduced RNA or cRNA yields have to be discarded since they likely have issues with RNA integrity (see below).

3. Experimental technique: Improper handling of samples can result in changes in gene expression because of factors unrelated to the experiment, including animal handling, tissue dissection, and sample preparation *(28,29)*. For example, a change in an animal's environment or handling can increase biological variance or noise and thus greatly complicate experiment analysis. The impact of laboratory environment of ethanol behavioral studies is well documented *(30)*. Because whole-genome microarrays are extremely sensitive to environmental factors, extraneous variables must be reduced so that observed changes in gene expression reflect experimental perturbations and not unrelated environmental factors. Thus, factors such as lab chow, light-dark cycle, and animal handler should remain constant during the course of an experiment as much as possible. For cell culture experiments, vendor and lot of reagents should be maintained constant and all cultures must be handled identically with very rapid approaches used for going from cell culture dish to frozen cell pellet or RNA extraction. Seemingly trivial factors such as position within the incubator or having a mildly prolonged incubation on ice before freezing or homogenization can produce systematic patterns of gene expression changes when monitoring the expression of more than 40,000 transcripts.

4. RNA quality control: High-quality total RNA is critical for successful labeled-cRNA synthesis and array hybridizations. As with any RNA work, a RNAse-free environment must be maintained at each step from tissue extraction through labeled-cRNA synthesis. This can be assisted by liberal application of anti-RNAse reagents such as RNAseZap (Ambion, Austin, TX) on bench surfaces, gloves, pipettes and sample tubes and the use of rapid protocols for going from intact tissue/cells to homogenization in denaturing RNA isolation buffers. Traditionally, RNA quality has been accessed by 260/280 UV absorbance ratios and RNAse-free agarose gel electrophoresis (*see* **Note 2**). More advanced techniques are now available to characterize RNA quality. The Agilent BioAnalyzer and BioRad Experion microfluidics systems are automated and require much less material (approx 0.5 µg) than gel electrophoresis (approx 5 µg). We routinely perform a size distribution analysis on both total RNA and the product cRNA. For total RNA, all samples should show near 1:1 molar ratios for 28S and 18S ribosomal RNA. For cRNA synthesis, size distribution patterns should be very similar across all samples and the highest molecular weight products should extend above 3000 nucleotides unless autopsy material or two-cycle amplification protocols are being used.

5. Randomization: as referred to previously, expression profiling with DNA microarrays is highly sensitive for detecting variation in gene expression due to seemingly trivial environmental or experimental design features. Although the source of these effects is sometimes unclear, it is known that array background and gene expression can correlate with reagent lot, day/time of RNA isolation, order of labeled-cRNA synthesis and hybridization, sample age, and technician, among other factors. In addition to controlling experimental conditions as much as possible, the use of randomization schemes at multiple levels throughout the experimental design can be very useful method for minimizing systematic bias in microarray experiments. Thus, it is critical to use a supervised randomization of control and treatment samples at RNA isolation, cRNA synthesis and array hybridization steps to eliminate systematic variance due to uncontrolled environmental factors.

3.2 Protocol

Tissue isolation or cell harvesting are highly individualized protocols dependent on the particular experimental system at study. As a general rule, however, very rapid harvesting of tissue/cells must be used to reduce ischemic, traumatic or cold stress induced changes in gene expression. For example, the mouse brain dissection protocol utilized in our laboratory requires the combined efforts of three to four laboratory members and takes tissue from seven to eight brain region microdissections to being frozen in liquid nitrogen within 5–7 min per mouse *(15)*.

RNA isolation, labeled-cRNA synthesis, and array hybridization protocols are extensive and therefore not within the scope of this review. There are several total

RNA isolation kits available with similar guanidine denaturation/phenol-chloro-form extraction protocols; we have used the Stat60 reagent (Tel-Test, Inc., Friendswood, TX) with very reproducible results. More recent protocols for fatty tissue such as brain have been developed, for example the BioRad Aurum™ Total RNA Fatty and Fibrous Tissue Kit (BioRad Laboratories Inc., Hercules, CA). These newer kits offer advantages either in a reduced number of steps or column steps for nucleic acid recovery, as opposed to ethanol precipitation in traditional methods. The best approach is to practice each step until high quality RNA can be obtained in the laboratory. In general, except for points mentioned above regarding cRNA synthesis, Affymetrix protocols are generally strictly followed for cDNA and cRNA synthesis and microarray hybridization (obtained through NetAffx at http://www.affymetrix.com/index.affx).

3.3   Microarray Quality Control

After hybridization and scanning, arrays must be rigorously evaluated for quality and consistency across arrays. Arrays that fail these standards will confound downstream analysis efforts and that sample should either be re-hybridized or discarded from the analysis. Microarray quality control includes metrics specific to the Affymetrix GeneChip platform and approaches that are applicable to all array types.

3.3.1   Affymetrix-Specific Quality Control Metrics

These parameters are shown on the metrics tab of the Affymetrix software and can be exported to a text file. These should be uniform across arrays and within thresh-old guidelines described below.

1. Scaling factor (SF): This is a measure of background and depends on the target average intensity (TGT) used during scanning. For TGT 190, used for older arrays, SF should be ≤3. For newer arrays TGT 500 is recommended; SF should be ≤10. Consistency within 3 SD units of the mean SF across all arrays of a batch is used as a cutoff for outliers
2. Statistical difference threshold (SDT): This is a measure of noise and is calculated from SF and RawQ. Standards for SDT are also dependent on TGT used for scan-ning. For TGT = 190 this value should be in the range 12–30; for TGT = 500 this value may be considerably greater. The critical feature of SDT is that it be consist-ent over all arrays and outliers are excluded as mentioned above.
3. Percent-present: The Affymetrix MAS and GCOS software will calculate the percentage of probesets (genes) present in the sample based on probe intensities corrected for background. Arrays with 40–60% present are generally accepted. Arrays outside this range are generally due to errors in cRNA synthesis or array hybridization. Again, array-to-array consistency is the most crucial factor and outliers are excluded as for the SF and SDT measures.

4. 3′-5′ GAPDH Control Expression Ratios: arrays should only be accepted that have signals for the 3′- and 5′-end of the gene for glyceraldehyde phosphate dehydrogenase (GAPDH) within an accepted range. The Affymetrix chips have probesets with oligonucleotides addressing the 5′-end or 3′-end of GAPDH. Since the cRNA probes are synthesized starting at the 3′-end, it is normal to have lower signal produced against 5′-end probes but an excessive 3′/5′ ratio is indicative of RNA or cRNA degradation or synthesis failure. GAPDH 3′/5′ ratios near or below 3.0 are acceptable but these are generally below 2.0 for good-quality RNA/cRNA.

3.3.2   Generalized Quality Control Measures

These descriptive statistical studies are standard approaches for assessing the quality of data from any array platform.

1. Scatter plot: Scatter plots of the log transformation of probe or probeset intensities between two arrays should be visually linear. Plots of arrays that curve at high or low intensity values, or other abnormal non-linear shapes, reflect abnormal saturation or background in one or more arrays. The shape of the scatter plot is a better indicator of array quality than Pearson correlations, which should be high (R > 0.96). Excessive scatter between closely related samples is also indicative of problems with one of the arrays and is accompanied by decreases in the Pearson correlation coefficients. Pairwise scattergrams between all arrays within an experiment should appear very similar. An alternative to the straight scatter plot is the so-called "M vs. A" plot. This transformation is visually somewhat easier to interpret than the normal scatter plot. In this case the x-axis is the average log intensity for each gene/probe across the two arrays being compared. The y-axis displays the log of the ratio of intensities between the two chips. Thus, if a gene shows no change, the ratio should be 1 with a log value of 0. The M vs. A plot will display a cigar-shaped scatter of points centered around 0 on the y-axis. Deviations from linearity at the high or low abundance range can be easily detected.
2. Box-and-whisker plot: the box plot function of the Bioconductor Affy Package will plot the median, 25th and 75th percentiles, and extremes of probe intensities. This function is useful to compare the relative background and scale of multiple arrays in an experiment, and may indicate which arrays should be removed prior to normalization and summarization.

3.4   Primary Analysis

Primary analysis refers to the production of intensity values for genes from the raw fluorescence measurements derived from scanning microarrays. For Affymetrix oligonucleotide arrays, this expression intensity determination is not as straightforward as one might think. The short oligonucleotides on these arrays are prone to nonspecific hybridization or lack of signal because of failed hybridization kinetics. Fortunately, the very high density of these arrays allowed the designers to incorporate two features aimed at circumventing the difficulties of using short oligonucleotides. First, each specific oligonucleotide on the array (termed a "perfect match" or PM probe) is paired with a "mismatch" or MM probe that has a single base substitution in the center of the 25 nucleotide oligonucleotide, aimed at disrupting specific hybridization. These MM probes are meant to detect non-specific hybridization. Secondly, each gene on the Affymetrix arrays is represented by a "probeset," which encompasses 8–15 different PM probes and their paired MM probes. The exact number of PM/MM probe pairs depends on the particular array design. Ideally, each different PM for a given gene should have the same hybridization intensity, but in actuality this rarely occurs due to variable oligonucleotide synthesis kinetics, hybridization kinetics, alternative splicing and a number of other factors.

Thus, the first step in low-level analysis of Affymetrix arrays is to derive a single intensity value from all the PM/MM probes making up a given probeset. A very large number of low-level analysis methods have evolved that produce robust estimates of gene expression from the individual probe data. The manufacturer's most recent algorithm, entitled MAS5 or GCOS, uses a Tukey bi-weight estimate of modeled oligonuceotide signals with greatly improved performance compared to the initial algorithms, and is particularly useful for low abundance genes. However, more advanced and robust gene expression summarization methods, for instance RMA *(31)*, MBEI *(32)*, and PDNN *(33)*, have been developed (*see* **Note 3**), and the use of at least one of these methods is often preferred. Some reports have suggested the use of multiple algorithms with the contention that different algorithms work better for different genes or that the use of multiple algorithms for statistical filtering produces a more robust set of selected genes *(13)*.

An additional approach developed by our laboratory involves a ratio method, the S-score algorithm, that compares expression of two arrays and produces a summary statistic reflecting the probability of their being a difference in expression for a given gene (probeset) between the two arrays compared *(34–37)*. This method is unique in that it compares individual probe intensities between two arrays rather than comparing probeset summary statistics. By performing S-score analyses across biological replicates and using statistical approaches to find reproducible changes in expression (see below), we have found the S-score algorithm to be very specific and sensitive for detecting small changes in expression *(15,37)*. An additional advantage of the S-score is the use of this approach for experiments having small numbers of replicates *(34,37)*.

3.5   *Statistical Filtering*

As mentioned in the discussion about replicates, statistical analysis of microarray data is complicated by multiple-testing errors affecting calculation of statistical significance. As a first step before the actual statistical analysis of any array platform, a filtering is performed to remove genes that are "absent" or expressed at levels below the linear range of detection for the microarray. Several algorithms exist to make decisions about which genes are absent, but a common approach is to simply define a cutoff as 3 standard deviation units above the background signal (defined either in an area with no probes or with cross-species probes known not to cross-hybridize (e.g., bacterial genes). Affymetrix arrays have a decision metric built into the low-level analysis software that defines a gene as "absent," "present," or "maybe." Some investigators choose to require that all genes are called "present" on *all* arrays for inclusion in statistical analysis. Our laboratory takes a more inclusive approach, to avoid generating excessive false negatives (see discussion below), by just eliminating genes judged to be "absent" in *all* samples.

Detailed discussions of statistical approaches for analysis of microarray data have been previously published *(38,39)*. Although there are vocal arguments about the best statistical approach to take with microarray data, some investigators feel that "statistical filtering" is not meant as a "final analysis" but rather, only a method for reducing false positives. Because multivariate analysis (*see* Subheading 3.6.) is frequently performed after statistical filtering, this secondary "network level" analysis serves to further reduce false positives. Such an approach, with a somewhat relaxed statistical filtering followed by multivariate analysis, serves to minimize the false negative rate and thus aids potential network analysis. It is difficult to study gene networks if you eliminate all but a handful of genes through a rigorous statistical filtering.

The most widely used primary approaches for statistical filtering with microarray data currently involve some type of permutation or re-sampling analysis to produce a false-discovery rate calculation to identify genes whose expression change is likely to not have occurred by chance alone. Many types of software are available for such statistical filtering. Perhaps the most popular application is the Significance Analysis of Microarrays (SAM) tool which can be used as a plug-in application for Excel spreadsheets on Windows based PCs *(40)* or within the TIGR MultiExperiment Viewer (TMEV) multi-functional analysis platform *(41)*. Other popular statistical approaches include ANOVA models (also present in the TMEV software) and the Timecourse package within the BioConductor suite of array analysis programs (www.bioconductor.org). This latter algorithm and the SAM Excel Plug-in software are useful particularly for analysis of time-course data.

3.6   *Multivariant Analysis*

Multivariant analysis groups genes with statistically similar expression patterns across all samples in an experiment. Such data reduction algorithms have proven to be exceedingly useful in analysis of microarray data for two major reasons *(4,6)*. First, grouping genes into "clusters" serves to reduce the complexity of the analysis and make overall interpretation easier. Second, use of techniques such as hierarchical and k-means clustering of microarray data has been shown to group the genes together in functionally meaningful ways *(6)*. A cluster can thus provide hints as to the biological function of the gene regulatory events themselves. For example, upregulation of a group of genes involved in glycolysis might implicate a demand for energy by whatever biological process is being studied. Additionally, clustering of an unknown or unannotated gene tightly with a group of known genes all having a coherent biological function can provide clues as to the function of the unknown genes.

The TMEV software again provides a comprehensive selection of multivariate analysis tools. These include additional clustering methods such as self-organizing maps, and post-hoc association methods such as Pearson-correlation template matching. Each clustering algorithm has unique features and a number of adjustable parameters resulting in a very large number of options for this approach. Generally, several methods are used and the results evaluated for biological relevance (*see* **Note 4**).

Thus, for a standard microarray analysis, after running quality control studies on the microarray data and eliminating genes consistently called "absent," we will import the RMA or S-score data from an experiment into TMEV and perform statistical filtering with SAM or ANOVA using permutation analysis. The statistical cutoff depends upon the source and quality of the data. Genes passing statistical filtering will be studied by multivariate analysis using the TMEV hierarchical or k-means clustering. The latter algorithm requires "guessing" the number of clusters and we frequently use principal components analysis to first estimate the number of clusters. The individual clusters are then subjected to bioinformatics analysis as described in Subheading 3.7.

3.7   Bioinformatics and Network Analysis

Bioinformatics methods are used to identify key regulated biological networks from lists of regulated genes. The biological significance or mechanism of regulation of selected genes can be explored by three general approaches: 1) Over-representation analysis: the identification of a common gene ontology, functional group assignment, or biological pathway for an abnormally large number of genes within a gene list or cluster *(42–45)*; 2) literature association analysis: the identification of pairings of genes within the biodmedical literature either directly or indirectly in a network of literature associations; and 3) biological network mining such as finding common transcription factor binding sites within a large number of genes from a microarray gene list or cluster.

All of these approaches take advantage of specialized software, forms of which are often found as open source freeware. In particular, we generally perform over-representation analysis of gene ontology or other functional groupings with web-based tools such as DAVID (http://david.abcc.ncifcrf.gov/) or WebGestalt (http://bioinfo. vanderbilt.edu/webgestalt/) *(46)*. For literature association analyses we generally use free-ware such as Chilibot (http://www.chilibot.net) *(47)* or the Agilent Literature plug-in for Cytoscape (http://www.cytoscape.org). Networks constructed from functional and regulatory relationships between genes can be studied using commercial programs such as BiblioSphere (http://www.genomatix.de) or the Ingenuity Pathway Analysis platform (http://www.ingenuity.com).

Finally, a number of academic databases can provide extremely important respources for microarray analysis. The WebQTL tool in GeneNetwork (http://www.genenetwork.org) is such a resource that is very frequently used by our laboratory in array analysis. GeneNetwork provides a searchable compilation of behavioral and gene expression QTL data from a variety of rodent models and other species *(48)*. This allows the identification of loci controlling the expression of target genes either in cis or trans, the correlation of gene expression across rich recombinant inbred strain batteries, and the correlation of gene expression and behavior across these same rodent lines (generally mouse). A host of other analysis tools and links makes WebQTL a powerful tool for identify gene networks and gene-behavior interactions from microarray data.

3.8 *Validation of Candidate Genes*

When microarray analysis was first introduced, it was generally deemed necessary to confirm array results in-bulk by identical experiments using more accepted forms, or arguably more reliable or more sensitive methods of mRNA quantification, that is, "gold standard" methods such as Northern blotting and quantitative real-time rtPCR (QPCR). It was not uncommon for a publication of microarray data to be accompanied by QPCR data for dozens of genes, irrespective of the significance of such genes within the results and conclusions of the report. With the acceptance of microarray hybridization as a reliable technology, expression validation solely for the sake of reproducing data and increasing confidence in array results no longer seems a necessity (except for an occasional obstinate reviewer!). However, in practice we tend to validate genes that are representative of gene networks networks or gene groupings selected by the bioinformatics analysis above. If sufficient materials exist, we prefer Northern blotting but QPCR confirmation can obviously also be used.

Acknowledgments This work was supported by NIAAA Grants RO1 AA014717 to MFM and F32 AA014726 to RTK.

4   Notes

1. Detailed descriptions and instructions regarding spotted cDNA arrays are available at the website of the Brown Lab at Stanford (http://cmgm.stanford.edu/pbrown/mguide/). Popular commerical oligonuclotide array vendors include

Affymetrix, Agilent, and Illumina and more detailed information can be obtained at their corporate web sites.

2. We usually dilute 1 μL of RNA sample in 99 μL of TE buffer for UV measurements. Ratios 260, 280 ≥ 2.0 and ≤ 2.2 are best. Alternatively, if using ddH$_2$O as diluent, ratios should be between 1.8 and 2.0.

3. A very large number of primary analysis algorithms for the Affymetrix platform have been developed. Many of these are available through the Bioconductor packages for the R programming environment (www.bioconductor.org).

4. Software is readily available for normalization, primary analysis, and both generating and viewing cluster diagrams. These programs include TIGR Multi-Experiment Viewer (http://www.tigr.org), the Bioconductor package (http://www.bioconductor.org) for the R statistical computing environment, and cluster analysis and visualization tools Cluster and Treeview (http://rana.lbl.gov/EisenSoftware.htm).

References

1. Nestler, E. J., and Aghajanian, G. K. (1997) Molecular and cellular basis of addiction. *Science* **278,** 58–63.
2. Schena, M., Shalon, D., Heller, R., Chai, A., Brown, P. O., and Davis, R. W. (1996) Parallel human genome analysis: microarray-based expression monitoring of 1000 genes. *Proc. Natl. Acad. Sci. USA* **93,** 10614–10619.
3. Lockhart, D. J., Dong, H., Byrne, M. C., et al. (1996) Expression monitoring by hybridization to high-density oligonucleotide arrays. *Nat. Biotechnol.* **14,** 1675–1680.
4. Quackenbush, J. (2001) Computational analysis of microarray data. *Nat. Rev. Genet.* **2,** 418–427.
5. Eisen, M. B., Spellman, P. T., Brown, P. O., and Botstein, D. (1998) Cluster analysis and display of genome-wide expression patterns. *Proc. Natl. Acad. Sci. USA* **95,** 14863–14868.
6. Hughes, T. R., Marton, M. J., Jones, A. R., et al. (2000) Functional discovery via a compendium of expression profiles. *Cell* **102,** 109–126.
7. Thibault, C., Lai, C., Wilke, N., et al. (2000) Expression profiling of neural cells reveals specific patterns of ethanol-responsive gene expression. *Mol. Pharmacol.* **58,** 1593–1600.
8. Hassan, S., Duong, B., Kim, K. S., and Miles, M. F. (2003) Pharmacogenomic analysis of mechanisms mediating ethanol regulation of dopamine beta-hydroxylase. *J. Biol. Chem.* **278,** 38860–29969.
9. Rimondini, R., Arlinde, C., Sommer, W., and Heilig, M. (2002) Long-lasting increase in voluntary ethanol consumption and transcriptional regulation in the rat brain after intermittent exposure to alcohol. *FASEB J.* **16,** 27–35.
10. Saito, M., Smiley, J., Toth, R., and Vadas, C. (2002) Microarray analysis of gene expression in rat hippocampus after chronic ethanol treatment. *Neurochem. Res.* **27,** 1221–1229.
11. Tadic, S. D., Elm, M. S., Li, H. S., et al. (2002) Sex differences in hepatic gene expression in a rat model of ethanol-induced liver injury. *J. Appl. Physiol.* **93,** 1057–1068.
12. Daniels, G. M., and Buck, K. J. (2002) Expression profiling identifies strain-specific changes associated with ethanol withdrawal in mice. *Genes Brain Behav,* **1,** 35–45.
13. Tabakoff, B., Bhave, S. V., and Hoffman, P. L. (2003) Selective breeding, quantitative trait locus analysis, and gene arrays identify candidate genes for complex drug-related behaviors. *J. Neurosci.* **23,** 4491–4498.
14. Treadwell, J. A., and Singh, S. M. (2004) Microarray analysis of mouse brain gene expression following acute ethanol treatment. *Neurochem. Res.* **29,** 357–369.

15. Kerns, R. T., Ravindranathan, A., Hassan, S., et al. (2005) Ethanol-responsive brain region expression networks: implications for behavioral responses to acute ethanol in DBA/2J versus C57BL/6J mice. *J. Neurosci.* **25,** 2255–2266.

16. Lewohl, J. M., Wang, L., Miles, M. F., Zhang, L., Dodd, P. R., and Harris, R. A. (2000) Gene expression in human alcoholism: microarray analysis of frontal cortex. Alcohol. *Clin. Exp. Res.* **24,** 1873–1882.

17. Mayfield, R. D., Lewohl, J. M., Dodd, P. R., Herlihy, A., Liu, J., and Harris, R. A. (2002) Patterns of gene expression are altered in the frontal and motor cortices of human alcoholics. *J. Neurochem.* **81,** 802–813.

18. Mulligan, M. K., Ponomarev, I., Hitzemann, R. J., et al. (2006) Toward understanding the genetics of alcohol drinking through transcriptome meta-analysis. *Proc. Natl. Acad. Sci. USA* **103,** 6368–6373.

19. Chesler, E. J., Wang, J., Lu, L., Qu, Y., Manly, K. F., and Williams, R. W. (2003) Genetic correlates of gene expression in recombinant inbred strains: a relational model system to explore neurobehavioral phenotypes. *Neuroinformatics* **1,** 343–357.

20. Schena, M., Shalon, D., Davis, R. W., and Brown, P. O. (1995) Quantitative monitoring of gene expression patterns with a complementary DNA microarray. *Science* **270,** 467–470.

21. Barnes, M., Freudenberg, J., Thompson, S., Aronow, B., and Pavlidis, P. (2005) Experimental comparison and cross-validation of the Affymetrix and Illumina gene expression analysis platforms. *Nucleic. Acids Res.* **33,** 5914–5923.

22. Van Gelder, R. N., von Zastrow, M. E., Yool, A., Dement, W. C., Barchas, J. D., and Eberwine, J. H. (1990) Amplified RNA synthesized from limited quantities of heterogeneous cDNA. *Proc. Natl. Acad. Sci. USA* **87,** 1663–1667.

23. Kacharmina, J. E., Crino, P. B., and Eberwine, J. (1999) Preparation of cDNA from single cells and subcellular regions. *Methods Enzymol.* **303,** 3–18.

24. Yang, Y. H., and Speed, T. (2002) Design issues for cDNA microarray experiments. *Nat. Rev. Genet.* **3,** 579–588.

25. Lee, M. L., Kuo, F. C., Whitmore, G. A., and Sklar, J. (2000) Importance of replication in microarray gene expression studies: statistical methods and evidence from repetitive cDNA hybridizations. *Proc. Natl. Acad. Sci. USA* **97,** 9834–9839.

26. Dudoit, S., Yang, Y. H., Callow, M. J., and Speed, T. J. (2000) Statistical methods for identifying differentially expressed genes in replicated cDNA microarray experiments. Technical Report. Stanford, CA: Stanford University School of Medicine; Report No. 578.

27. Luo, L., Salunga, R. C., Guo, H., et al. (1999) Gene expression profiles of laser-captured adjacent neuronal subtypes. *Nat. Med.* **5,** 117–122.

28. Sandberg, R., Yasuda, R., Pankratz, D. G., et al. (2000) Regional and strain-specific gene expression mapping in the adult mouse brain. *Proc. Natl. Acad. Sci. USA* **97,** 11038–11043.

29. Geschwind, D. H. (2000) Mice, microarrays, and the genetic diversity of the brain. *Proc. Natl. Acad. Sci. USA* **97,** 10676–10678.

30. Wahlsten, D., Metten, P., Phillips, T. J., et al. (2003) Different data from different labs: lessons from studies of gene-environment interaction. *J. Neurobiol.* **54,** 283–311.

31. Irizarry, R. A., Bolstad, B. M., Collin, F., Cope, L. M., Hobbs, B., and Speed, T. P. (2003) Summaries of Affymetrix GeneChip probe level data. *Nucleic Acids Res.* **31,** e15.

32. Li, C., and Wong, W. H. (2001) Model-based analysis of oligonucleotide arrays: expression index computation and outlier detection. *Proc. Natl. Acad. Sci. USA* **98,** 31–36.

33. Zhang, L., Miles, M. F., and Aldape, K. D. (2003) A model of molecular interactions on short oligonucleotide microarrays. *Nat. Biotechnol.* **21,** 818–821.

34. Zhang, L., Wang, L., Ravindranathan, A., and Miles, M. F. (2002) A new algorithm for analysis of oligonucleotide arrays: Application to expression profiling in mouse brain regions. *J. Mol. Biol.* **317,** 225–235.

35. Kerns, R. T., Zhang, L., and Miles, M. F. (2003) Application of the S-score algorithm for analysis of oligonucleotide microarrays. *Methods* **31,** 274–281.

36. Kennedy, R. E., Kerns, R. T., Kong, X., Archer, K. J., and Miles, M. F. (2006) SScore: an R package for detecting differential gene expression without gene expression summaries. *Bioinformatics* **22,** 1272–1274.

37. Kennedy, R. E., Archer, K. J., and Miles, M. F. (2006) Empirical validation of the S-score algorithm in the analysis of gene expression data. *BMC Bioinformatics* 7:154.
38. Kim, S. Y., Lee, J. W., and Sohn, I. S. (2006) Comparison of various statistical methods for identifying differential gene expression in replicated microarray data. *Stat. Methods Med. Res.* **15,** 3–20.
39. Reimers, M. (2005) Statistical analysis of microarray data. *Addict. Biol.* **10,** 23–35.
40. Tusher, V. G., Tibshirani, R., Chu, G. (2001) Significance analysis of microarrays applied to the ionizing radiation response. *Proc. Natl. Acad. Sci. USA* **98,** 5116–5121.
41. Saeed, A. I., Sharov, V., White, J., et al. (2003) TM4: a free, open-source system for microarray data management and analysis. *BioTechniques* **34,** 374–378.
42. Consortium TGO. (2001) Creating the gene ontology resource: design and implementation. *Genome Res.* **11,** 1425–1433.
43. Doniger, S. W., Salomonis, N., Dahlquist, K. D., Vranizan, K., Lawlor, S. C., and Conklin, B. R. (2003) MAPPFinder: using Gene Ontology and GenMAPP to create a global gene-expression profile from microarray data. *Genome. Biol.* **4:**R7.
44. Hosack, D. A., Dennis, G., Jr., Sherman, B. T., Lane, H. C., and Lempicki, R. A. (2003) Identifying biological themes within lists of genes with EASE. *Genome Biol.* **4:**R70.
45. Pavlidis, P., Qin, J., Arango, V., Mann, J. J., Sibille, E. (2004) Using the gene ontology for microarray data mining: A comparison of methods and application to age effects in human prefrontal cortex. *Neurochem. Res.* **29,** 1213–1222.
46. Zhang, B., Kirov, S., and Snoddy, J. (2005) WebGestalt: an integrated system for exploring gene sets in various biological contexts. *Nucleic Acids Res.* **33**(Web Server issue), W741–W748.
47. Chen, H., and Sharp, B. M. (2004) Content-rich biological network constructed by mining PubMed abstracts. *BMC Bioinformatics* **5,** 147.
48. Chesler, E. J., Lu, L., Wang, J., Williams, R. W., and Manly, K. F. (2004) WebQTL: rapid exploratory analysis of gene expression and genetic networks for brain and behavior. *Nat. Neurosci.* **7,** 485–486.

Index